What role does viscosity play in accretion discs? How do you calculate [the ...]
of a pulsar? And can strong shocks account for the energy spectrum [...]
Galaxy? These are just some of the exciting questions that Profes[sor ...]
develop the physics needed by the astronomer and high energy astrophysicist.

The highly acclaimed first edition of *High energy astrophysics* instantly established itself
as a classic in the teaching of contemporary astronomy. Reflecting the immense interest
and developments in the subject, Professor Longair has developed the second edition into
three texts; in this second volume, he provides a comprehensive discussion of the high
energy astrophysics of stars, the Galaxy and the interstellar medium. He develops an
understanding for the essential physics with an elegance and infectious enthusiasm for
which his teaching is internationally renowned, and he illustrates the issues throughout
with results from forefront research.

This book takes the student with a knowledge of physics and mathematics at the
undergraduate level – but not necessarily with training in astronomy – to the point where
current astronomical research can be understood.

High energy astrophysics

Volume 2
Stars, the Galaxy and the interstellar medium

High energy astrophysics

Volume 2
Stars, the Galaxy and the interstellar medium

Second edition

M. S. Longair

Jacksonian Professor of Natural Philosophy,
Cavendish Laboratory,
University of Cambridge

CAMBRIDGE
UNIVERSITY PRESS

Published by the Press Syndicate of the University of Cambridge
The Pitt Building, Trumpington Street, Cambridge CB2 1RP
40 West 20th Street, New York, NY 10011-4211, USA
10 Stamford Road, Oakleigh, Melbourne 3166, Australia

First published 1981

Second edition 1994

A catalogue record for this book is available from the British Library

Library of Congress cataloguing in publication data

(Revised for vol. 2)
Longair, M. S., 1941–
  High energy astrophysics.
  Includes bibliographical references and index.
  Contents: v. 1. Particles, photons and their detection – v. 2. Stars, the galaxy,
and the interstellar medium.
  1. Nuclear astrophysics. 2. Particles (Nuclear physics)   I. Title.
QB464.L66 1992   523.01'97   92-191337
ISBN 0-521-38374-9
ISBN 0-521-38773-6 (pbk.)

ISBN 0 521 43439 4   hardback
ISBN 0 521 43584 6   paperback

TAG

For Deborah

Contents

Preface

There is little to add to the remarks which I made in the preface to Volume 1 of *High energy astrophysics*. The process of revising and updating the first edition has resulted in a very major expansion in the length of the text, so that it could not be contained within a single volume. I had hoped to complete the work in two volumes, but, as the work progressed, it became apparent that the second volume would have become unwieldy, and so this book is Volume 2 of what will be a three-volume work. In this volume, I concentrate upon the high energy astrophysics of our own Galaxy, and Volume 3 will be devoted to extragalactic high energy astrophysics.

It is worthwhile explaining the point of view which I have adopted in introducing the various topics contained in Volumes 2 and 3. As in the first edition, my aim has been to produce self-contained texts which include most of the essential astronomy and astrophysics needed to understand the context as well as the content of high energy astrophysics. For this reason, the present volume begins with descriptions of the current picture of the large-scale distribution of matter and radiation in the Universe, as well as a broad survey of relevant astrophysics, before getting down to studying the high energy astrophysics in detail. In my view, it is no longer possible, if it ever was, to consider the high energy processes independently of the astrophysical environments within which they take place. I have based the contents of these chapters upon my review article 'The new astrophysics', which appeared in the book *The new physics* (ed. P.C.W. Davies, Cambridge University Press, 1989) and upon other review articles which I have presented during the last year, specifically my 1992 Milne Lecture 'Modern cosmology – a critical assessment' and 'The frontiers of modern astrophysics and cosmology' prepared for the 1992 OECD Expert Astronomy Meeting on Megaprojects held in La Laguna, Tenerife.

Although the text is intended to be read as a continuously developing narrative, the separate chapters of all three volumes can be used in a modular fashion, for a variety of courses in modern astrophysics, with different emphases. I have also designed the contents so that topics can be treated at different levels of difficulty by including qualitative as well as more detailed quantitative treatments of a number of important topics. For example, the interstellar gas may be studied at different depths in Chapter 17, and similarly synchrotron radiation can be tackled either from the physical point of view or in a more quantitative fashion in Chapter 19.

The subject continues to develop at a remarkable pace, and many topics are in a state of flux. As in Volume 1, my aim is to concentrate upon those aspects of high energy astrophysics which are likely to be of lasting importance and which provide useful tools which can be applied in a wide range of different astrophysical circumstances.

Malcolm Longair
Cambridge, UK.

June 1993.

Acknowledgements

Many people have influenced the contents of this volume. Some of the initial spade work was carried out whilst I was at the Royal Observatory, Edinburgh, but most of the major rewriting and updating has been carried out within the last year. Many colleagues have generously provided their advice and data for use in this volume. Among them, it is a special pleasure to thank Tom Faber, Andy Fabian, Douglas Gough, David Green, Alan Heavens, Anthony Hewish, Jim Pringle and Martin Rees. I am also very grateful to Mrs Pam Hicks for retyping many of my drafts into TₑX format to speed the process of publication.

As usual, all errors of fact, interpretation, balance and bias are entirely my responsibility.

A couple of lines are quite inadequate to express how much I owe to the unfailing love and support of my family, Deborah, Mark and Sarah.

13

The contents of the Universe – the grand design

At last, after the 12 chapters which comprise the whole of Volume 1, we leave the Solar System behind and enter the astronomical domain, where we can no longer detect high energy particles directly but can only infer their presence from the radiations they emit. High energy processes are now known to be important in essentially all classes of astronomical object, and so we begin our study with a survey of the contents of the Universe – this will provide the astrophysical context for our study. This will be a very broad-brush description, and should be supplemented by the more specialised texts listed in the Further reading and references section.

13.1   The large-scale distribution of matter and radiation in the Universe

The modern picture of how matter and radiation are distributed in the Universe on a large scale is derived from a wide variety of different types of observation.

13.1.1   The isotropy of the Universe as a whole

On the very largest scale, the best evidence for the overall isotropy of the Universe comes from measurements of the *cosmic microwave background radiation*. This is the intense diffuse radiation observed in the centimetre and millimetre wavebands discovered by Penzias and Wilson in 1965. It is wholly convincing that this radiation is the cooled remnant of the very hot early phases of the Big Bang. The radiation decoupled from the matter when the Universe was only about 1/1000 of its present size, and provides direct evidence for the isotropy of the matter and radiation content of the Universe.

The most important observations of the cosmic microwave background radiation have been made by the Cosmic Background Explorer (COBE), which was launched in November 1989 and which has measured the spectrum and

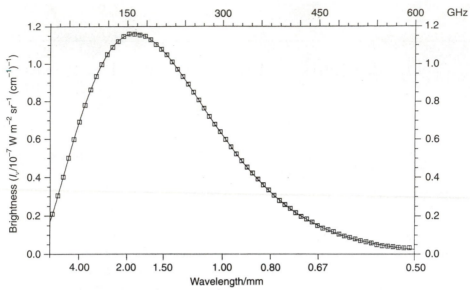

Figure 13.1. The spectrum of the Cosmic Microwave Background Radiation as measured by the COBE satellite in the direction of the North Galactic Pole. Within the quoted errors, the spectrum is that of a perfect black body at radiation temperature 2.735± 0.06 K. (From J.C. Mather, E.S. Cheng, R.E. Eplee Jr., R.B. Isaacman, S.S. Meyer, R.A. Shafer, R. Weiss, E.L. Wright, C.L. Bennett, N.W. Boggess, E. Dwek, S. Gulkis, M.G. Hauser, M. Janssen, T. Kelsall, P.M. Lubin, S.H. Moseley, Jr., T.L. Murdock, R.F. Silverberg, G.F. Smoot and D.T. Wilkinson (1990). *Astrophys. J.*, **354**, L37.)

isotropy of the background with very high precision. The spectrum of the background is very precisely of black-body form (Fig. 13.1) with radiation temperature $T = 2.735 \pm 0.06$ K. In fact, this is a conservative estimate of the error, the error boxes shown in Fig. 13.1 corresponding to 1% of the peak intensity. The more recent analyses reported by Mather (1994) have shown that the spectrum is a perfect black body at wavelengths longer than 500 μm at intensity levels 0.03% of the peak intensity. This is the most beautiful example of which I am aware, of a naturally occurring black-body radiation spectrum.

A second major result of the COBE mission is the measurement of the isotropy of the Microwave Background Radiation. The radiation has the same intensity to better than 1 part in 1000 wherever one looks at the sky on all angular scales from about 1 arcmin to 360 degrees. Just below this sensitivity level, a global anisotropy of dipole form is observed so that the radiation intensity is slightly greater in the direction $\alpha = 11^{\text{h}}.3$; $\delta = -7^{\circ}.5$ and slightly less in the opposite direction (Fig. 1.8(*b*)). This dipole anisotropy can be wholly attributed to the Doppler effect associated with the motion of the Earth at a velocity of about 350 km s^{-1}, through a frame of reference in which the Cosmic Microwave Background Radiation would be 100% isotropic on the large scale.

One of the most important recent discoveries for cosmology has been the detection of tiny intensity fluctuations in the distribution of the Cosmic Microwave

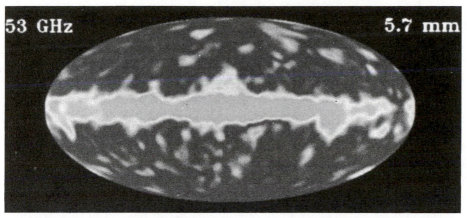

Figure 13.2. The map of the whole sky in galactic coordinates as observed in the millimetre waveband at a wavelength of 5.7 mm by the COBE satellite, once the dipole component associated with the motion of the Earth through the background radiation has been removed (for comparison, see Fig. 1.8(*b*)). The residual radiation from the plane of the Galaxy can be seen as a bright band across the centre of the picture. The fluctuations see at high galactic latitudes are noise from the telescope and the instruments, the rms value at each point being 36 μK, but, when statistically averaged over the whole sky at high latitudes, an excess sky noise signal of 30 $\pm$ 5 μK is observed (From G.F. Smoot, C.L. Bennett, A. Kogut, E.L. Wright, J. Aymon, N.W. Boggess, E.S. Cheng, G. DeAmici, S. Gulkis, M.G. Hauser, G. Hinshaw, C. Lineweaver, K. Loewenstein, P.D. Jackson, M. Janssen, E. Kaita, T. Kelsall, P. Keegstra, P. Lubin, J.C. Mather, S.S. Meyer, S.H. Moseley, T.L. Murdock, L. Rokke, R.F. Silverberg, L. Tenorio, R. Weiss and D.T. Wilkinson (1992). *Astrophys. L. Lett.*, **396**, L1.)

Background Radiation over the sky from careful analysis of the COBE intensity maps. Fig. 13.2 shows the COBE map of the whole sky at a wavelength of 5.7 mm, once the dipole component of the background has been subtracted – this map shows the intensity distribution at a sensitivity level of about 1/100000 of the total intensity. The bright band of radiation stretching along the centre of the diagram is emission from the plane of our own Galaxy, and is the bremsstrahlung of hot gas and the synchrotron radiation of high energy electrons in the disc of our Galaxy. Away from the Galactic plane, very low amplitude intensity fluctuations can be seen. It is important to realise that the intensity maxima and minima seen on this map are *not* fluctuations in the Cosmic Microwave Background Radiation associated with large-scale structures in the Universe, but rather are simply the noise fluctuations from the telescope and receiving system expected at low intensity levels. A positive signal of cosmological significance is found when these spatial fluctuations are averaged over the whole sky. When this is done, it is found that there is an 'excess noise' signal, over and above instrumental noise, at the level of 30$\pm$5 μK on an angular scale of 10°, corresponding to the detection of real cosmological fluctuations at a significance level of about 6σ – the fluctuations correspond to fractional intensity fluctuations $\Delta I/I \approx 10^{-5}$. An angular scale of

$10°$ corresponds to very large physical scales in the Universe now, about ten times larger than the largest scales discussed below. The detection of this signal is of the greatest interest for cosmology (see, for example, Longair 1989), but, for our present purposes, the important point is that the Universe as a whole is known to be the same in all directions on angular scales $10°$ and greater, with a precision of about one part in 10^5. This is a quite remarkable result for cosmology.

One important aspect of these studies for high energy astrophysics is that this isotropic background radiation corresponds to an energy density of radiation of $\epsilon_{rad} = aT^4 = 4.2 \times 10^{-14}$ J m^{-3} = 2.64×10^5 eV m^{-3}, present throughout the whole Universe at the present epoch. This energy density of radiation pervades the whole Universe, and provides by far the greatest contribution to the universal background radiation.

When we look at the distribution of *visible matter* in the Universe, we observe a very different picture. Fig. 13.3 shows the distribution of visible matter derived from counts of galaxies made by workers at the Lick Observatory in California, USA, in the 1960s. The processed data provide a view of the whole of the Northern Galactic Hemisphere, that is, the view looking up out of the plane of our own Galaxy (for a brief discussion of coordinate systems in astronomy, see Appendix A4). Towards the centre of this picture, we obtain a reasonably unobscured view of the distribution of galaxies. It can be seen that there are irregularities in this distribution, although, looked at on a large scale, one region looks very much like another. There is no question about the reality of a great deal of this structure. The dense knots correspond to regions of strong clustering of galaxies, for example, the bright region towards the centre of the picture corresponding to the well-known Coma cluster of galaxies.

What has provoked a great deal of interest is the nature of the other structures on large angular scales apparent in this picture. To the eye, there appear to be holes and filaments in the distribution of visible matter. One of the obvious problems in interpreting Fig. 13.3 is that it is a two-dimensional representation of the distribution of the galaxies, and much more would be learned if their distances were known as well. The most detailed picture of the distribution of galaxies in the nearby Universe has been provided by Geller, Huchra and their colleagues at the Harvard-Smithsonian Center for Astrophysics. They have almost completed a huge survey of a statistical sample of nearly 15000 galaxies, for which precise radial velocities (or redshifts) have been measured. These data have resulted in the best three-dimensional picture of the large-scale distribution of galaxies in the local Universe to date (Fig. 13.4). As described below, because of the expansion of the Universe, the distances of galaxies participating in the universal expansion can be found from Hubble's law, $v = H_0 r$. This has been used to plot the 'distances' of the galaxies in Fig. 13.4. Galaxies which are members of rich clusters of galaxies have large velocities within the cluster, the typical velocity dispersion being about 10^3 km s^{-1}. These systems can be recognised as the 'fingers' pointing radially towards our Galaxy, which is located at the centre of the diagram. The radius of the bounding circle is 150 Mpc, assuming

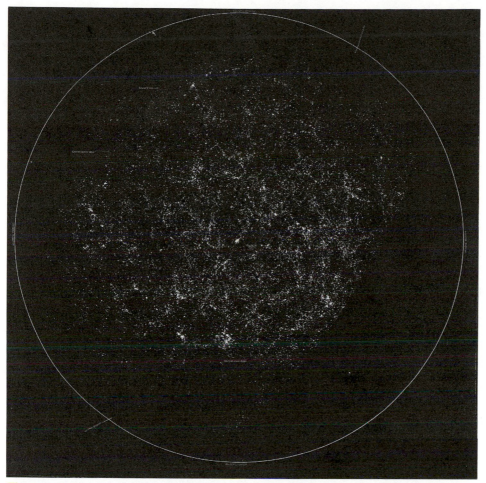

Figure 13.3. The distribution of galaxies in the Northern Galactic Hemisphere, derived from counts of galaxies undertaken by Shane, Wirtanen and their colleagues at the Lick Observatory in the 1960s. Over 10^6 galaxies were counted in their survey. The Northern Galactic Pole is at the centre of the picture, and the Galactic equator is represented by the solid circle bounding the diagram. The projection of the sky onto the plane of the picture is an equal area projection. This photographic representation of the galaxy counts was made by Peebles and his colleagues. The large sector missing from the lower right-hand corner of the picture corresponds to an area in the Southern Celestial Hemisphere, which was not surveyed by the Lick workers. The decreasing surface density of galaxies towards the circumference of the picture, that is, towards the galactic equator, is due to the obscuring effect of dust in the interstellar medium of our own Galaxy. The prominent cluster of galaxies close to the centre of the picture is the Coma cluster. (From M. Seldner, B. Siebars, E.J. Groth and P.J.E. Peebles (1977). *Astron. J.*, **82**, 249.)

Hubble's constant to be 100 km s^{-1} Mpc^{-1}. The striking features of Fig. 13.4 are the gross large-scale irregularities. On the one hand, there are filaments and 'sheets' of galaxies, including the feature known as the 'Great Wall', which runs

Right ascension

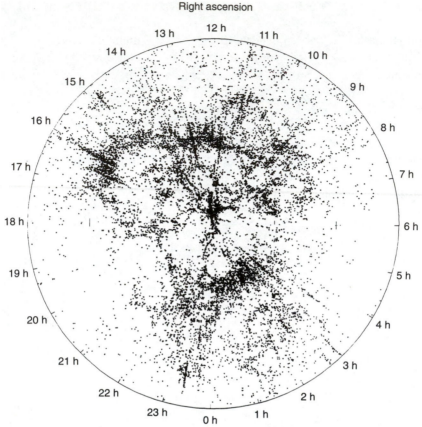

Figure 13.4. The distribution of galaxies in the nearby Universe, as derived from the Harvard-Smithsonian Center for Astrophysics survey of galaxies. The map contains over 14 000 galaxies, which form a complete statistical sample around the sky between declinations $\delta = 8°.5$ and $44°.5$. All the galaxies have recession velocities less than 15 000 km s^{-1}. Our Galaxy is located at the centre of the map, and the radius of the bounding circle is 150 Mpc, assuming Hubble's constant to be 100 km s^{-1} Mpc^{-1}. The galaxies within this slice around the sky have been projected onto a plane to show the large-scale features in the distribution of galaxies. Rich clusters of galaxies which are gravitationally bound systems, with internal velocity dispersions of about 10^3 km s^{-1}, appear as 'fingers' pointing radially towards our Galaxy. It can be seen that the distribution of galaxies is highly irregular, with huge holes, filaments and clusters of galaxies occurring throughout the local Universe. (Courtesy of Margaret Geller, John Huchra and the Harvard-Smithsonian Center for Astrophysics (1992).)

from 9^h to 17^h, roughly halfway to the bounding circle; on the other hand, there are huge voids in which the number density of galaxies is significantly depressed relative to the mean value. The clusters and groups of galaxies are always found within the sheets and filaments. The voids have sizes up to about 100 Mpc, and yet the filaments can be very thin. The random velocities within the sheets and filaments can be as small as 300 km s^{-1}. The filaments and sheets of galaxies are the largest known physical structures in the Universe, but they are still only about one-tenth of the scale on which the fluctuations in the Cosmic Microwave Background Radiation have been detected by the COBE satellite.

It is useful to have a physical picture for the nature of these structures. The topological studies of Gott and his colleagues show that we can think of the distribution of galaxies as being sponge-like (Gott, Melott and Dickinson 1986; Melott, Weinberg and Gott 1988). The material of the sponge represents the location of the galaxies, and the holes in the sponge represent the large voids seen in the distribution of galaxies. The intriguing aspect of this topology is that the material of the sponge is all joined together, and, in addition, the holes are all joined together too – this configuration is possible in three dimensions, but not in two. This form of topology has implications for the nature of the fluctuations out of which these structures must have formed. One of the biggest challenges in astrophysical cosmology is to reconcile the remarkable smoothness of the Cosmic Microwave Background Radiation (Fig. 13.2) with the gross irregularity in the large-scale distribution of visible matter (Fig. 13.4).

Further evidence on the isotropy of the distribution of galaxies on a very large scale is provided by the distribution of the brightest extragalactic radio sources over the sky (Fig. 13.5). These radio sources are almost exclusively associated with distant radio galaxies and quasars, and thus they sample a smaller but more distant population of galaxies than those mapped in Fig. 13.3. The sample of 31 000 bright radio sources in Fig. 13.5 covers most of the Northern Celestial Hemisphere, and consists of radio galaxies and quasars typically at redshifts about one. This means that these objects sample the distribution of galaxies on the largest scales possible at roughly the present epoch in the Universe. The distribution of points in Fig. 13.5 shows no statistical departure from uniformity on any scale except for a small excess associated with the Galactic plane; the precision with which this statement can be made is simply limited by the number of sources counted. Notice that this measure of the isotropy of the Universe refers to discrete objects. In contrast, the Cosmic Microwave Background provides information about the isotropy of the pregalactic plasma and the distribution of radiation on the last scattering surface, which occurred at a redshift of about 1000 when the Universe was of the order of $1/30 000$ of its present age.

On finer scales, the clustering of galaxies appears to occur on a very wide variety of scales, from pairs and small groups of galaxies, like the Local Group of galaxies, to giant *clusters* of galaxies, such as the Coma or Pavo clusters which contain many thousands of members (Fig. 13.6). The rich regular clusters are self-gravitating bound systems, but there are also irregular clusters which have an irregular, extended appearance, and it is not so clear that these are bound systems (see Volume 3).

The term *supercluster* is used to define structures on scales larger than those of clusters of galaxies. Superclusters may consist of associations of clusters of galaxies, or a rich cluster with associated groups and an extended distribution of galaxies. Some authors would classify the 'stringy' structures seen in Figs. 13.3

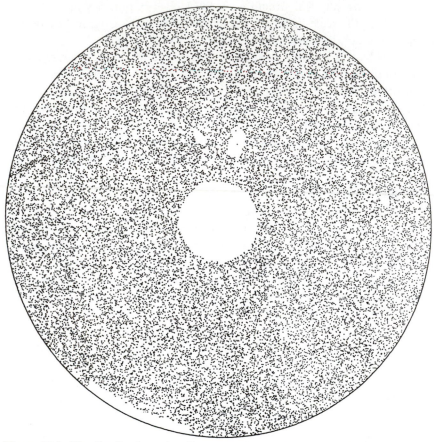

Figure 13.5. The distribution of radio sources in the Greenbank Catalogue of radio sources at 6 cm. (From P.C. Gregory and J.J. Condon (1991). *Astrophys. J. Suppl.*, **75**, 101.) The picture includes 31 000 radio sources. In this equal area projection, the North Celestial Pole is in the centre of the diagram and the Celestial Equator around the solid circle. The area about the North Celestial Pole was not surveyed. There are 'holes' in the distribution about the bright sources Cygnus A and Cassiopeia A and a small excess of sources associated with the Galactic plane. Otherwise, the distribution does not display any significant departure from a random distribution. (From P.J.E. Peebles (1993). *Principles of physical cosmology.* Princeton: Princeton University Press.)

and 13.4 as superclusters, or supercluster cells. From the physical point of view, the distinction between the clusters and the superclusters is whether or not they are gravitationally bound. Even in the rich, regular clusters of galaxies, which have had time to relax to a state of statistical equilibrium, there has only been time for individual galaxies to cross the cluster up to about ten times during the lifetime of the Universe, and so, on larger scales, there is scarcely time for the systems to become gravitationally bound. Our own Galaxy, and the Local Group of galaxies to which it belongs, are members of what is known as the *Local Supercluster*. This is a huge flattened distribution of galaxies centred on the

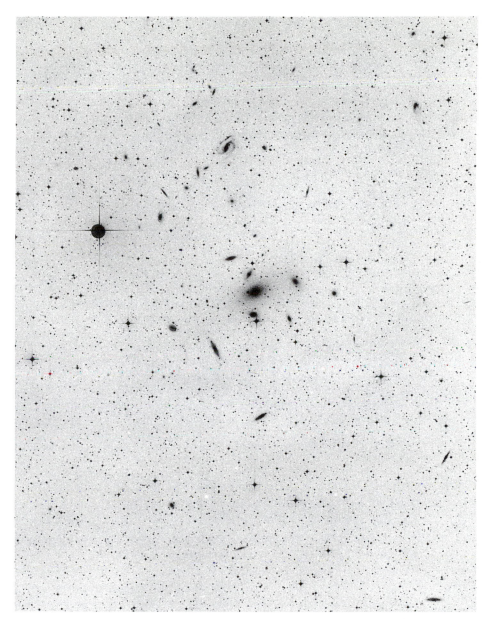

Figure 13.6. The rich cluster of galaxies in the constellation of Pavo in the southern hemisphere. The central galaxy is a supergiant, or cD galaxy, which is very much brighter than all the other galaxies in the cluster. It is located close to the dynamical centre of the cluster. (Courtesy of the Royal Observatory, Edinburgh.)

Virgo cluster, which lies at a distance of 15–20 Mpc from our own Galaxy. It can be seen very prominently in maps of the distribution of bright galaxies, running more or less perpendicular to the plane of the Galaxy, and is the feature close to the centre of Fig. 13.4.

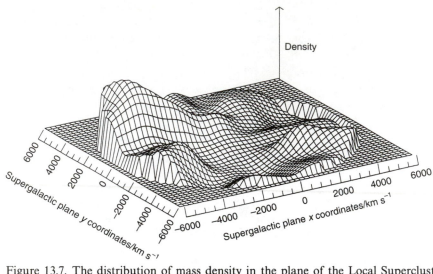

Figure 13.7. The distribution of mass density in the plane of the Local Supercluster, reconstructed using the procedures developed by E. Bertschinger, A. Dekel and their colleagues. The coordinates are distances in the supergalactic plane measured in kilometres per second, corresponding to their recessional velocities in a uniform Hubble flow. Our Galaxy and the Local Group of galaxies lie at the centre of the diagram. The large hump to the left of the diagram is the mass excess associated with the 'Great Attractor', which is associated with the Hydra–Centaurus supercluster. The fractional excess density (or density contrast) in the Great Attractor corresponds to $\delta\rho/\rho = 1.2 \pm 0.4$, if the density parameter $\Omega = 1$. ρ is the average density of matter throughout the whole region. (E. Bertschinger, A. Dekel, S.M. Faber, A. Dressler and D. Burstein (1990). *Astrophys. J.*, **364**, 370.)

Notice that, in this discussion of the large-scale structure of the Universe, I have been careful to refer to the distribution of *visible* matter, because it is important to distinguish between this and the actual distribution of all the gravitating mass in the Universe. Measurements of the masses of giant spiral and elliptical galaxies, as well as of clusters of galaxies such as the Pavo cluster seen in Fig. 13.6, show that there must be about 10–20 times more mass present in these systems than would be inferred from their visibile light – this is the well-known *dark matter problem*.

A remarkable analysis of this problem for the local Universe has been under-taken by E. Bertschinger, A. Dekel and their colleagues. Using only the velocities and distance estimates for local galaxies, they have devised a numerical scheme for finding self-consistent solutions for the local mass and velocity distributions on a large scale. An example of their solutions for the local density distribu-tion is shown in Fig. 13.7, which is a two-dimensional cut through the Local Supergalactic plane. What is particularly elegant about this approach is that the derived density distribution is independent of the number density of galaxies – the galaxies simply act as tracers of the velocity field in the local Universe. It turns out that the large-scale features seen in the density distribution in Fig. 13.7

can be associated with the large-scale superclustering of galaxies. For example, the large density enhancement to the left of the figure is associated with the *Great Attractor*, and the other density peaks can be associated with other regions of superclustering. The regions where there are large holes or voids in the galaxy distribution agree well with low density regions in Fig. 13.7. In general terms, the conclusion of this work is that the total gravitating mass seems to follow quite closely the distribution of visible mass, but there must be about 10 to 50 times more mass present than would be predicted from the optical light of galaxies. The presence of so much dark matter on a large scale has profound implications for the theory of the origin and stability of galaxies, clusters and other large-scale structures, and quite possibly for fundamental physics.

There is no agreement about the nature of the dark matter. The observational problem is that any forms of matter which emit or absorb very little radiation are very difficult to detect. Thus, some of the dark matter could well be in *baryonic* form – rocks, planets, or 'brown dwarfs', which are stars which are not massive enough, and consequently not hot enough in their interiors, to convert hydrogen into helium (see Section 13.3). Limits to the amount of such baryonic matter which can be present in the Universe can be found from arguments concerning primordial nucleosynthesis. The dark matter could be in some much more exotic form – for example, black holes, massive neutrinos, or ultraweakly interacting particles as yet unknown to particle physics. The last idea is of particular interest to the particle physicists who use the early Universe as a laboratory in which theories of elementary particles can be tested. This theme of the dark matter haunts much of modern astrophysics and cosmology.

13.1.2   *Hubble's law and the expansion of the Universe*

In 1929, Hubble made his great discovery that the Universe of galaxies is not static but is in a state of uniform expansion (Hubble 1929). The basis for this result was the observation that all galaxies are receding from our own Galaxy, and that the further away a galaxy is from us, the greater its velocity of recession v, that is,

$$v = H_0 r \tag{13.1}$$

where r is the distance of the galaxy and H_0 is a constant, appropriately known as *Hubble's constant*. The observational basis for this relation is shown in Fig. 13.8; the galaxies plotted are the brightest galaxies in clusters which are found empirically to have more or less the same intrinsic luminosities. Because of this constancy of their intrinsic luminosities, the observed flux densities (or apparent magnitudes) provide distance measures for the galaxies. The velocity is measured from the shift of the spectral lines in the galaxy's spectrum to longer wavelengths, this shift being interpreted as a Doppler velocity shift. It is known as the *redshift*, z, of the galaxy, and is defined by

$$z = \frac{\lambda_{\mathrm{obs}} - \lambda_{\mathrm{em}}}{\lambda_{\mathrm{em}}} \tag{13.2}$$

where λ_{em} is the emitted wavelength of the line and λ_{obs} is the wavelength at

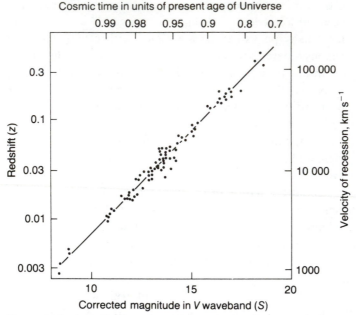

Cosmic time in units of present age of Universe

Figure 13.8. A modern version of the velocity–distance relation (or Hubble diagram) for the brightest galaxies in clusters. (After A.R. Sandage (1968). *Observatory*, **88**, 99.) In this logarithmic plot, the corrected apparent magnitudes (that is −2.5 times the logarithm of the flux density S) of the brightest galaxies in clusters are plotted against their redshifts, z. The straight line shows what would be expected if $S \propto z^{-2}$. This correlation indicates that the brightest galaxies in clusters have remarkably standard properties, and that the distances of the galaxies are proportional to their redshifts, which, for small redshifts, implies that velocity is proportional to distance.

which it is observed. For velocities much less than the velocity of light, $v \ll c$, the relativistic formula for the redshift $(1 + z) = [(1 + v/c)/(1 - v/c)]^{1/2}$ reduces to $z = v/c$. Fig. 13.8 is one of the most remarkable correlations in astronomy – the points lie closely clustered about the line $v \propto r$.

Redshifts can be measured very accurately, and hence, because of Hubble's law (13.1), relative distances of extragalactic objects can also be measured accurately. One of the great problems of observational cosmology is the measurement of accurate distances which are *independent of redshift*. This is necessary in order to calibrate the velocity–distance relation, that is, to determine the value of Hubble's constant, H_0. There is no general agreement about the best value of Hubble's constant to use in cosmological calculations. There are two schools of thought, one favouring values of H_0 in the range 45–60 km s^{-1} Mpc^{-1}, the other adopting somewhat larger values in the range 75–100 km s^{-1} Mpc^{-1}. The discrepancy results from the fact that different methods of estimating distances are employed. My own view is that the discrepancies simply reflect the considerable observational difficulty of measuring accurate distances. In view of this uncertainty, it is common practice to write Hubble's constant as $H_0 = 100h$ km s^{-1} Mpc^{-1}, where

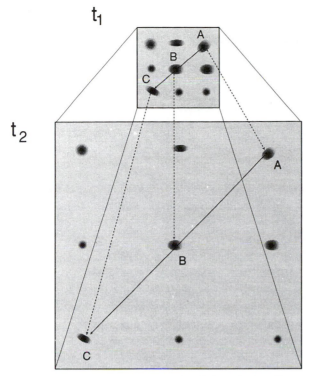

Figure 13.9. Illustrating the origin of the velocity–distance relation for an isotropically expanding distribution of galaxies. The distribution of galaxies expands uniformly between the epochs t_1 and t_2. If, for example, we consider the motions of the galaxies relative to the galaxy A, it can be seen that galaxy C travels twice as far as galaxy B between the epochs t_1 and t_2 and so has twice the recession velocity of galaxy B relative to A. Since C is always twice the distance of B from A, it can be seen that the velocity–distance relation is a general property of isotropically expanding Universes.

the variable $h = 1$ if one adopts $H_0 = 100$ km s^{-1} Mpc^{-1} and $h = 0.5$ if $H_0 = 50$ km s^{-1} Mpc^{-1}. For the purposes of illustration, I will normally adopt $h = 0.5$.

The expansion of the distribution of galaxies as a whole follows directly from Hubble's law and the overall isotropy of the Universe discussed in Section 13.1.1. If the distribution of galaxies is to remain isotropic and a velocity–distance relation observed, the distribution of galaxies as a whole has to expand uniformly. Let us demonstrate this important conclusion by a simple calculation. Consider a uniformly expanding system of points (Fig. 13.9). Then, the definition of a uniform expansion is that the distances to any two points should increase by the same factor in a given time interval, that is, we require

$$\frac{r_1(t_2)}{r_1(t_1)} = \frac{r_2(t_2)}{r_2(t_1)} = \dots = \frac{r_n(t_2)}{r_n(t_1)} = \dots = \alpha = \text{constant}$$

for any set of points. The recession velocity of galaxy 1 from the origin is therefore

$$v_1 = \frac{r_1(t_2) - r_1(t_1)}{t_2 - t_1} = \frac{r_1(t_1)}{t_2 - t_1}\left[\frac{r_1(t_2)}{r_1(t_1)} - 1\right] = \frac{r_1(t_1)}{t_2 - t_1}(\alpha - 1) = H_0 r_1(t_1)$$

Similarly,

$$v_n = \frac{r_n(t_1)}{t_2 - t_1}(\alpha - 1) = H_0 r_n(t_1)$$

Thus, a uniformly expanding distribution of galaxies automatically results in a velocity–distance relation of the form $v \propto r$.

13.1.3   *World models*

This is certainly *not* a text-book on cosmology, but it is helpful to derive some of the useful results which we will return to at various stages in our story. Fortunately, a number of important results can be derived from simple arguments using Newton's theory of gravity rather than Einstein's general theory of relativity. The reason this can be done is that the natural starting point for the development of the standard world models is the construction of uniform, isotropically expanding universes, as we deduced in Section 13.1.2. Because of the assumption of isotropy, the physics of the expansion must be the same everywhere in the Universe, including within local volumes. Therefore, local physics also becomes global physics.

To derive the dynamics of the standard world models, we therefore carve out of the Universe a uniformly expanding sphere of matter, and consider its dynamics. We locate our observer at the centre of the sphere, so that the sphere expands uniformly with $v \propto r$ about that point in all directions (Fig. 13.10). Because the sphere is uniform and finite, we can work out the deceleration of a test particle or galaxy of mass m_g located at the surface of the sphere using Gauss's theorem. This allows us to replace the sphere of matter by a point mass $M = \frac{4\pi}{3}r^3\rho$ at the origin, where r is the radius of the sphere and ρ is the density of matter in the sphere. Then, the force acting upon the galaxy or test particle is

$$m_g\frac{d^2 r}{dt^2} = -\frac{GMm_g}{r^2} = -\frac{4\pi Gm_g\rho}{3}r \tag{13.3}$$

We note that the mass of the test particle m_g cancels out on either side of the equation, so that this equation tells us something about the dynamics of the Universe as a whole and not about test particles.

Now, the Universe is expanding as a whole, and so let us change our radial variables by writing distances relative to their values at some reference epoch t_0. We write $r = R(t)r_0$, where r_0 is the distance of the galaxy at our reference epoch, which we take to be the present epoch t_0, and $R(t)$ is known as the *scale factor* which describes how the relative distance between any two points in the Universe changes with time. Now, the density of the Universe decreases as it expands, and so, if ρ_0 is its density at the present time t_0, at some other epoch t, the density is $\rho = \rho_0 R^{-3}$. Making these substitutions into equation (13.3), we find

$$\frac{d^2 R}{dt^2} = -\frac{4\pi G\rho_0}{3}R^{-2} \tag{13.4}$$

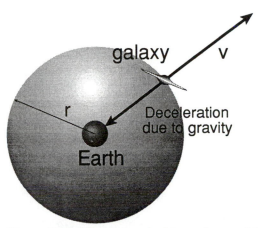

Figure 13.10. Illustrating the Newtonian model for the dynamics of the classical Friedman models of general relativity. The velocity of recession of a galaxy at distance r can be considered to be decelerated by the gravitational attraction of the matter within distance r of our own Galaxy. Because of the assumption of isotropy, an observer on any galaxy participating in the expansion of the Universe would carry out exactly the same calculation.

Let us take the first integral of this equation by multiplying by dR/dt and integrating with respect to time:

$$\left(\frac{dR}{dt}\right)^2 = \frac{8\pi G\rho_0}{3}\frac{1}{R} + C \tag{13.5}$$

where C is a constant. The solutions of this equation describe the standard Friedman models of general relativity, which are illustrated in Fig. 13.11. We can understand these solutions by considering the behaviour of equation (13.5) as the scale factor R tends to infinity. If C is positive, the Universe expands to infinity and (dR/dt) remains positive; if C is negative, the Universe does not expand to infinity, but (dR/dt) goes to zero at some finite value of R, after which collapse ensues, the dynamics being the exact mirror image of the expansion phase. The case $C = 0$ is known as the *critical model*, in which the expansion velocity tends to zero as the time tends to infinity, that is, the Universe has exactly its own escape velocity.

Let us work out the dynamics and density for the critical model. The solution of equation (13.5) for the case $C = 0$ is very simple. First, we note that, if we substitute $r = R(t)r_0$ into Hubble's law, $v = H_0 t$, we find

$$\frac{dr}{dt} = H_0 r \qquad r_0\frac{dR}{dt} = H_0 r_0 R \qquad H_0 = \frac{1}{R}\frac{dR}{dt}$$

Since we have normalised $R(t)$ to be unity at the present epoch $t = t_0$, we can

Scale factor

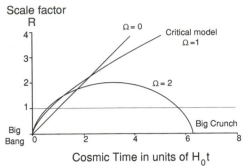

Cosmic Time in units of $H_0 t$

Figure 13.11. The dynamics of the classical Friedman models parameterised by the density parameter $\Omega = \rho/\rho_{\text{crit}}$. If $\Omega > 1$, the Universe collapses to $R = 0$ as shown; if $\Omega < 1$, the Universe expands to infinity and has a finite velocity of expansion as R tends to infinity. In the case $\Omega = 1$, $R = (t/t_0)^{2/3}$, where $t_0 = \frac{2}{3}H_0^{-1}$. The time axis is given in terms of the dimensionless time $H_0 t$. At the present epoch, $R = 1$, and in this presentation the three curves have the same slope of unity at $R = 1$, corresponding to a fixed value of Hubble's constant. If t_0 is the present age of the Universe corresponding to $R = 1$, then, for $\Omega = 0$, $H_0 t_0 = 1$; for $\Omega = 1$, $H_0 t_0 = \frac{2}{3}$; and, for $\Omega = 2$, $H_0 t_0 = 0.57$.

write $H_0 = (dR/dt)_{t_0}$. The solution of equation (13.5) if $C = 0$ is then

$$R = \left(\frac{t}{t_0}\right)^{2/3}$$

where $t_0 = \frac{2}{3}H_0^{-1}$ is the age of the Universe for this model. The density of the Universe is also uniquely defined in this model. Substituting for $(dR/dt)_0$ into equation (13.5) at the present epoch, $t = t_0$, $R = 1$, we find that the *critical density* is

$$\rho_{\text{crit}} = \frac{3H_0^2}{8\pi G} \tag{13.6}$$

This is a very useful reference density for the Universe. The only problem is that it depends upon the value of Hubble's constant, which is not accurately known. For all our purposes, it will be adequate to adopt a value of Hubble's constant of $H_0 = 50$ km s^{-1} Mpc^{-1}, for which $\rho_{\text{crit}} = 5 \times 10^{-27}$ kg m^{-3}; more generally, $\rho_{\text{crit}} = 2 \times 10^{-26} h^2$ kg m^{-3}. The corresponding particle number density, if the Universe were all composed of hydrogen, is $10 h^2$ m^{-3}.

Another very useful piece of cosmological terminology is to define cosmic densities relative to the critical density. Thus, suppose the Universe actually has density ρ. It is then convenient to define a *density parameter* Ω which is the ratio of the density of the Universe to the critical value $\Omega = \rho/\rho_{\text{crit}}$. It is also convenient to introduce density parameters to define the density of different components of the Universe relative to the critical density. For example, the density of visible matter amounts to about 1–2% of the critical cosmological density, and so the density parameter for visible matter is $\Omega_{\text{vis}} \approx 0.01$–0.02. Inspection of equation (13.4) shows that the deceleration of the Universe is directly proportional to the density of matter, and hence is proportional to Ω.

Identical results come out of general relativity, with the additional bonus that the geometry of space is included in the constant C. If C is positive, the spatial geometry of the Universe is hyperbolic, so that, if we sum the angles of a triangle over cosmological distances, the answer would be less than 180°. If C is negative, the geometry is spherical, and the angles of a triangle add up to more than 180°. Only in the case of the critical Universe, $C = 0$, is the geometry flat. Formally, the curvature of the geometry is given by $\kappa = -C/c^2 = 1/\mathscr{R}^2 = (\Omega - 1)/(c/H_0)^2$, where $\mathscr{R}$ is the radius of curvature of the spatial sections. If $\Omega = 1$, $\mathscr{R} = \infty$, and the geometry is flat; if $\Omega > 1$, κ is positive, and the geometry is spherical; whereas if $\Omega < 1$, κ is negative, and the geometry is hyperbolic. These results are discussed in all the standard text-books – my own version is given in *Theoretical concepts in physics* (Longair 1992).

The critical Universe is of particular interest for a number of reasons. First of all, it is the only 'natural' solution to come out of the field equations. Secondly, it turns out that it is the only 'stable' solution of the standard Friedman world models, in the sense that, if Ω differs from one at any stage, the dynamics diverge rapidly from the case $\Omega = 1$, and so the Universe would be very far from the critical model now. Since our Universe is probably within about a factor of ten of the critical model, it must have been remarkably fine-tuned to a value very close to the critical model in its earliest stages. For this reason, theorists argue that the Universe must have $\Omega = 1$. The *inflationary model* of the early evolution of the Universe is an example of how the Universe might evolve naturally into the critical model. It expands exponentially in the early stages, and so evolves very rapidly to a state in which the geometry is flat, corresponding to the critical model.

We will often need typical cosmological distances and timescales. If we adopt $H_0 = 50$ km s^{-1} Mpc^{-1}, a typical cosmological timescale is $1/H_0 = 20 \times 10^9$ years, that is 20 billion years. In the case of the critical world model, its age is just two-thirds of this value, $T_0 = 13 \times 10^9$ years. A typical cosmological distance is $l = c/H_0 \approx 3000/h$ Mpc $= 1.8 \times 10^{26}$ m, if $h = 0.5$.

The density parameter and the deceleration of the Universe are not well determined. As we will discuss in Volume 3, the visible matter in galaxies corresponds to $\Omega_{\text{vis}} \approx 0.01$–$0.02$. There is probably about ten times as much dark matter present in giant galaxies and in clusters of galaxies, which would increase the value of Ω to about 0.1–0.2. It is not known whether or not Ω tends to unity on the very largest scales.

13.2   The galaxies

13.2.1   *Normal galaxies*

Galaxies are the basic building blocks of the Universe. Much of their visible mass is in the form of stars. It is the gravitational pull of the stars on one another which holds a galaxy together, although we have also to take account of the dark matter present in the outer regions of galaxies.

Figure 13.12 (*a*), (*b*). Examples of different types of galaxies. (*a*) The normal spiral galaxy M31 or the Andromeda nebula. (Courtesy of the Mount Palomar Observatory.) (*b*) The barred spiral galaxy NGC 1365. (Courtesy of David Malin and the Anglo–Australian Observatory.)

Figure 13.12 (c), (d). Examples of different types of galaxies. (c) The elliptical galaxies NGC 1399 and 1404, the brightest members of the Fornax cluster of galaxies. Both galaxies show a prominent distribution of globular clusters in their outer regions. (Courtesy of David Malin and the Anglo–Australian Observatory.) (d) The S0 or lenticular galaxy NGC 3115. (Courtesy of David Malin and the Anglo–Australian Observatory.)

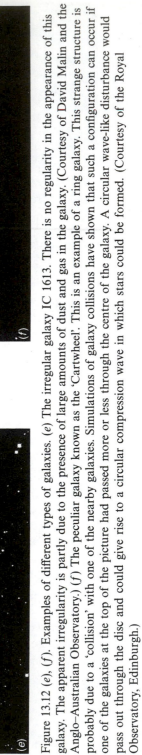

Figure 13.12 (e), (f). Examples of different types of galaxies. (e) The irregular galaxy IC 1613. There is no regularity in the appearance of this galaxy. The apparent irregularity is partly due to the presence of large amounts of dust and gas in the galaxy. (Courtesy of David Malin and the Anglo–Australian Observatory.) (f) The peculiar galaxy known as the 'Cartwheel'. This is an example of a ring galaxy. This strange structure is probably due to a 'collision' with one of the nearby galaxies. Simulations of galaxy collisions have shown that such a configuration can occur if one of the galaxies at the top of the picture had passed more or less through the centre of the galaxy. A circular wave-like disturbance would pass out through the disc and could give rise to a circular compression wave in which stars could be formed. (Courtesy of the Royal Observatory, Edinburgh.)

Many different types of galaxy have been identified, but the basic distinction which is apparent on any photograph is between spiral and elliptical galaxies. A small fraction of galaxies are known as irregular galaxies. The *spiral galaxies* (Fig. 13.12(a)) have a disc shape with a central bulge, like our own Galaxy. The relative sizes of the disc and bulge vary from galaxy to galaxy. Their masses range from systems up to 100 times more massive than our own Galaxy which has mass about 100 billion times the mass of the Sun (10^{11} $M_\odot$), to dwarf systems, which are only about 10 million times the mass of the Sun (10^7 $M_\odot$). The discs of spiral galaxies rotate, and this is what gives them their characteristic flattened shapes. There are *normal* spirals and also *barred* spiral galaxies, in which the central bulge is elongated and the spiral arms trail from the ends of the 'bar' (Fig. 13.12(b)). According to recent analyses, it is likely that the central bulge of our own Galaxy is in the form of an ellipsoidal bar rather than a pure spheroid. The spiral arms are defined by the most massive and luminous stars, which have the shortest lifetimes, and also by the ingredients of regions of star formation, that is, by gas clouds and dust. There is continuing star formation in the arms of spiral galaxies. The *elliptical galaxies* (Fig. 13.12(c)) have much smoother profiles, with little evidence for dust, gas or spiral arms in general. They are spheroidal in shape and can have masses from about 100 times the mass of our Galaxy to only about 10^7 $M_\odot$. They are self-gravitating systems in which the random velocities of the stars prevent the galaxy collapsing.

There are galaxies which appear to be intermediate between the spirals and the ellipticals, which are known as *SO* or *lenticular galaxies* (Fig. 13.12(d)). They look rather like spiral galaxies which have been stripped of their spiral structure, that is, they possess stars in a disc and have a central bulge but have little evidence, or at best vestigal evidence, for spiral arms. Many of them are found in clusters of galaxies.

The *irregular galaxies* (Figs. 13.12(e) and (f)) are generally less massive than typical spiral and elliptical galaxies. They have an irregular structure and often have large amounts of gas and dust. In a number of cases, in particular those referred to as 'peculiar' or 'interacting' galaxies, it is likely that the irregular appearance is due to the galaxy having had a recent strong gravitational interaction or 'collision' with a nearby galaxy.

13.2.2   *Galaxies with active nuclei*

There are a number of special classes of galaxy, most of which are very much less common than the above types. From the point of view of high energy astrophysics, the most interesting are the galaxies with *active galactic nuclei*. The first class of galaxies with active nuclei was recognised in the 1940s by Seyfert, and these are appropriately known as *Seyfert galaxies*. They appear to be spiral galaxies, but they possess star-like nuclei (Figs. 13.13(a) and (b)). When Seyfert took the spectra of the nuclei of these galaxies, he found that the emission lines are very broad and strong, unlike those found in emission line regions in normal galaxies, such as the regions of ionised hydrogen associated with young stars or planetary nebulae. Once it was realised in the 1960s that these distinctive features are the signatures of active galactic nuclei, many more galaxies of this type were discovered, and they are among the most important classes of active nuclei because they are relatively common. Because their nuclei are relatively bright, they are ideal objects for detailed spectroscopic studies.

The next class of galaxies with active nuclei to be discovered were the *radio galaxies*. By the mid-1950s, it was established that these galaxies must be sources of vast fluxes of high energy particles and magnetic fields. A few of these galaxies had prominent star-like nuclei and were called *N-galaxies*. They are similar to the Seyfert galaxies, and also had strong broad emission lines in their optical spectra, but the relation between these phenomena was not clear at that time.

In the early 1960s, the first *quasars* were discovered. Among the optical objects which were associated with the bright radio sources were a few 'stars' which had quite unintelligible spectra. The radio source 3C 273 (Fig. 1.9) was one of those which had been found in early radio surveys of the sky. The optical counterpart of 3C 273 looked exactly like a star on a photographic plate, and, in addition, it was found to be variable in brightness. The remarkable discovery made by Maarten Schmidt in 1962 was that 3C 273 lies at a distance similar to those of the most distant galaxies for which distances could be readily measured at that time. The object was called a *quasi-stellar object* (or *quasar*) because it looked like a star but clearly could not be any normal sort of star at this very great distance. 3C 273 is more than 1000 times more luminous than a galaxy such as our own, and its optical luminosity varies over a period of one to ten years or less. Following this discovery, many more quasars were found, all of them characterised by stellar appearance and very great distances. In addition to those quasars which are strong radio sources, *radio-quiet quasars* were discovered in 1965 – these are just as remarkable objects optically, but are very weak sources of radio emission indeed relative to the *radio-loud quasars*. The quasars are among the most extreme examples of active galactic nuclei known. That they actually are the nuclei of galaxies has been demonstrated by the observation of the underlying galaxies in some of the more nearby examples.

Possibly the most extreme examples of active galactic nuclei are the objects known as BL Lacertae or *BL-Lac objects* and the *optically violently variable (OVV) quasars* or *blazars*. These are similar to the quasars, but they differ from them in two respects. First, they vary extremely rapidly in luminosity, variations being detected on timescales of days or less; they must therefore be very compact. Secondly, in the case of the BL-Lac objects, their optical spectra are normally featureless and the continuum radiation is strongly polarised. It is plausible that, in the BL-Lac objects, we observe more or less directly the primary source of energy in active nuclei.

It should be emphasised that the most extreme examples of active galactic nuclei are very rare indeed. It is because they are so luminous and have such distinctive properties that they can be discovered relatively easily. The astrophysical problem is to understand how they can produce such vast amounts of

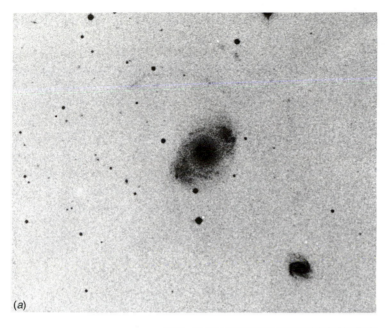

(a)

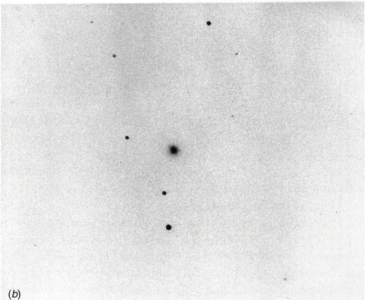

(b)

Figure 13.13. Two images of the nearby Seyfert galaxy NGC 4151. This was one of the galaxies noted by Karl Seyfert in his first catalogue of galaxies with star-like nuclei and broad emission lines. (a) An image of NGC 4151 in a long exposure, showing the inner structure of the disc of the galaxy. In an even longer exposure, spiral arms are clearly visible in the outer regions. (b) A short exposure image of NGC 4151, showing the star-like nucleus of the galaxy. (Courtesy of the Royal Greenwich Observatory.)

energy from very compact volumes. The current consensus of opinion is that these different types of active nuclei are all basically the same. There exists some ultracompact source of energy in these objects, and precisely what one observes depends upon the environment of the compact object and the way in which it obtains its fuel. Even rather quiescent nuclei, like the centre of our own Galaxy, may possess small-scale versions of the phenomena observed in quasars and other active nuclei, although this is still a controversial issue. It is likely that there is a continuity in activity which runs from systems such as the centre of our own Galaxy, through the Seyfert galaxies and N-galaxies, to the quasars and BL-Lac objects. The most convincing models for generating the energy needed to power active galactic nuclei involve the presence of supermassive black holes with masses about 10^6 to 10^{10} $M_\odot$. The many ramifications of these ideas will form a central theme of this and the following volume.

13.3   Stars and stellar evolution

The principal components of normal galaxies are stars, gas and dust. The stars provide much of the mass of the galaxy, and hence are responsible for the self-gravitating forces that bind it into a stable association of stars. The space between the stars is far from empty, and is referred to as the *interstellar medium.* The gas and dust play a key role in providing the material out of which new generations of stars form, and also in providing a temporary resting place for the mass ejected from stars during the course of their evolution and for the debris expelled when stars die. Studies of stars and stellar evolution are central to all astronomy since stars are long-lived objects and are responsible for most of the visible light we observe from normal galaxies. Much of the mass of the Universe is tied up in stars, although precisely what fraction is not precisely known because of uncertainty about the nature and total amount of dark matter present in the Universe.

The study of the stars begins with measurements of the total amount of radiation emitted by a star (its luminosity, L) and also its surface temperature, T. What makes the study of the structure and evolution of stars one of the most exact of the astrophysical sciences is the fact that, although a wide range of combinations of surface temperature and luminosity are found among the stars, most of them lie along certain well-defined loci or branches in the luminosity–temperature diagram. If we plot quantities which are equivalent to luminosity and surface temperature against one another, it is found that the stars occupy quite specific regions of this luminosity–temperature diagram (Fig. 13.14), which is known as a *Hertzsprung–Russell* (or *H–R*) *diagram* or, equivalently, as a *colour–magnitude diagram.* (A brief description of astronomical measures of distance, mass, flux density, luminosities, magnitudes, colours and so on is given in Appendices A1, A2 and A3.) Fig. 13.14 shows where most stars lie. The majority of stars lie along a locus which runs from the bottom right to the top left of this diagram, and which is known as the *main sequence.* What distinguishes stars along this sequence is their mass. The most massive stars lie at the top left

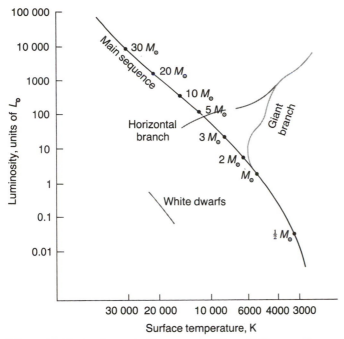

Figure 13.14. A schematic Hertzsprung–Russell diagram for stars. The diagram is also
known as a colour–magnitude diagram or a luminosity–temperature diagram. Most stars
lie in the regions of the diagram as indicated. The names of the various sequences
are shown. The masses and luminosities of stars are quoted in terms of the mass and
luminosity of the Sun, which are written $M_\odot$ and $L_\odot$, respectively.

of the main sequence, and the lowest mass stars are at the bottom right. For stars
with masses in the range 1–$10M_\odot$, this relation can be written $L \propto M^\alpha$, where
$\alpha \approx 4$. The exponent α is smaller for stars with masses greater than $10M_\odot$ and
also for stars less massive than the Sun. Our Sun lies at about the middle of the
main sequence and is a very ordinary star.

Extending from about the location of the Sun on the luminosity–temperature
diagram towards the top right is what is known as the *giant branch*. The stars
in this region of the diagram are large and cool. Extending across the diagram
from the giant branch are stars of what is known as the *horizontal branch*. There
are, in addition, stars which lie significantly below the main sequence; these are
very faint, blue, compact stars known as *white dwarfs*. Fig. 13.15 shows the H–R
diagram for nearby stars determined from recent parallax observations made by
the *Hipparcos* astrometry satellite of the European Space Agency. This satellite
was specifically designed to measure the positions, distances and spatial motions
of 120 000 stars with apparent magnitudes less than about ninth magnitude
with roughly a ten-fold improvement in accuracy as compared with ground-based
measurements. The stars plotted in Fig. 13.15 can be considered a random sample
of stars in the solar neighbourhood. The main sequence can be clearly seen, as

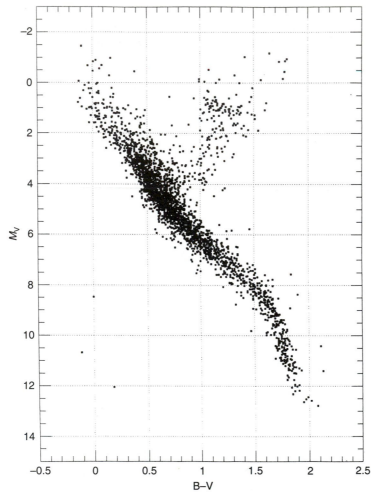

Figure 13.15. The Hertzsprung–Russell diagram for 2927 nearby stars, as determined by the *Hipparcos* astrometric satellite from the first year's data. All stars plotted have parallaxes determined to better than 10 % . Many of these stars had no ground-based photometry, but had their accurate colours and magnitudes first determined by *Hipparcos.* The colour scale is the Tycho photometric scale: $(B–V)_{Tycho} \approx 1.2 \times (B–V)_{Johnson}$. (Courtesy of Dr M.A.C. Perryman and the *Hipparcos* Science Team.)

can a few stars belonging to the giant branch and a few white dwarfs. This diagram is of special importance because these are the best distances available for nearby stars, and they provide a fundamental calibration of the luminosities of nearby stars of different types. The giant branch is much more clearly defined in old star clusters, an example of which is shown in Fig. 14.8.

One of the main goals of the theory of stellar structure and evolution is to understand why it is that stars appear only in certain regions of this luminosity–temperature diagram, and how they evolve from one part to another. Gentle introductions to these ideas are given by Tayler (1974), Shu (1982), Karttunen *et al.* (1987) and Audouze and Israël (1988). Comprehensive discussions of stars and stellar evolution are given by Böhm-Vitense (1989–92) and by Kippenhahn and Weigert (1990), the latter book giving a particularly illuminating discussion of the physical processes which are important at each stage in a star's evolution.

A star is an object in which the force of gravity, which would cause it to collapse, is balanced by the pressure gradient of the hot gas within the star. In all stable stars, this hydrostatic equilibrium is very precisely maintained, the source of energy to maintain the pressure gradient for stars on the main sequence, and the giant and horizontal branches, being nuclear energy generation occurring in their centres.

We consider first the main sequence stars. The most common element in the Universe is hydrogen, and the next most abundant is helium-4 (^{4}He), with a cosmic abundance of about 24% by mass. The abundance of all the heavier elements, including species such as carbon, nitrogen, oxygen and iron, amount to only about 1 to 2% by mass of that of hydrogen. Except for the oldest stars in the Galaxy, most stars have more or less the same chemical abundances. In the centres of main sequence stars, the temperature is sufficiently high for hydrogen to be converted into helium, releasing in the process about 0.7% of the rest mass energy of the hydrogen, which is just the nuclear binding energy of helium. This is the energy source which is ultimately responsible for the light radiated through their surfaces by main sequence stars. Theoretical models of stars, using the best available nuclear reaction rates and opacities of typical stellar material for radiation escaping through the body of the star, can give a convincing quantitative explanation for the observed properties of main sequence stars.

The nuclear reactions by which hydrogen is converted into helium depend upon the temperature. If the central temperature of the star is less than about 2×10^7 K, the proton–proton (p–p) chain reaction is the primary energy source for the star (see Section 14.1); if the temperature is greater than this value, the reaction cycle known as the carbon–nitrogen–oxygen (CNO) cycle is the dominant process (see Section 14.2). The internal structure of the star is crucially dependent upon which of these processes is dominant. For example, if the helium is synthesised by the p–p chain, energy is transported through the central regions by radiation; in contrast, the rate of the CNO cycle is so sensitive to temperature that the energy generating regions are unstable to convection when it is the dominant process, and so the energy is transported by convection to the outer regions of the star.

In the Sun, the central temperature is about 1.6×10^7 K, and the region within which the p–p nuclear chain reactions take place occupies roughly the central 10% of the Sun by radius. The energy diffuses outwards from this central region and is transferred by radiation within the central 70% of the Sun by radius, a process known as radiative diffusion. In the outer 30% of the Sun, energy transfer is by convection rather than by radiation, and these convective motions are believed to be responsible for the remarkable forms of activity observed on the Sun's surface (Fig. 12.5). In stars more massive than about 1.5 $M_\odot$, the CNO cycle is dominant

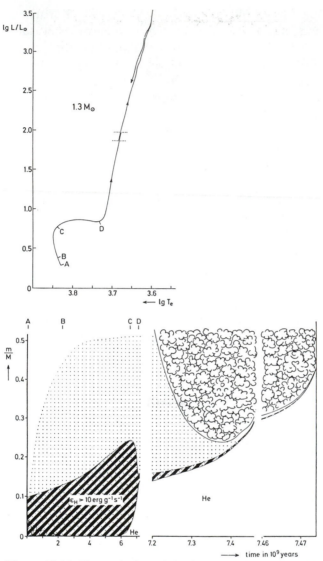

Figure 13.16. The evolution of the internal structure of a $1.3M_\odot$ star, showing how it evolves from the main sequence to the giant branch. The letters A, B, C and D show the structure of the star and the corresponding location of the star on the H–R diagram. The main region of hydrogen burning is indicated by the hatched areas, and the 'cloudy' areas indicate regions which are convective. Notice that the star evolves to the giant branch when roughly 10% of the hydrogen is converted into helium. The diagram illustrates clearly the formation of the extensive outer convective zone when the star moves up the giant branch. Notice also the changing timescale along the abscissa, which illustrates that the star spends most of its lifetime close to the main sequence. The scale on the ordinate is a radial coordinate given in terms of the fractional mass contained within a given radius. (From R. Kippenhahn and A. Weigert (1990). *Stellar structure and evolution*, p. 314. Berlin: Springer-Verlag.)

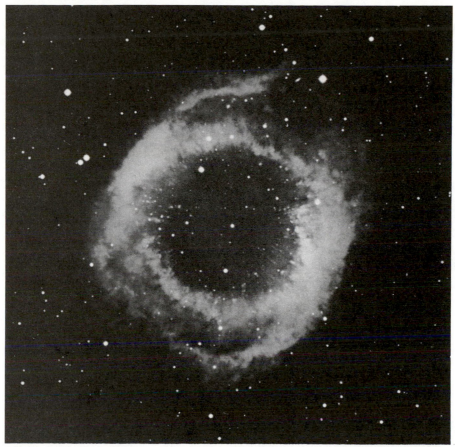

Figure 13.17. The planetary nebula NGC 7293, also known as the Helix Nebula. The shell of hot gas is excited by the compact hot star in the centre. The star is so hot that it is a strong emitter of ultraviolet radiation, which is responsible for the heating, ionisation and excitation of the shell of gas. (Courtesy of David Malin and the Anglo–Australian Observatory.)

and the stars are convective in their centres. The process of converting hydrogen into helium in stars is remarkably stable. Once a star starts burning hydrogen into helium on the main sequence, it remains at roughly that point on the main sequence for almost the whole of its lifetime. A star like the Sun is expected to remain at its location on the main sequence for about ten billion years (10^{10} years).

When the star has converted about 10% of its hydrogen into helium, a limit known as the *Schönberg–Chandrasekhar limit*, it becomes unstable. The core contracts, and its envelope expands to become a giant star. During this process, the star moves rapidly from left to right across the H–R diagram until it reaches a locus known as the *Hayashi limit* at which the star would become totally convective and unstable. A deep convective zone forms in the outer envelope of the star, which then evolves gradually up the giant branch. This instability involves a number of changes in the internal structure of the star. All the evolutionary stages after the star moves off the main sequence take place very much more rapidly than its evolution whilst on the main sequence. This result enables us to work out an approximate age, $t_{\odot}$, for the Sun. The total energy liberated in converting m kg of hydrogen into helium is $0.007mc^2$, and hence, when 10% of the mass of the Sun has been converted into helium, the total energy release is $7 \times 10^{-4} M_{\odot} c^2$. Since the Sun's luminosity has remained roughly constant over its lifetime $t_{\odot}$ on the main sequence, $L_{\odot} t_{\odot} = 7 \times 10^{-4} M_{\odot} c^2$. Inserting $L_{\odot} = 3.9 \times 10^{26}$ W and $M_{\odot} = 2 \times 10^{30}$ kg, we find $t_{\odot} = 10^{10}$ years, in agreement with more detailed calculations.

We can extend this simple calculation to work out the ages of main sequence stars of different masses. Repeating the above calculation for a star of mass M, we find $t(M) \propto L/M \propto M^{-(\alpha-1)}$. Thus, stars with mass $2M_{\odot}$ are expected to have lifetimes about 1.3×10^9 years, whereas stars with mass about five times the mass of the Sun have much shorter lifetimes, about 8×10^7 years. These estimates are in good agreement with much more detailed calculations (see, for example, Fig. 13.19). It is of particular importance that the most massive stars, $M \geq 10 M_{\odot}$, have very short lifetimes, $t(M) \leq 10^7$ years, compared with the age of galaxies, which are of the order of 10^{10} years.

In a star like the Sun, once the nuclear fuel in the core is exhausted, hydrogen continues to be burned in a shell about the helium core. The various stages involved in the evolution of the internal structure of a $1.3 M_{\odot}$ star are illustrated in Fig. 13.16, which also shows how these changes are reflected in the location of the star on the H–R diagram. The star eventually ends up at the top right of the giant branch, an area occupied by long–period variable and unstable stars. When the whole star eventually becomes unstable at this late stage in its evolution, the outer layers are blown off, producing the characteristic *planetary nebula* phase of the star's evolution (Fig. 13.17), and the core of the star collapses to form a very hot helium star, which then cools and ends up as a white dwarf, to the lower left of the main sequence. This evolution is shown schematically in Fig. 13.18.

For stars on the main sequence, the central temperature is roughly proportional to the mass of the star, and so, in higher mass stars, the CNO cycle dominates. In Fig. 13.19, the diagram corresponding to Fig. 13.16 is shown but now for a metal-rich $5M_{\odot}$ star. The heavily hatched areas indicate the regions in which there is large nuclear energy production. It can be seen that, after the star leaves the main sequence, hydrogen burning continues in a shell about the helium core, and then the triple-α reaction begins to convert helium into carbon in the core. This is followed by helium shell burning, which takes place about an inert carbon–oxygen core. As the different types of nuclear burning are initiated, the star makes excursions from the giant branch to higher surface temperatures and then returns to the giant branch. Notice the formation of the huge convective envelope about the region of helium shell burning. In more massive stars, the nuclear processes in the core can proceed through to carbon and oxygen burning to produce elements such as silicon, which can then be burned through to iron,

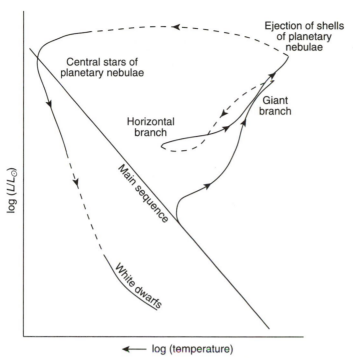

Figure 13.18. A schematic evolutionary track for a low-mass star belonging to a globular cluster ($M \approx M_\odot$). Regions of rapid evolution which may involve mass loss, and in which the evolution is not so well defined, are indicated by dashed lines. Following evolution onto the giant branch, and possibly an excursion onto the horizontal branch, the star ends up at the tip of the giant branch. At this point, strong mass loss takes place and a planetary nebula is formed. The central regions contract to become a hot helium star, which then evolves through to a white dwarf. The observed luminosities and colours of the central stars of planetary nebulae lie along the evolutionary sequence indicated by the solid line at the top left of the diagram. (After D. Michalas and J. Binney (1981). *Galactic astronomy: structure and kinematics*, p. 152. New York: W.H. Freeman and Co.)

the chemical element with the greatest nuclear binding energy. More details of these processes are given in Section 14.3. Theoretical evolutionary tracks for stars of different main sequence masses are shown in Fig. 13.20. Notice that, in constructing these models, it is assumed that there is no mass loss from the surface of the star. We will return to this problem in Section 14.4.

   For stars with masses less than that of the Sun, the central temperatures are lower, and there is a temperature below which the nuclear reactions which burn hydrogen into helium are unable to generate sufficient energy to provide pressure support for the star. The theory of stellar structure suggests that this lower mass limit for hydrogen burning stars is about $0.08 M_\odot$. The search for intrinsically faint stars is thus an important problem for the theory of stellar evolution. On the one hand, it is important to know whether or not this prediction of theory is correct; on the other hand, there is the important question of what happens to

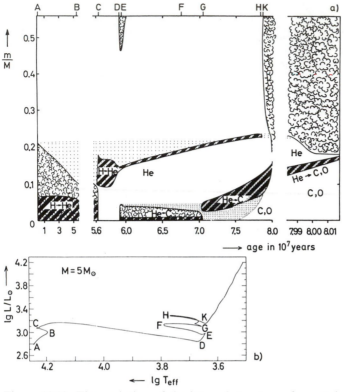

Figure 13.19. The evolution of the internal structure of a metal-rich star of $5M_\odot$, illustrating the synthesis of carbon and oxygen in the core of the star. The abscissa shows the age of the model star after the ignition of hydrogen in units of 10^7 years. The ordinate shows the radial coordinate in terms of the mass m within a given radius relative to M, the total mass of the star. The cloudy regions indicate convective zones. The corresponding positions of the star on the H–R diagram at each stage in its evolution are shown in the bottom diagram. (From R. Kippenhahn and A. Weigert (1990). *Stellar structure and evolution*, p. 294. Berlin: Springer-Verlag.)

those 'stars' which have masses less than $0.08M_\odot$, if indeed they exist at all. We know that planetary sized bodies exist within our Solar System, Jupiter having mass about $0.001M_\odot$. The big problem is that it is very difficult to observe what must be intrinsically very faint, cool objects. They would have to be very close to the Solar System to be observable. Because of their indeterminate colours, these very low mass, cool objects are collectively referred to as *brown dwarfs*. Recently, a few very low mass, cool stars have been discovered, and these are possible candidates for brown dwarfs. Their lifetimes and sources of energy are, however, poorly understood. Brown dwarfs, by which we mean gravitationally bound objects with mass less than about $0.08M_\odot$, are one of the more important types of cold matter which can make a significant contribution to resolving the dark matter problem, to which we return in Volume 3.

We know rather precisely the types of objects which can be formed at the

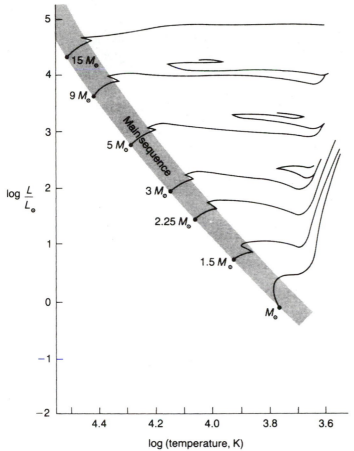

Figure 13.20. Evolutionary tracks for stars of different initial masses on the Hertzsprung–Russell diagram. These models assume that there is no mass loss. Most of the lifetime of the star is spent on the main sequence, and the subsequent phases of evolution on the giant branch take place over much shorter timescales. (After D. Michalas, and J. Binney (1981). *Galactic astronomy: structure and kinematics*, p. 139. New York: W.H. Freeman and Co.)

end of a star's lifetime. There are three types of 'dead star'. In all three cases, there is no longer any nuclear generation of energy. One form of dead star is a *white dwarf*, in which internal pressure support is provided by electron degeneracy pressure; a white dwarf has mass roughly equal to the mass of the Sun or less. A second possible end point is as a *neutron star*. In this case, internal pressure support is provided by neutron degeneracy pressure. These stars are very compact indeed, having masses about the mass of the Sun and radii about 10 km. They have been found in two ways. In the first, they are the parent bodies of radio *pulsars*, which are rotating, magnetised neutron stars. These are observed to emit very intense pulses of radio emission once per rotation period (which is about one second). In the second case, they are the compact 'invisible' secondary stars in *binary X-ray sources*, in which the X-rays are produced by matter falling from the normal primary star onto the neutron star, a process known as *accretion*. The third possibility is that the star collapses to a *black hole*. We will show in Chapter 15 that white dwarfs and neutron stars cannot have masses greater than about $3M_\odot$, and that, for greater masses, the only stable configuration is as a black hole. Thus, it is expected that, if a massive stellar core with mass $M \geq 3M_\odot$ collapses, a black hole will form, unless there is some means by which it can lose mass effectively so that a stable neutron star or white dwarf can be formed. White dwarfs are the end point of the evolution of the cores of stars with masses $M \sim M_\odot$. Stars with masses greater than about 3–4 $M_\odot$ probably do not end up forming white dwarfs but end catastrophically in supernova explosions, in which neutron stars or black holes are formed. It is likely that black holes are the remnants of the most massive stars with $M \geq 10M_\odot$. We will discuss these arguments and the properties of these forms of dead star in Chapter 15 – they are central to high energy astrophysics.

13.4   The interstellar medium
13.4.1   *The general interstellar medium*

One of the areas of astrophysical research which has changed out of all recognition over the last 20 years has been the study of the medium between the stars – the interstellar medium. Originally, it was thought that the interstellar medium was a rather simple, quiescent medium, but it is now clear, from a wide range of different types of observation, that there are many phases and components in the interstellar medium. The medium has four main constituents: gas in all its phases (that is, atomic, molecular and ionised gas), very high energy particles, dust and magnetic fields. It is not stationary, but is constantly being buffetted by winds blowing from stars, by stellar explosions and by large-scale perturbations, such as the gravitational influence of spiral arms or interactions with companion galaxies. It is continually being depleted by the formation of stars and replenished by the various forms of mass loss described in Section 14.3. Let us review briefly the various components of the interstellar medium; we will return to these in more detail in Chapter 17.

Gas The coolest components are the *giant molecular clouds*, which are present throughout the interstellar medium. These are observed by telescopes operating at millimetre and sub-millimetre wavelengths through their intense molecular line emission (Fig. 13.21). Many molecular species have been detected in the interstellar gas in giant molecular clouds. The commonest emission processes are dipole transitions of molecules which have non-zero electric dipole moments, such as carbon monoxide, CO. Almost 100 different molecular species have been detected, most of them containing a few atoms, but molecules with up to 13 atoms have been detected. A number of molecules have only been observed in the rarefied conditions of the interstellar medium because they are destroyed by collisions in

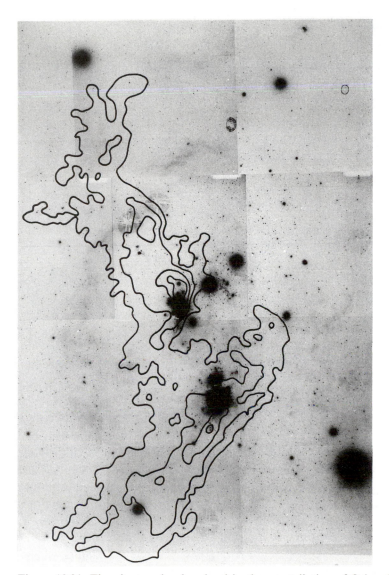

Figure 13.21. The giant molecular cloud in the constellation of Orion. The enormous extent of the clouds can be appreciated by noting the positions of the stars of the constellation of Orion. The contours show the intensity distribution of carbon monoxide, which is the commonest molecular species after molecular hydrogen. At higher angular resolution, it is found that the giant molecular clouds contain many dense knots, which may be the sites of new generations of stars. Unfortunately, molecular hydrogen does not emit millimetre line radiation because the molecule does not possess a dipole moment. The Orion nebula can be seen halfway down Orion's sword. (After R.J. Maddalena, M. Morris, J. Moscowitz and P. Thaddeus (1986). *Astrophys. J.*, **303**, 375. Photograph of the constellation of Orion kindly produced by the Photolabs, Royal Observatory, Edinburgh, from a mosaic of 12 UK Schmidt Telescope plates.)

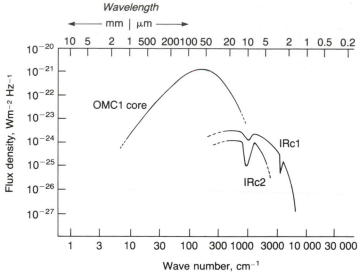

Figure 13.22. The far infrared spectrum of the central region of the Orion Nebula. The emission is the reradiation of heated dust grains. In the case of the source labelled 'OMC1 core', the dust is heated to typical temperatures of about 100 K. The discrete sources labelled IRc1 and IRc2 are hotter compact sources exhibiting strong absorption features at 3.1 and 9.7 μm, which are associated with ice and silicate grains, repsectively. The sources may be very young stars or protostars within the Orion complex. (From C.G. Wynn-Williams (1981). *Sci. Am.*, **245**, 35.)

laboratory experiments. Evidently, a great deal of organic and inorganic chemistry must be taking place in the interstellar medium. The typical giant molecular cloud has a mass about $10^6 M_\odot$ and a temperature in the range 10–100 K (Fig. 13.22). Most of the mass is in the form of molecular hydrogen, which is shielded from the intense interstellar flux of ionising radiation by dust and by self-shielding by the molecular hydrogen itself. Molecular line and infrared observations have shown that there exist within the large clouds much denser regions, and it is within these that stars are formed. Hot molecular gas is observed close to the regions where stars have already formed, the gas being heated by the radiation and by outflows from newly formed stars as they push back the surrounding medium.

Hot ionised gas at temperatures of about 10000 K is observed around young stars and the hot dying stars in the centres of planetary nebulae. The beautiful diffuse structures which are seen in young objects, such as the Orion Nebula and the Horsehead Nebula (Fig. 13.23), and in old systems, such as the shells of planetary nebulae (Fig. 13.17), are due to the excitation of strong emission lines in this ionised gas. The hot stars are the source of intense ultraviolet radiation, which ionises and heats the surrounding gas clouds and causes them to radiate intense line emission.

In addition to these hot components associated with the regions where stars form and die, there is very hot gas which is ejected in supernova explosions. These

Figure 13.23. (*a*) The Orion Nebula. In this picture, special photographic processes have been used to bring out faint features. This region of ionised hydrogen is excited by four young bright stars close to the centre of the nebula, known as the Trapezium stars. The light of the nebula is mostly due to the emission lines of hydrogen and oxygen. (Courtesy of David Malin and the Anglo–Australian Observatory.) (*b*) The Horsehead Nebula in the constellation of Orion. The dark patches are regions of dense interstellar dust, in many of which protostars and young stars may be formed as revealed by molecular line and infrared observations. (Courtesy of the Royal Observatory, Edinburgh; photography by David Malin.)

violent events, in which a whole star can be disrupted, liberate huge amounts of energy, sufficient to heat up the gas in a large volume surrounding the explosion to very high temperatures, $\sim 10^7$ K. The overlapping of the explosions of different generations of supernovae results in a considerable fraction of the interstellar gas being heated to these high temperatures. This gas is detected through its X-ray emission at energies $\epsilon = h\nu \sim 0.1 - 1$ keV, and through the observation of absorption lines in the interstellar gas due to highly ionised species, such as five-times ionised oxygen, O^{+5} or OVI. Its temperature is typically about 10^6 K.

Between the very hot regions and the giant molecular clouds, there is diffuse gas consisting of a mixture of ionised and unionised gas. These different components are not stationary, but are shocked by supernova explosions and by strong stellar winds. These motions compress the gas, and may assist in the formation of giant molecular clouds and, in turn, star formation.

Dust The black areas seen in photographs such as those of the Orion and Horsehead Nebulae (Fig. 13.23) are due to obscuring dust. Dust is present throughout the interstellar gas, and it has the unfortunate effect of obscuring some of the most interesting regions we wish to study. Fortunately, dust becomes transparent in the infrared waveband, and it is then possible to observe deep inside the regions where stars are forming by observations at these wavelengths (Fig. 8.17). In addition, the dust is heated by the radiation it absorbs, and therefore it is a strong emitter at the temperatures to which it is heated, which are typically in the range from about 30 to 500 K. Indeed, in the far infrared waveband, the regions containing the largest quantities of dust become the most intense emitters rather than the most obscuring objects, as can be seen in the far infrared image of the star forming regions in Orion (Fig. 13.24). The dust radiates at temperatures less than about 1000 K because at higher temperatures it would be evaporated and so would no longer exist. Dust plays a crucial role in the processes by which stars form – first, it helps protect the fragile molecules from being dissociated by the intense interstellar radiation field, and, secondly, it acts as an efficient means by which the protostar can radiate away its gravitational binding energy. Wherever stars are forming, the regions are characterised by very intense emission in the far infrared waveband.

High energy particles In addition to gas and dust, there are very high energy particles present throughout the interstellar medium. These are probably accelerated in supernova explosions and, possibly, in pulsars, and are then dispersed throughout the interstellar medium. The electron component is detected by its radio emission, which is due to the synchrotron radiation of ultrahigh energy electrons spiralling in the interstellar magnetic field. High energy protons are also observed through the gamma-rays (γ-rays), which they emit when they collide with ordinary matter in the interstellar gas. In these high energy collisions, pions of all types are produced, and the neutral pions decay almost instantaneously into γ-rays. These very high energy particles are very rare, but they are so en-

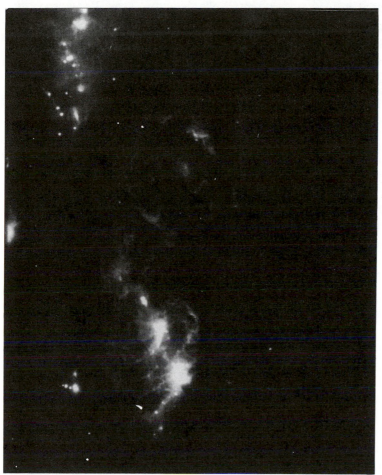

Figure 13.24. The region of the Orion giant molecular cloud, as observed by the Infrared Astronomical Satellite (IRAS). In the wavebands 10–100 μm, the region around the Orion Nebula and the giant molecular clouds are very intense far infrared emitters, the total far infrared luminosity of the region being about $10^6 L_\odot$. (Courtesy of NASA and the Jet Propulsion Laboratory.)

ergetic that the pressure of this relativistic gas has an important influence upon the dynamics of the interstellar medium. The study of these particles and their significance for high energy astrophysics will be a central theme of this volume.

The Galactic magnetic field Finally, there are a number of separate observations which indicate that there is a weak, large-scale magnetic field present in the interstellar gas. The evidence comes from observations of radio synchrotron emission of high energy electrons in the interstellar medium, and from the polarisation of the radio emission and of the light of nearby stars. Zeeman splitting of the 21-cm line of neutral hydrogen and of molecular radio lines has also been measured in a number of interstellar clouds. Typical values of the strength of the interstellar magnetic fields are $B \approx 10^{-8} - 10^{-10}$ T. Although apparently rather weak, this magnetic field has an important influence upon the dynamics of the different components of the interstellar medium.

13.4.2   Star formation

The understanding of processes by which stars form is one of the most important unsolved problems of the astrophysics of stars and galaxies. There are many uncertainties concerning the processes by which stars form from an observational point of view. We do not know how many massive and how many light stars are formed in any particular molecular cloud. We do not know how the rate at which stars form depends upon physical conditions within the cloud. We do not know how the molecular clouds themselves form. These problems can now be addressed by infrared and millimetre observations.

Theoretically, the position is not very much better. The fundamental problem to be understood is the sequence of events which takes place between the formation of a giant molecular cloud, with a typical density of about 10^9 molecules m^{-3}, and the formation of a new star, which has a density billions of billions of times greater. The denser regions of molecular clouds collapse under their own gravity, and, as they collapse, they heat up. Eventually, the cloud becomes opaque to its own radiation, and there has to be some means by which the heat generated in the collapse can be removed from the cloud. This is where the dust plays a crucial role. The radiation from the collapsing core is absorbed by the dust and is reradiated at far infrared wavelengths, at which the cloud is transparent. This accounts for the observation that regions where active star formation is proceeding are intense sources of far infrared radiation. Thus, the binding energy of the star is reradiated away by the heated dust. The slow collapse and heating up continues until the protostar becomes so hot in its centre that nuclear burning of hydrogen into helium begins, and the star starts its life as a main sequence star. Although this sounds a very straightforward picture, there are many theoretical uncertainties. For example, it is not clear how any rotation or magnetic fields present in the giant molecular cloud, which tend to prevent collapse of the cloud, are expelled from the protostar.

These studies have implications far beyond their immediate relevance for the formation of stars. For example, in order to understand the evolution of galaxies, we have to know the rate at which new stars are formed under a wide range of different astrophysical conditions. This field of study is of the greatest importance for essentially all branches of contemporary astronomy and cosmology.

14

Aspects of stellar evolution relevant to high energy astrophysics

14.1   Introduction

In this chapter, we look in a little more detail into some aspects of stellar evolution which will be important for studies of high energy astrophysical processes in our Galaxy and in extragalactic systems. For example, we need to know how far we can trust the theory of stellar structure and evolution; we need to know more details of the processes of nucleosynthesis in stars in order to understand the origin of the chemical composition of the interstellar gas and of the cosmic rays; we need to study what is known about the processes of mass loss from stars and the processes which can lead to the formation of dead stars; we need to investigate binary star systems in order to contrast their properties with those of X-ray binary systems containing neutron stars and black holes, and to discuss how such close binary stars can be formed. This survey is in no sense complete, and reference should be made to the texts recommended at the end of the book for more details.

14.2   The Sun as a star

Granted the outline of stellar evolution presented in Section 13.3, how well can the theory account for the properties of our own Sun? As by far the brightest star from our location in the Universe, it can be studied in much more detail than any other star, and is a benchmark for the theory of stellar structure and evolution. Until the last 20 years, the study of the Sun and the stars was largely confined to the interpretation of their surface properties. High resolution spectroscopic observations of stars provided measurements of the temperatures and chemical compositions of their surface layers as well as measurements of their surface gravities and motions in their atmospheres. Masses can only be determined for those stars which are members of binary systems, and only for a few giant stars have angular diameters been measured by optical interferometric techniques. Given these limitations, it is remarkable how well the theory of stellar structure and evolution can account for the populations of stars observed on the H–R diagram and how the stars evolve from one region to another (see, for example, Kippenhahn and Weigert 1990).

Over the last 20 years, two powerful new techniques have provided the means for studying the internal structure of the Sun and the stars. These are: the measurement of the neutrinos released in the nuclear reactions taking place in the centre of the Sun, and the measurement of the modes of oscillation of the Sun, the discipline known as *helioseismology*. Let us look at these important advances in a little detail since they are crucial for understanding stellar structure and evolution.

14.2.1 *Observations of solar neutrinos*

During evolution on the main sequence, the energy source for all stars is the conversion of hydrogen into helium nuclei. For stars less massive than about $1.5 M_\odot$, such as the Sun, the proton–proton (or p–p) chain provides the main route for the synthesis of helium, whereas, for more massive stars, the carbon–nitrogen–oxygen (or CNO) cycle is the principal energy source. The nuclear reactions involved in the p–p chain are as follows (see Kippenhahn and Weigert 1990). The first stages in the chain involve the formation of deuterium (^{2}H) and ^{3}He:

$$\text{p} + \text{p} \rightarrow {}^2\text{H} + \text{e}^+ + \nu_\text{e} \quad : \quad {}^2\text{H} + \text{p} \rightarrow {}^3\text{He} + \gamma \tag{14.1}$$

The first reaction, in which deuterium is formed, has a very small cross-section since it is a weak interaction involving β^+ decay. In fact, the reaction rate for this process has never been measured experimentally at the energies of interest for nucleosynthesis in the Sun, and so the reaction rates are based upon theoretical estimates. This reaction is also the principal source of neutrinos from the Sun, but they are of low energy, the maximum energy being 0.420 MeV. There are then three alternative routes for the formation of helium. The most straightforward is the pp1 branch:

$$\text{pp1} : \quad {}^3\text{He} + {}^3\text{He} \rightarrow {}^4\text{He} + 2\text{p} \tag{14.1a}$$

The other routes involve the formation of ^{7}Be as a first step:

$$ {}^3\text{He} + {}^4\text{He} \rightarrow {}^7\text{Be} + \gamma \tag{14.1b}$$

Then ^{7}Be can either interact with an electron (the pp2 branch) or a proton (the pp3 branch) to form two ^{4}He nuclei:

$$\text{pp2} : \quad {}^7\text{Be} + \text{e}^- \rightarrow {}^7\text{Li} + \nu \quad : \quad {}^7\text{Li} + \text{p} \rightarrow {}^4\text{He} + {}^4\text{He} \tag{14.1c}$$

$$\text{pp3} : \quad {}^7\text{Be} + \text{p} \rightarrow {}^8\text{B} + \gamma \quad : \quad {}^8\text{B} \rightarrow {}^8\text{Be}^* + \text{e}^- + \nu_\text{e}$$

$$ {}^8\text{Be}^* \rightarrow 2\,{}^4\text{He} \tag{14.1d}$$

Which of these reaction chains is dominant is temperature sensitive, pp1 being most important at low temperatures ($T < 10^7$ K) and the other chains at higher temperatures. Notice that the pp2 and pp3 chains depend upon there being ^{4}He present to begin with, but, since about 24% of the mass of ordinary matter in the Universe is expected to be in the form of ^{4}He as a result of primordial nucleosynthesis, there is a considerable amount of helium present even in unprocessed stellar material.

The p–p chain is the primary source of energy in the Sun, and one of the key tests of the theory of stellar nucleosynthesis is provided by the flux of electron neutrinos, v_e, generated in the p–p chain. Because of their very small cross-sections for interaction with matter, neutrinos escape essentially unimpeded from their point of origin within the central 10% of the Sun by radius, and thus the detection of the flux of solar neutrinos provides a direct test of the processes of nucleosynthesis. The predicted intensity spectrum of neutrinos from the various branches of the p–p chain, as well as contributions from the CNO cycle, is shown in Fig. 7.25, according to the standard solar model of Bahcall and Ulrich (Bahcall 1989). It can be seen that the greatest neutrino fluxes are expected from the first reaction in the p–p chain (14.1), but their maximum energy is only $\epsilon_v = 0.420$ MeV. Until recently, detectors sensitive to these low energy neutrinos have not been available. In 1958, however, Cameron and Fowler suggested that the much rarer but higher energy neutrinos from the pp3 chain might be detectable.

The electron neutrinos emitted in the decay of the ^{8}B nucleus have maximum energy 14.06 MeV (see Fig. 7.25), and can be detected on Earth through the nuclear transmutations which they induce in chlorine nuclei bound in a form of 'cleaning fluid', perchloroethylene, C_2Cl_4:

$$^{37}Cl + v_e \rightarrow \,^{37}Ar + e^- \tag{14.2}$$

for which the threshold energy is 0.814 MeV. The argon created in this reaction is radioactive, and, when it is flushed out of the tank, the amount produced can be measured from the number of radioactive decays of the ^{37}Ar nuclei (see Section 7.5). This flux provides a direct measurement of the rate of production of neutrinos in the Sun.

The first detailed predictions of the solar neutrino flux were made by Bahcall in 1964, and at about the same time the famous *solar neutrino experiment* was begun by Dr Raymond Davis and his colleagues using a 100000 gallon tank of C_2Cl_4 located at the bottom of the Homestake gold-mine in South Dakota. Over the subsequent years, a significant flux of neutrinos has been detected, but it corresponds to only about one-quarter of the flux predicted by the standard solar model (Fig. 7.27). This discrepancy is the famous *solar neutrino problem*. The results quoted by Bahcall (1989) are:

$$\text{observed flux of neutrinos} = 2.1 \pm 0.9 \text{ SNU}$$

$$\text{predicted flux of neutrinos} = 7.9 \pm 2.6 \text{ SNU}$$

where 1 SNU = 1 solar neutrino unit = 10^{-36} absorptions per second per ^{37}Cl nucleus. The errors quoted are formal 3σ errors for both the observations and the predictions.

The discrepancy indicated by the above figures is based upon the standard solar model described by Bahcall, and independent workers have now evaluated the expected flux of high energy neutrinos on the basis of their own models of

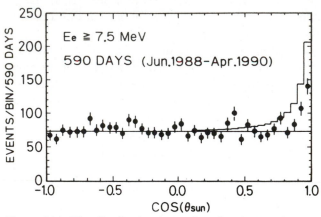

Figure 14.1. The distribution in $\cos\theta_{Sun}$ for the 590-day sample for $E_e \geq 7.5$ MeV. θ_{Sun} is the angle between the momentum vector of an electron observed at a given time and the direction of the Sun. The isotropic background, which is roughly 0.1 events day^{-1} bin^{-1}, is due to spallation products induced by cosmic ray muons, γ-rays from outside the detector and radioactivity in the detector water. The angular resolution of the detector system has been taken into account in calculating the expected distribution of arrival directions of the neutrinos from the Sun. (K.S. Hirata, K. Inoue, T. Katija, T. Kifune, M. Kihara, M. Nakahata, K. Nakamura, S. Ohara, N. Sato, Y. Suzuki, Y. Totsuka, Y. Yaginumi, M. Mori, Y. Oyama, A. Suzuki, K. Takahashi, M. Yamada, M. Koshiba, T. Suda, T. Tajima, K. Miyano, H. Miyata, H. Takei, Y. Fukuda, E. Kodera, Y. Nagashima, M. Takita, K. Kaneyuki, T. Tanimori, E.W. Beier, L.R. Feldscher, E.D. Frank, W. Frati, S.B. Kim, A.K. Mann, F.M. Newcomer, R. Van Berg and W. Zhang (1990). *Phys. Rev. Lett.*, **65**, 1297).

the Sun. Turck-Chièze *et al.* (1988) predict a slightly smaller flux of high energy neutrinos of 5.8 ± 1.3 SNU, compared with Bahcall and Ulrich's 7.9 SNU, which would reduce the discrepancy between the results of the ^{37}Cl experiment and the expectations of the solar models.

Confirmation that the flux of high energy neutrinos indeed originates within the Sun has been provided by the Kamiokande II experiment. As discussed in Section 7.5, the objective of the original Kamiokande experiment was the detection of fast electrons and positrons liberated in the decay of nucleons to measure the half-life of the proton. Since the original experiment, great efforts have been made to reduce the number of background events within the detector volume, so that the system would be sensitive enough to detect solar neutrinos. The great advantage of the Kamiokande II experiment is that the arrival directions of the incoming neutrinos can be measured. The lower energy limit of the experiment for the detection of electrons and positrons was reduced to 7.5 MeV, and so solar neutrinos with energies in the range $7.5 \leq \epsilon_v \leq 14$ MeV could be detected. The important interaction is electron–neutrino scattering, and so the scattered electrons, and consequently the incoming neutrinos, must have $\epsilon_v \geq 7.5$ MeV to register a signal in the detector system. Since the rest mass energy of the electron is only 0.512 MeV, the scattered electrons are relativistic and are strongly beamed in the forward direction in a narrow cone about the arrival direction of the neutrino, as a simple relativistic calculation shows. According to Bahcall's calculation, 90% of the electrons with energies greater than 7.5 MeV lie within about 10° of the direction of the incident neutrino (see Bahcall 1989, Table 8.14). The Cherenkov detectors which line the walls of the Kamiokande II experiment can measure the direction of travel of the scattered electrons, and thus the arrival directions of the neutrinos can be found.

The beautiful results derived from 590 days of observation from the Kamiokande II experiment are shown in Fig. 14.1. The solid histogram shows the predicted angular distribution of scattered electrons expected from the standard solar model, and the points show the numbers of electrons arriving from different directions with respect to the direction of the Sun. It can be seen that there is a small but significant excess flux of neutrinos coming from the direction of the Sun, but it is less than that expected from the standard solar model. As the Kamiokande II team state: 'These provide unequivocal evidence for the production of ^{8}B by fusion in the Sun.' (See the reference in the caption to Fig. 14.1.) The results quoted by the Kamiokande II team are:

$$\frac{\text{measured flux of neutrinos}}{\text{predicted flux of neutrinos}} = 0.46 \pm 0.13 \text{ (stat)} \pm 0.08 \text{ (syst)} \qquad (14.3)$$

where (stat) refers to the statistical errors and (syst) to the systematic errors. The actual flux of neutrinos measured in the experiment was 2.65×10^{10} m^{-2} s^{-1}. The expected flux from the models of Turck-Chièze *et al.* (1988) would be about 4.2×10^{10} m^{-2} s^{-1}, and so the discrepancy between the observations and the models would be considerably reduced.

A key test of the solar models is the detection of the low energy neutrinos from the first interaction (14.1) of the p–p chain, $p + p \rightarrow {}^2H + e^+ + \nu_e$, since this is the essential first step in the synthesis of helium and is much more directly related to the observed total (or bolometric) luminosity of the Sun than the high energy neutrinos. The predicted flux of low energy neutrinos is therefore less sensitive to physical conditions in the centre of the Sun than the flux of high energy neutrinos. The best approach for measuring the much more plentiful low energy neutrinos is to use gallium as the detector material and to measure the neutrino flux from the number of radioactive germanium nuclei created by the neutrino interactions:

$$\nu_e + {}^{71}Ga \rightarrow e^- + {}^{71}Ge \qquad (7.14)$$

The principles involved in the two experiments currently being carried out to search for these neutrinos were described in Section 7.5; they are known as the SAGE and GALLEX projects. The first results of the SAGE experiment found no neutrinos, and have provided an upper limit of 79 SNU at the 90% confidence level (Abazov *et al.* 1991).

In June 1992, the first results of the GALLEX experiment were reported. A positive detection of the flux of low energy neutrinos has been made, the reported flux being 83 ± 19 (stat) ± 8 (syst) SNU compared with a flux of $132 \, ^{+20}_{-17}$ SNU expected from the standard solar models.

Thus, the solar neutrino problem is far from resolved, but my impression is that the discrepancy between the observed and predicted fluxes of neutrinos may well be less severe than it was once thought to be. Nonetheless, it is crucial to resolve this problem because it is a direct test of the processes of nucleosynthesis inside the Sun. It may be that there is some aspect of the nuclear physics which has not been included in the reaction chains (14.1), and this would have important consequences for particle physics. An example of the type of modification to the standard theory which might be required is the Mikheyev, Smirnov, Wolfenstein (MSW) effect, which Bahcall and Bethe (1990) have shown could resolve the present discrepancy if it turns out to be real (see also Bahcall 1989). We must wait and see how this story develops.

14.2.2 *Helioseismology*

By the 1970s, much of the overall picture of stellar evolution described in Section 13.3 was falling into place, largely through the application of high speed computers to the detailed study of the physical processes taking place inside the stars. Besides the detection of solar neutrinos, however, no other method was available for comparing the internal structures of the stars with the expectations of theory. The picture changed dramatically with the discovery of *solar oscillations* (see Deubner and Gough 1984). The first detections were made in the early 1960s by Leighton and his colleagues, who discovered 'five-minute' oscillations in their studies of the velocity field of the solar atmosphere. The full significance of these observations was only fully appreciated in the 1970s. The 'five-minute' oscillations are standing acoustic waves, which are largely confined to the outer layers of the Sun. In the mid-1970s, other oscillations were discovered with much longer periods of 20–60 minutes and 160 minutes, which had quite different properties from the 'five-minute' oscillations. Some of these modes probe deep inside the Sun, and it is these which have provided very powerful tools for testing the theory of stellar structure and evolution. It is no exaggeration to say that the whole subject of the astrophysics of the Sun and the stars has been completely rejuvenated by these discoveries. The internal structure of the Sun has been the subject of the most intensive study, but it is now feasible to extend these techniques to bright nearby stars.

It is simplest to think of the Sun as a resonant sphere which, when perturbed, oscillates at frequencies corresponding to its normal modes of oscillation. The convective envelope of the Sun provides a natural source of excitation, which can stimulate the Sun to resonate in its normal modes. In terrestrial seismology, the resonance modes of the Earth can be found by tracing the paths of sound waves inside the Earth, and exactly the same procedure can be employed to study physical conditions inside the Sun. For this reason, these studies are referred to as *helioseismology*.

There are two principal methods for measuring the solar oscillations, both of

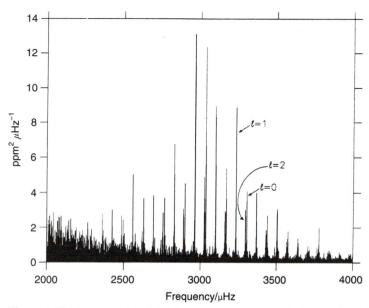

Figure 14.2. An example of the frequency spectrum of solar oscillations, showing some of the normal modes of oscillation, of the Sun. These data were derived from 160 days of observation by the IPHIR experiment on board the PHOBOS spacecraft. This power spectrum of the low degree *p*-modes shows an alternating pattern of double and single peaks; the double peaks are the $l = 0, 2$ modes and the single peaks are the $l = 1$ modes. (T. Toutain and C. Frölich (1992). *Astron. Astrophys.*, **257**, 287.)

which are technically very challenging. In one approach, the brightness of the Sun is measured with very high precision, so that variations as small as one part in 10^6 of the total intensity can be measured. In the other approach, very precise measurements of the Doppler shifts of the solar atmosphere can be made – the techniques must be precise enough to measure velocity differences of about 1 m s^{-1} or less. Both approaches have now been successfully used to measure the resonant modes of the Sun, those sensitive to physical conditions in the core of the Sun being of particular interest for the solar neutrino problem. An example of the quality of the data now available from these studies is shown in Fig. 14.2. This power spectrum of the total luminosity of the Sun was obtained from the IPHIR experiment on board the PHOBOS space probe whilst in transit between the Earth and Mars. The power spectrum displays both large and small splittings of the spectral lines.

The theory of the modes of oscillation of the Sun is a beautiful example of the power of classical theoretical physics applied to an astrophysical problem. It is remarkable that much of the pioneering analysis is contained in the sixth edition of Lamb's classical text, *Hydrodynamics* (1932). A clear introduction to the key physical ideas is provided by Deubner and Gough (1984). The modes of oscillation of the Sun can be thought of as standing waves resulting from the interference of oppositely directed propagating waves. In the simplest approximation, the Sun can be considered to be spherically symmetric, and so the natural representation of the perturbations is in terms of associated Legendre functions, similar to those used to describe the amplitude of the wave function inside the hydrogen atom. Following Deubner and Gough, if ξ is the vertical component of the fluid displacement, the decomposition into normal modes can be written as:

$$\xi(r, \theta, \phi, t) = \Re \left[R(r)\, P_l^m(\cos\theta) \, {\cos \atop \sin} m\phi \, \exp(i\omega t) \right] \qquad (14.4)$$

where the separation of variables consists of the associated Legendre function $P_l^m(\cos\theta)$ describing the angular variation of the amplitude of the displacement and the radial function $R(r)$. $\Re$ indicates that the real part of the function should be taken. The adopted terminology for the Sun is that l is called the *degree*, n the *order* and m the *azimuthal order* of a particular mode.

The modes of oscillation consist of two types: acoustic or p-modes, in which the restoring force is provided by pressure fluctuations; and gravity or g-modes, for which the restoring force is buoyancy. The modes of greatest interest for the study of the internal structure of the Sun are the acoustic modes of small degree l, since they can probe into its central regions. For a mode of given degree, there are many different orders n which measure the vertical component of the wave number. As in the hydrogen atom, n is related to the number of nodes in the solutions of the radial wave equation. Rather than investigate the formal mathematical results, which are discussed by Deubner and Gough, Fig. 14.3 shows graphical representations of three of the normal modes of oscillation of the Sun for a standard solar model. It can be seen that low degree (l) modes propagate into the central regions of the Sun, whereas the high l modes are confined to the outer regions.

The power spectrum of the Sun shown in Fig. 14.2 displays the low degree p-modes, and there are two types of separation of the resonant frequencies. The 'large' separations, corresponding to frequency differences $\Delta\nu_0$ of about 60 μHz, correspond to modes of the same degree l but of order n differing by one. There are also 'small' differences δ_{nl} associated with alternate resonances, and these are associated with the difference in frequency between modes with 'quantum numbers' (n, l) and those with $(n-1, l+2)$. The physical significance of $\Delta\nu_0$ is that it is associated with the average sound speed throughout the Sun. For low values of l, the modes are identical in the outer regions of the Sun but differ in the central regions. Thus, δ_{nl} is sensitive to physical conditions in the core of the Sun. The plot of δ_{nl} against $\Delta\nu_0$ is known as the *asteroseismological H–R diagram*, and is a sensitive measure of the central hydrogen content of the star and its mass (Fig. 14.4).

One of the most striking examples of the power of these techniques is illustrated by the work of the Birmingham (UK) group, who have used their global network of observing stations to measure precisely the small separations δ_{nl} for the low degree modes. Values of δ_{nl} have been measured for the $l = 0, 1$ modes for values of n in the range $15 \leq n \leq 27$ (Elsworth *et al.* 1990). They find that their observations are in excellent agreement with the standard solar models, but are not consistent with non-standard solar models which involve large amounts of core mixing, rapid rotation of the core of the Sun or the presence of weakly interacting massive particles in the core of the Sun. In consequence, we can

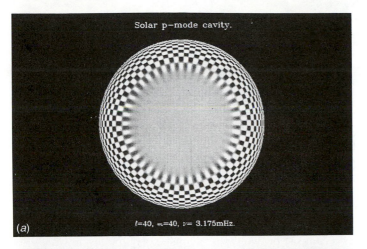

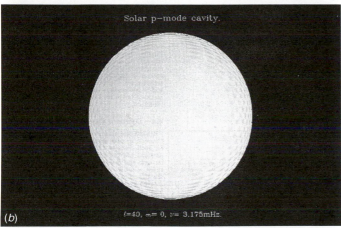

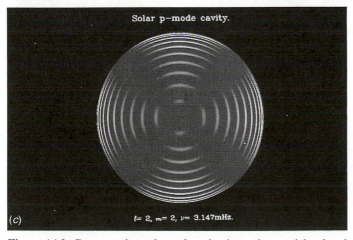

Figure 14.3. Cross-sections through pulsating solar models, showing how different modes of oscillation probe different regions within the Sun. The *p*-modes can probe physical conditions in the central regions of the Sun. (From T. Appourchaux, C. Catala, S. Catalano, S. Frandsen, A. Jones, P. Lemaire, O. Pace, S. Volonte and W. Weiss (1991). *Commun. Asteroseismol.*, number 35.)

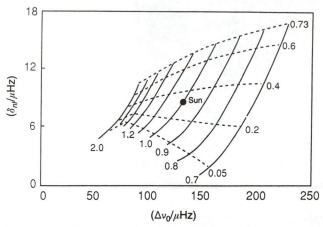

Figure 14.4. The asteroseismological H–R diagram, in which the fine structure splitting δ_{nl} is plotted against $\Delta \nu_0$. The stellar models are sensitive to the mass of the star and to the central hydrogen content. The solid lines show the predicted loci for different stellar masses; the dashed lines show the loci for different helium abundances by mass. The location of the Sun on this diagram is indicated. (From J. Christensen-Dalsgaard (1988). In *Advances in helio- and asteroseismology*, eds J. Christensen-Dalsgaard and S. Fransden, p. 295. Dordrecht: D. Reidel Publishing Company.)

have confidence that the astrophysics of the standard solar model is a very good description of the internal structure of the Sun right into its core. If there is a solar neutrino problem, its resolution probably lies with the nuclear physics rather than with the astrophysics.

A measure of the quality of data which can now be provided by observations of solar oscillations is shown in Fig. 14.5, which was kindly prepared by Prof. Douglas Gough. By analogy with terrestrial seismology, the frequencies of the standing waves depend upon the variation of the speed of sound as a function of depth within the Sun. The modes of different order l sample different volumes of the solar interior, and so the solar oscillation data can be inverted to determine the speed of sound throughout the Sun. Thus, the physical quantity which is most directly related to the power spectrum of the solar oscillations is the variation of the speed of sound with radius in the Sun. In Fig. 14.5(a), the square of the speed of sound as a function of radius within the Sun is shown for a number of different standard solar models. When the helioseismological data are inverted to derive the square of the sound speed, it turns out that the theory is consistent with the model labelled $t = 4.6 \times 10^9$ years and $Y_0 = 0.28$, to within the thickness of the line. In Fig. 14.5(b), the detailed comparison between the theory and the observations is shown, and it can be seen that they agree to better than 1% throughout the Sun. In the most recent analyses, the internal rotation of the Sun has been determined from the rotational splitting of vibrational modes, and the rotational velocity as a function of depth and polar angle has been determined.

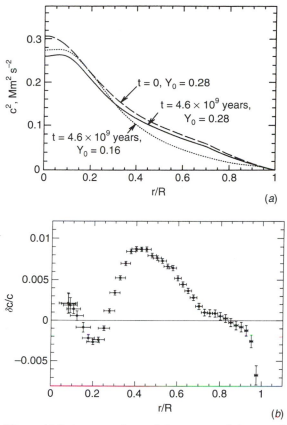

Figure 14.5. A comparison of the square of the sound speed inside the Sun as a function of radial distance from the centre, according to the theory of stellar structure and from inversions of the solar oscillation data. (*a*) The variation of the square of the sound speed with radius for different assumptions about the age of the Sun, t, and its helium abundance by mass, Y_0. The observations are in agreement with the line labelled $t = 4.6 \times 10^9$ years and $Y_0 = 0.28$, to within the thickness of the line. (*b*) A comparison of the theory and the observations in terms of the fractional deviations of the sound speed inferred from observation relative to the best fitting standard model of the interior structure of the Sun. The agreement is better than 1%. (Courtesy of Prof. Douglas Gough.)

This is crucial information for understanding how the solar dynamo operates, and this in turn is related to the origin of the 22-year solar cycle.

To obtain more stringent tests of the solar models, it is necessary to carry out these studies over very long time periods. To obtain the necessary continuity of observation, they must be carried out using a world-wide network of observing stations, or from space, the SOHO mission of the European Space Agency being the ideal type of mission for this purpose. Although our Sun is the prime target for these studies, the brightest nearby stars can also be investigated using similar techniques, and the European Space Agency has developed a satellite project known as PRISMA which would be dedicated to seismological studies of the nearby stars. In these cases, the velocities and intensities would be averaged over the whole surface of the star, a procedure similar to that carried out in some of the solar experiments. Fortunately, the low degree modes, which are preferentially observed when 'whole star' observations are made, are also the modes which provide most information about the internal structure throughout the star.

14.3   Nucleosynthesis in stars

In a classic paper, Burbidge *et al.* (1957) described a wide range of nuclear processes which are important in synthesising the chemical elements during the course of stellar evolution. As might be expected, the physical conditions of density, temperature and chemical composition are crucial in determining which are the dominant processes.

With increasing mass, the central temperature of the star increases and then the carbon–nitrogen–oxygen cycle (CNO cycle) becomes the dominant energy source because of the very steep dependence of the energy generation rate ϵ upon temperature, $\epsilon \propto T^{17}$, compared with $\epsilon \propto T^4$ for the p–p chain. The CNO cycle uses ^{12}C as a catalyst in the formation of helium through the following sequence of reactions:

$$^{12}\text{C} + \text{p} \rightarrow {}^{13}\text{N} + \gamma \; ; \; {}^{13}\text{N} \rightarrow {}^{13}\text{C} + \text{e}^+ + \nu_e \; ; \; {}^{13}\text{C} + \text{p} \rightarrow {}^{14}\text{N} + \gamma$$
$$^{14}\text{N} + \text{p} \rightarrow {}^{15}\text{O} + \gamma \; ; \; {}^{15}\text{O} \rightarrow {}^{15}\text{N} + \text{e}^+ + \nu_e \; ; \; {}^{15}\text{N} + \text{p} \rightarrow {}^4\text{He} + {}^{12}\text{C} \quad (14.5)$$

The cycle proceeds through the successive addition of protons to the ^{12}C nucleus, with two intermediate inverse β-decays, which have the effect of converting two of the protons into neutrons.

The expected products of the p–p chain and the CNO cycle provide important tests of the role of these processes of nucleosynthesis in stars. In the case of the CNO cycle, the abundances of ^{13}C, ^{14}N, ^{15}N and ^{15}O are enhanced relative to their initial abundances. Indeed, in the steady state CNO cycle, large ^{13}C/^{12}C, He/H and N/C abundances are expected. The problem is therefore to find stars in which it is possible to study the products of nucleosynthesis within the cores of stars, rather than within their surface layers in which the relative abundances are directly measurable.

The problem is a complex one because the observed abundances depend upon many factors. For example, what was the initial chemical composition from which the star formed? Was the star formed from a pure hydrogen–helium plasma, or was it already enriched by the products of the previous cycle of star formation? What is the role of convection and diffusion in dredging up material from the interior of the star to the surface layers? Has the star lost its outer layers, revealing the chemical elements in the interior? Despite these complications, there is good evidence for the types of abundance trends expected from the CNO cycle and the p–p chain in those classes of star where it is reasonable to expect the products of nucleosynthesis to be observable. Particularly interesting cases are those in which anomalously high values of $^{13}C/^{12}C$ and N/C abundance ratios are observed.

An outline of the evolution of massive stars was given in Section 13.3. Post-main sequence evolution proceeds by the process of successive core and shell burning of the elements to produce nuclei with higher and higher binding energies. For the most massive stars, the sequence of burning runs through helium burning to produce carbon and oxygen, carbon and oxygen burning to produce silicon, which can eventually be burned to create iron peak elements. These processes can be written as:

$$\left.\begin{aligned} ^{12}C + {}^{12}C &\rightarrow {}^{24}Mg + \gamma \\ &\rightarrow {}^{23}Mg + n \\ &\rightarrow {}^{23}Na + p \\ &\rightarrow {}^{20}Ne + {}^{4}He \\ &\rightarrow {}^{16}O + 2\,{}^{4}He \end{aligned}\right\} \quad T \geq 5 \times 10^8 \text{ K} \quad (14.6)$$

$$\left.\begin{aligned} ^{16}O + {}^{16}O &\rightarrow {}^{32}S + \gamma \\ &\rightarrow {}^{31}P + p \\ &\rightarrow {}^{31}S + n \\ &\rightarrow {}^{28}Si + {}^{4}He \\ &\rightarrow {}^{24}Mg + 2\,{}^{4}He \end{aligned}\right\} \quad T \geq 10^9 \text{ K} \quad (14.7)$$

In the case of silicon burning, which begins at a temperature of about 2×10^9 K, the reactions proceed slightly differently because the high energy γ-rays remove protons and ^{4}He particles from the silicon nuclei, and the heavier elements are synthesised by the addition of ^{4}He nuclei through reactions which can be schematically written as:

$$\left.\begin{aligned} ^{28}Si + \gamma' \text{ s} &\rightarrow 7\,{}^{4}He \\ ^{28}Si + 7\,{}^{4}He &\rightarrow {}^{56}Ni \end{aligned}\right\} \quad (14.8)$$

It is therefore expected that, in the final stages of evolution of very massive stars, the star will take up an 'onion-skin' structure, with a central core of iron peak elements and successive surrounding shells of silicon, carbon and oxygen, helium and hydrogen (Fig. 14.6). Calculations of the explosive burning of shells of carbon, oxygen and silicon have shown good agreement with the observed abundances of the heavy elements up to the iron peak. There is good evidence that these types of event occur in Type II supernovae (Section 15.2). The important point is that these processes of nucleosynthesis lead to the production of elements up to the iron peak, and these possess the greatest binding energy of the chemical elements.

Iron is the most tightly bound of the chemical elements, and therefore the process of nuclear burning to reach lower energy states cannot proceed beyond iron. To proceed further, two processes are important, both involving neutron reactions with iron peak elements. In these reactions, a neutron is absorbed and the subsequent products depend upon whether or not the nucleus formed has

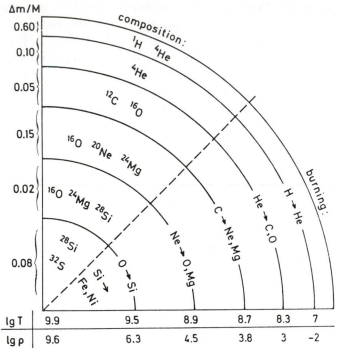

Figure 14.6. A schematic illustration of the 'onion-skin' picture of the interior structure of a highly evolved star. Typical values of the mass, density (in g cm^{-3}) and temperature (in K) of the different shells are indicated along the axes. $\Delta m/M$ is the fractional mass of the star contained in each shell. (After R. Kippenhahn and A. Weigert (1990). *Stellar structure and evolution*, p. 329. Berlin: Springer-Verlag.)

time to decay before the addition of further neutrons takes place. The case in which the decay always occurs first is referred to as the *slow* or *s-process*, and that in which several neutrons are added before β-decay terminates the sequence is known as the *rapid* or *r-process*. The latter is likely to be important in the extreme conditions during explosive nucleosynthesis, where very high densities and temperatures are attained and large fluxes of neutrons are produced by the inverse β-decay process (see Section 15.2). This is believed to be the process which is responsible for the synthesis of neutron-rich species such as the heaviest isotopes of tin, ^{122}Sn and ^{124}Sn.

The products of the s-process are estimated by calculations in which iron, by far the most abundant of the elements heavier than oxygen, is irradiated by neutrons. The products are sensitive to the irradiation time, but it has been shown that, if it is assumed that there is a range of irradiation times, the Solar System abundances of the elements heavier than iron can be accounted for. This theory has been particularly successful in accounting for the anomalously high abundances of heavy elements such as barium and zirconium, and, in particular, for the unstable element technetium, Tc, the longest lived isotope of which has a lifetime of only

2.6×10^6 years. These considerations are of importance for studies of the chemical composition of the very heavy elements in the cosmic rays. It is likely that these are accelerated in supernova explosions, and hence the chemical composition of the very heavy elements provides a diagnostic test for the relative importance of the r- and s-processes in synthesising the elements beyond the iron peak.

14.4   Stellar evolution in globular clusters – the age of the Galaxy

One of the most difficult problems in observational astronomy is the determination of the distances of stars (see Appendix A1), and for this reason star clusters are of special importance in testing the theory of stellar evolution because all the stars can be assumed to be at the same distance. It can also be assumed that all the stars in a cluster formed at more or less the same time in the distant past from material of the same chemical composition. The ages of star clusters range from very young systems, probably no more than about 10^6 years old, such as those found in the vicinity of the Orion Nebula, to very old clusters, the globular clusters, which are among the oldest stellar systems in the Galaxy, with ages of about 10^{10} years. One of the goals of the study of star clusters of different ages is to determine their luminosity–temperature diagrams and to compare these with the predictions of the theory of stellar evolution. The H–R diagrams for a number of clusters of different ages are shown in composite form in Fig. 14.7. It can be seen that these have a general shape, which can be understood in terms of the theory outlined in Section 12.3. A key problem for observational astronomy is the measurement of accurate colours and magnitudes for the stars in clusters, so that the comparison of theory and observation can be made as precise as possible.

The theoretical predictions are derived by working out the loci of stars of different masses on a theoretical luminosity–temperature diagram, as illustrated in Fig. 13.20, and then populating the different regions according to how long the stars remain in different parts of the diagram. The calculations depend upon the assumed form of the *initial mass function*, that is, the numbers of stars which begin their lives with different masses along the main sequence. These diagrams thus contain crucial information, not only about stellar evolution but also about the initial distribution of masses of the stars in the cluster when it formed. Of special importance for many aspects of astrophysics and cosmology is the determination of the ages of globular clusters from the location and shape of the distribution of stars near the *main sequence termination point*. As described in Section 13.3, the main sequence termination point is a measure of the age of the globular clusters, and the oldest of these provide a measure of the age of our Galaxy and the Universe itself.

As an example, the H–R diagram for the globular cluster 47 Tucanae is shown in Fig. 14.8, and it can be seen how well the comparison between theory and observation can be made if great care is taken in measuring the magnitudes and colours of the stars. The goodness of fit of the theory to the observations depends upon the metal abundance as well as the age of the globular cluster.

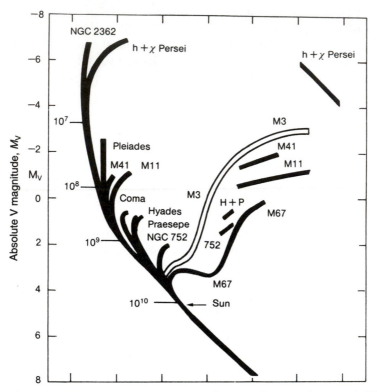

Figure 14.7. The Hertzsprung–Russell diagrams for star clusters of different ages. The differences in their H–R diagrams can be attributed to their different ages. The youngest cluster is NGC 2362 and the oldest is M67. (From D. Michalis and J. Binney (1981). *Galactic astronomy: structure and kinematics*, p. 105. New York: W.H. Freeman and Co.)

For the cluster 47 Tucanae, Hesser *et al.* (see caption to Fig. 14.8) find an age of $1.2 - 1.4 \times 10^{10}$ years and a metal abundance corresponding to 20% of the solar value. This is a crucial datum for cosmology because it provides a lower limit to the age of the Universe, as well as important constraints on Hubble's constant and the deceleration parameter (see Section 13.1.3).

Because the globular clusters are so old, they provide important information about some of the earliest generations of stars which formed in the Galaxy. In the oldest globular clusters, the heavy elements are much less abundant than they are in stars forming at the present day. This is consistent with the general picture, in which the cosmic abundance of the elements has been built up over the last 10^{10} years through nucleosynthesis inside stars. It appears, however, that, even in the oldest globular clusters, there are none which are totally devoid of heavy elements. Indeed, it is very difficult to find stars which have abundances of elements heavier than helium less than about 1/100 of the solar value. This is an important fact for studies of the formation of galaxies, since it means that, even in the oldest systems we can study directly, there must already have been some synthesis of the elements. Whether this can have taken place before or after the galaxy formed is

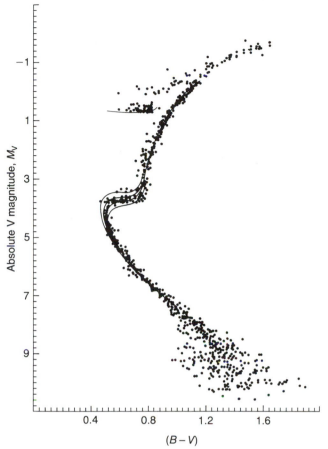

Figure 14.8. The Hertzsprung–Russell diagram for the globular cluster 47 Tucanae. The scatter in the points increases at faint magnitudes because of the increase in observational error associated with the photometry of faint stars. The solid lines show the best fits to the data using theoretical models of the evolution of stars from the main sequence onto the giant branch due to VandenBerg. The best-fit isochrones have ages in the range $1.2 - 1.4 \times 10^{10}$ years and the cluster is metal-rich relative to the other globular clusters, the metal abundance corresponding to about 20% of the solar value. (From J. E. Hesser, W. E. Harris, D. A. VandenBerg, J. W. B. Allright, P. Shott and P. Stetson (1989). *Publ. Astron. Soc. Pacific*, **99**, 739.)

an intriguing and important question since it bears upon the precise sequence of events which must have taken place when galaxies first formed.

14.5   Post-main-sequence evolution and mass loss

Stars lose mass from their surfaces throughout much of their lives. One of the most important discoveries of the Einstein X-ray Observatory was that essentially all classes of stars emit X-rays, the radiation generally originating from hot stellar coronae or stellar winds. Thus, coronae similar to that observed about our own Sun must be common to most classes of star. The immediate consequence is that there must be stellar winds, and hence mass loss, associated with essentially all classes of star (see Figs. 10.1 and 12.3). Stellar coronae are believed to be heated by hydromagnetic waves or shock waves originating in the convective layers close to the surface of the star, and this energy is dissipated above the photosphere, leading to strong heating of the lower density gas in the immediate vicinity of the Sun (see Section 12.2). The gas in the solar corona is heated to temperatures in excess of 10^5 K, so that it is no longer bound to the Sun, and a stellar wind, in our case the *Solar Wind*, is created. This may be termed *quiescent mass loss*, since it occurs in all classes of normal star.

There are, however, much more violent forms of mass loss, which are believed to be associated with the various evolutionary changes which stars undergo, both when they are on the main sequence and after they have left it. Some of the evidence is directly observational; other evidence is derived from theoretical arguments concerning the types of star which appear in different parts of the H–R diagram.

Direct observational evidence comes from a variety of sources. One of the most direct is the observation of the profiles of emission lines of *P-Cygni* type (Fig. 14.9). In this type of profile, the emission line originates in the stellar atmosphere, but the short wavelength side of the line is strongly modified by absorption by the same type of ions responsible for the emission line in the outflowing material, which is moving along the line of sight towards the observer. The outflowing material absorbs not only the emission line radiation but also the underlying continuum of the star. Observations of this type have been made with particular success in the ultraviolet waveband by the International Ultraviolet Explorer (IUE) because the resonance lines of a number of the common elements fall in this waveband. In the example of the Wolf–Rayet star HD 93131, shown in Fig. 14.9, P-Cygni profiles are associated with the ions of NIII, NIV and NV, HeII and CIV. As a result, mass-loss rates have been determined for many classes of hot star. Similar techniques have been used to evaluate mass-loss rates from cooler stars. The profiles of the MgII line at 280 nm show similar but much less pronounced asymmetric P-Cygni profiles.

Another piece of direct evidence for mass loss comes from the observation of expanding shells about highly evolved stars. This is particularly the case for *planetary nebulae*, in which roughly spherical shells of gas are observed moving outwards from the central star, the velocities being typical of the escape velocity from the surface of a star belonging to the giant branch (Fig. 13.17).

Another important technique is the observation of dust shells around giant stars. These are detected by their far infrared emission, either in the wavebands accessible from the ground at 10 and 20 μm, or from space infrared telescopes such as the Infrared Astronomy Satellite (IRAS). Dust particles condense in the cooling outflows from giant stars, and these are then heated up by the stellar radiation from the giant star. The dust is heated to temperatures in the range 100–1000 K, and this is readily detected as intense far infrared radiation.

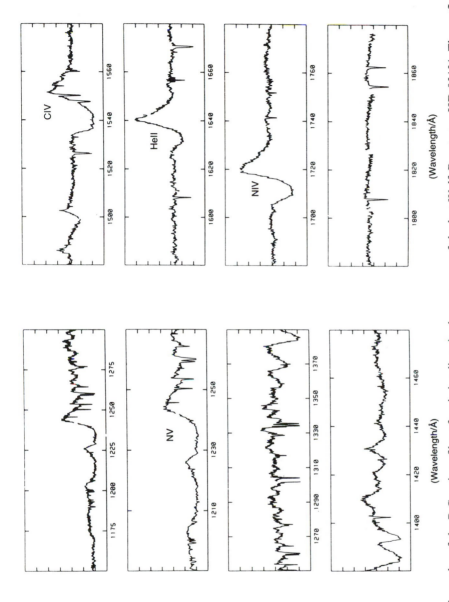

Figure 14.9. Examples of the P-Cygni profiles of emission lines in the spectrum of the hot Wolf–Rayet star HD 93131. The outflow of gas in the form of a wind causes absorption of both the line and continuum radiation to the short wavelength side of the line. In this spectrum there are many strong emission lines, and P-Cygni profiles are observed in the lines of NIV and NV, HeII and CIV, as indicated. (After A.J. Willis, K.A. van der Hucht, P.S. Conti and C.D. Garmany (1986). *Astron. Astrophys. Suppl.*, **63**, 417.)

These observations are clear evidence for mass loss from stars on the main sequence and on the giant branch. There are also theoretical reasons for believing that mass loss must be important. First of all, the sequence of nuclear burning within stars as they move up the giant branch does not proceed continuously but as a discontinuous series of 'jumps'. For example, once the hydrogen in the core of a star is exhausted, hydrogen burning continues, but now in a shell about an isothermal helium core. As evolution proceeds, the core contracts until the temperature is high enough for helium burning in the core to begin. As illustrated in Fig. 13.19, there is a sequence of nuclear burning phases within the core and in shells about the core. The evolution of the position of the stars on the H–R diagram is also illustrated in Fig. 13.19. During these transitions from one form of burning to another, the star has to reorganise its internal structure, and in the process the outer layers of the star may be expelled.

One reason for expecting that this must occur during evolution on the giant branch is derived from models of *horizontal branch stars*, which indicate that they have masses of about $0.5M_\odot$ or less. Further evidence on the internal structure of the horizontal branch stars and their masses comes from the variable *RR-Lyrae* stars, which are only found within a certain *instability strip* on the H–R diagram (Fig. 14.10). The models of these stars, which can account for their light curves, their luminosities and surface properties, suggest that their masses are about $0.5M_\odot$. Since the main sequence termination point for even the oldest stars in the Galaxy has just reached one solar mass, the horizontal branch stars must have lost mass from their outer layers. Thus, it is entirely reasonable to suppose that the horizontal branch stars result from stars which originate on the giant branch and which lose mass during the various convulsions undergone by the star as it evolves up the giant branch. Models of horizontal branch stars indicate that they evolve back towards the giant branch, and ultimately towards the tip of the giant branch.

As the star moves towards the tip of the giant branch, it reaches the region occupied by long period variables and unstable stars. These are stars in the very final phases of evolution, and there appears to be a continuity in properties between the various classes of objects found in this region of the diagram. The long period variables and the OH/IR stars appear to form a continuous sequence with increasingly long periods, leading ultimately to a region of the H–R diagram populated by unstable stars. In these stars, collapse of the core takes place with the expulsion of the envelope of the giant star, leading to the *planetary nebula* phase of evolution. Evidence that the cores of these stars collapse is provided by the very high temperatures of the central stars in planetary nebulae, and also by the chemical abundances of the elements necessary to explain their surface properties. They appear to be essentially helium stars, the implication being that most of the outer layers of the stars have already been expelled. These very hot compact stars follow a sequence on the H–R diagram which indicates that they will end up as white dwarf stars (Fig. 13.18).

We have concentrated, in the above discussion, on the evolution of stars with masses roughly that of the Sun. Higher mass stars evolve much more rapidly

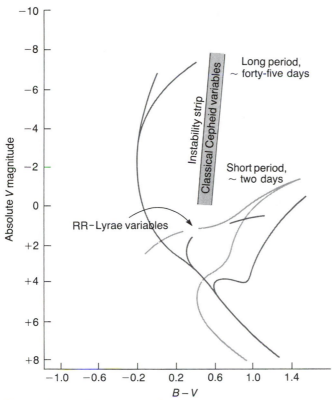

Figure 14.10. A Hertzsprung–Russell diagram illustrating the region occupied by Cepheid and RR-Lyrae variable stars. This region is known as the *instability strip*. The latter stars provide important information about the masses and properties of stars which lie on the horizontal branch. (After D. Michalis and J. Binney (1981). *Galactic astronomy: structure and kinematics*, pp. 125, 155. New York: W.H. Freeman and Co.; and F.H. Shu (1982). *The physical universe: an introduction to astronomy*, p. 169. Mill Valley, CA: University Science Books.)

and follow the evolutionary paths shown in Fig. 13.20. These evolutionary tracks assume that mass loss is unimportant, but it is now clear that, for the most luminous stars, mass loss plays a major role in their evolution. There is an upper limit to the luminosities of stars, which is set by a vibrational instability. Stars with masses greater than about $60M_\odot$ are subject to a radial pulsational instability, which becomes non-linear and ejects layers of gas from the surface of the star until its mass is reduced to about $60M_\odot$. Many of the massive stars with masses up to this limiting value exhibit enormous mass-loss rates, values as large as $10^{-4}M_\odot$ year^{-1} being common, and the extreme object η-Carinae having a mass-loss rate of about $10^{-2}M_\odot$ year^{-1}. These stars lose mass at such a high rate that they can lose their hydrogen envelopes during what would normally be the main sequence phase of evolution, exposing the helium cores created in their centres. In some cases, they may evolve towards the red giant region and then suffer further mass loss from their surfaces. Mass loss of this form may be the origin of the class of star known as *Wolf–Rayet* stars, which appear to be massive helium stars with high abundances of carbon or nitrogen. Typical mass-loss rates in these stars are about $3 \times 10^{-5} M_\odot$ year^{-1}. A number of the Wolf–Rayet stars are members of binary systems, and so Roche lobe overflow may also be an important mass-loss mechanism (see Section 14.6).

The Wolf–Rayet stars come in two varieties: the WC stars, which exhibit very strong carbon lines but no nitrogen; and the WN stars, which have strong nitrogen lines but are very deficient in carbon. It is likely that these differences reflect the different evolutionary status of the two types. The WN stars can be naturally associated with massive O stars in which the products of hydrogen burning through the CNO cycle are exposed due to the effects of strong mass loss from their surfaces. On the other hand, the WC stars can be naturally associated with stars which have proceeded through to helium burning in their cores. The triple-α process takes place at a higher temperature than the CNO cycle and has the effect, not only of creating ^{12}C, but also of destroying the nitrogen. Evidently, there must be considerable mixing and mass loss to make the products of nuclear processing apparent in the stellar atmosphere. These stars may be important in explaining some of the abundance anomalies observed in the cosmic rays. Examples of modified evolutionary tracks including mass loss are shown in Fig. 14.11. There is convincing evidence that this type of mass loss must have been important in the evolution of the progenitor star of the supernova known as SN1987A (see Section 15.2.2). The problem of understanding the ultimate fate of high-mass stars will be taken up in Chapter 15.

During post-main-sequence evolution, two basic processes take place. The star attempts to achieve a state of higher gravitational binding energy, the energy released providing the pressure support for the star. The second aspect is that, in achieving this more highly bound state, the star loses mass. If we sum over all forms of mass loss from stars in our own Galaxy, it is likely that about $1 - 10 \ M_\odot$ of material each year are returned to the interstellar medium. This is an important result because it means that the interstellar medium is constantly being replenished by stellar mass loss. Over a period of 10^{10} years, it is likely that a considerable fraction of the mass of the Galaxy will have been circulated through stellar interiors, which provides a plausible explanation for the fact that the abundances of the elements in stars seem to have a fairly universal character. What we have not addressed in this section is how the observed abundances of the elements are created. Obviously, many of the mass-loss processes described above involve the expulsion of the outer layers of the stars, and newly synthesised elements in their cores are not available for enriching the interstellar gas unless there is considerable mixing. It is likely that supernova explosions are responsible for much of the chemical enrichment, whereas the overall gaseous content of the interstellar gas is maintained by the more quiescent forms of mass loss described in this section (see also Section 15.2.1).

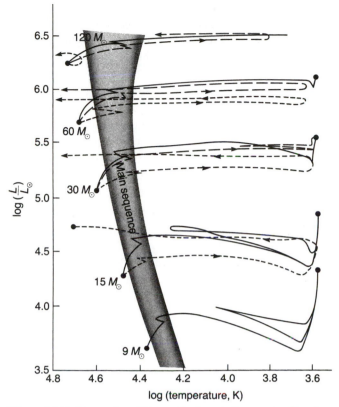

Figure 14.11. Examples of theoretical evolutionary tracks of massive stars once account is taken of the effects of mass loss during their evolution on the main sequence and in the red giant phase. The solid dot at the end of each track corresponds to the onset of carbon burning. The solid lines indicate the evolution of the stars if there is no mass loss. The dashed lines indicate the modification to the evolutionary tracks if there is moderate mass loss, and the dotted lines illustrate the case of strong mass loss. The effect of mass loss is to make the surface temperatures of the stars rather greater than if no mass loss took place. As a result, the evolutionary tracks of the stars are shifted to the left across the H–R diagram, and this is thought to be responsible for the absence of cool stars as luminous as the most luminous of the hot stars, as would be predicted by the evolutionary tracks shown in Fig. 13.15. (After A. Maeder (1981). *Astron. Astrophys.* **102**, 405–6.)

14.6   Binary stars and stellar evolution

The above picture of stellar evolution refers to single stars. We know, however, that many of the stars in our Galaxy are, in fact, double systems, and this can strongly influence their evolution, particularly if they are members of close binary systems. Double star systems can be detected either as *visual binaries* or, if spectroscopic observations show periodic variations of the radial velocities of one or both stars, as *spectroscopic binaries*. The periods of the binary stars can range from a few hours in the case of close binary systems to thousands of years, the upper limit almost certainly being set by the limited period over which precise observations have been carried out (Griffin 1985). More than half the stars in the Galaxy may be members of binary systems according to recent statistics. Part of the great importance of binary stars is that the determination of their orbits provides a direct method of estimating stellar masses.

So far as stellar evolution is concerned, the greatest interest concerns the close binary systems, in which the close proximity of the stars influences their evolution. The separations of close binaries can range from a few times the radii of the stars to systems in which the stars share a common envelope; these are known as *contact binaries*. Close binary systems are known containing a wide range of stellar types, from massive binaries containing O and B stars, through intermediate-mass binaries, to systems containing compact stars, either white dwarfs, neutron stars or black holes, to systems such as the binary pulsar, which consists of a pair of neutron stars. We will deal with systems containing compact stars in the next chapter. Here we consider how stars in binary systems can have very different evolutionary histories from isolated stars.

The most instructive way of understanding how binary stars evolve is through consideration of the gravitational equipotential surfaces of the rotating system. In a frame of reference rotating with the binary system, we add to the gravitational potential of each star a centrifugal potential associated with their binary motion. Thus, if we consider some point at radial distance r from the centre of mass of the binary system, the equipotential surfaces are given by

$$\phi = \frac{GM_1}{r_1} + \frac{GM_2}{r_2} - \Omega^2 r^2 = \text{constant} \qquad (14.8)$$

where r_1 and r_2 are the distances from the centres of the stars of masses M_1 and M_2 to the point at r. This results in the forms of equipotential surface shown in Fig. 14.12. There is a critical equipotential surface which encompasses both stars, which is referred to as the *Roche lobe* of the binary system. The equipotential surfaces within the Roche lobe show that the shapes of the stars are significantly distorted from spheres if they fill a significant fraction of their Roche lobe.

In the extreme case of contact binary systems, the common envelope of the binary lies outside the Roche lobe. These stars are recognised by their very short periods, which are less than about half a day, and their distinctive light curves, which approximate much more closely to sine waves than to typical eclipse light curves. Examples of this type of binary are those known as *W UMa-type binaries*. The common envelope leads to a very different type of internal stellar structure, for example, the stars having a common convective envelope. This results in a number of important differences as compared with the properties of isolated stars. For example, the mass–luminosity relation becomes $L \propto M$, rather than $L \propto M^4$ as found for isolated main sequence stars.

For non-contact close binaries, the stars do not fill their Roche lobes and the stars evolve more or less as normal stars. Interesting phenomena occur as the stars evolve off the main sequence. The more massive of the pair evolves off the main sequence at an earlier time than the less massive star, and, as it becomes a red giant, it expands to fill its Roche lobe. Matter always seeks the lowest gravitational

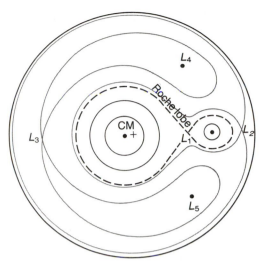

Figure 14.12. Illustrating the equipotential surfaces of a binary star system in the rotating frame of reference. The equipotential surfaces shown correspond to the sum of the equipotential surfaces for each star plus a centrifugal potential term to take account of the rotation of the stars about their common centre of mass (CM). In this example, the mass ratio of the stars is 10:1. These surfaces define the shapes of the stars in the binary system. The equipotential surface which connects both stars is known as the *Roche lobe* of the system, and the point at which the two lobes touch is called the inner Lagrangian point L_1. Other turning points in the value of the potential are labelled L_2, L_3, etc. In close binary stars, the common surface of the stars may well lie outside the Roche lobe. (After S.I. Shapiro and S.A. Teukolsky (1983). *Black holes, white dwarfs and neutron stars: the physics of compact objects*, p. 400. New York: Wiley-Interscience.)

potential, and this is achieved if matter passes through the Lagrangian point L_1 onto the secondary companion. In this way, the mass of what was initially the less massive star increases whilst the mass of the primary decreases. In the case of massive binaries, this can lead to the secondary component becoming more massive than primary. It is a sequence of events such as this which can lead to apparently anomalous situations in which a low-mass white dwarf is found as a companion to an intermediate or high-mass star.

There are many variations upon this theme of mass transfer in close binary systems (Fig. 14.13). The secondary star may now evolve into a red giant and the reverse process of mass transfer back onto the original primary can occur. The end point could be the formation of a binary white dwarf or a white dwarf–neutron star pair. Another intriguing variation occurs if one of the stars undergoes a supernova explosion. These explosions may be associated with either low-mass or high-mass stars, as described in Section 15.2. In the process, a considerable amount of mass is ejected from the system, and the binary may either remain bound or unbound. If the less massive star explodes, due, for example, to mass transfer onto a white dwarf, the system can remain bound, and this provides a means of creating a binary system containing a neutron star. If the more massive

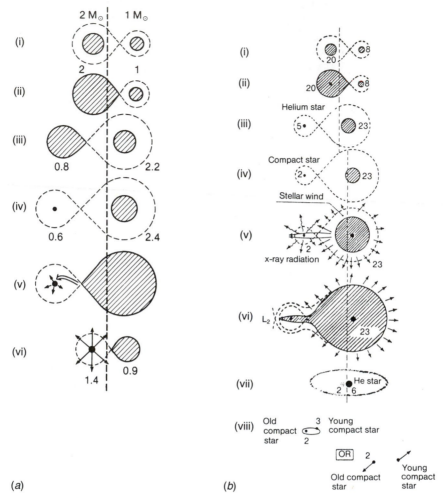

(a)   *(b)*

Figure 14.13. Examples of the evolution of stars in binary systems: (*a*) A low-mass binary. The stages of evolution are as follows. (i) Both stars on the main sequence; (ii) mass transfer from the more massive of the stars to the lighter once the former becomes a red giant; (iii) formation of a light sub-giant and a more massive main-sequence companion; (iv) formation of a white dwarf and a main-sequence star; (v) mass transfer onto the white dwarf when the second star becomes a giant; (vi) the white dwarf exceeds the Chandrasekhar mass and explodes as a Type I supernova. (*b*) A massive binary. (i) Main-sequence phase; (ii) more massive star fills Roche lobe and mass transfer begins; (iii) end of first mass transfer phase; the first Wolf–Rayet phase begins; (iv) the helium star has exploded leaving a $2M_\odot$ remnant, which may be a neutron star or a black hole; (v) the massive star becomes a giant star and mass is transferred onto the compact star which becomes a strong X-ray source; (vi) large-scale mass loss onto the secondary star begins; (vii) the second Wolf–Rayet phase, in which the more massive star has lost its envelope leaving a roughly $6M_\odot$ helium star; (viii) the $6M_\odot$ star explodes as a supernova. The binary may or may not be disrupted in the explosion, depending upon the mass of the remnant. (From H. Karttunen, P. Kröger, H. Oja, M. Poutanen and K.J. Donner (1987). *Fundamental astronomy*, p. 271. Berlin: Springer-Verlag.)

star explodes as a Type II supernova, the system is likely to be disrupted, and then the components disperse at very high velocities, roughly the velocities of the stars in the binary orbit. This is a plausible explanation for the fact that radio pulsars are observed to have large proper motions, corresponding to transverse velocities of typically 300 km s^{-1}.

Various classes of binary system have been associated with most of these permutations. For example, the *symbiotic stars* are believed to be associated with binaries in which mass transfer occurs between a red giant star and a compact companion which may be a dwarf main sequence star or a white dwarf. The *cataclysmic variables* probably consist of semi-detached binaries, in which the 'cataclysmic' variability is associated with an accretion disc about a white dwarf. The *X-ray binaries* consist of a main sequence star and a neutron star or black hole.

It will be noted that, as mass transfer takes place, the inner regions of the star are exposed and material can be deposited onto the surface of the secondary component. These processes provide an explanation for some of the abundance anomalies found in the surfaces of some binary stars. It is also a mechanism by which the *Wolf–Rayet* stars can lose their hydrogen envelopes, exposing the products of hydrogen and helium burning. The theme of binaries containing compact stars will be taken up in more detail in the next chapter and in Chapter 16 on accretion processes in astrophysics.

15

Dead stars

15.1   The formation of dead stars

The types of stars described in Chapters 13 and 14 are held up by the thermal pressure of hot gas, the source of energy to provide the pressure being nuclear energy generation in their cores. As evolution proceeds off the main sequence, up the giant branch and towards the final phases when the outer layers of the star are ejected, the nuclear processing continues further and further along the route to using up the available nuclear energy resources of the star. The more massive the star, the more rapidly it evolves and the further it can proceed along the path to the formation of iron, the most stable of the chemical elements. An intriguing question is whether or not the star is disrupted by the various 'flashes' which are expected to take place as new regimes of nucleosynthesis are switched on, for example, at the points E and G in Fig. 13.19. In the most massive stars, $M \geq 10 M_{\odot}$, it is likely that the nuclear burning can proceed all the way through to iron, whereas, in less massive stars, the oxygen flash, which occurs when core burning of oxygen begins, may be sufficient to disrupt the star. In any case, at the end of these phases of stellar evolution, the core of the star runs out of nuclear fuel, and it collapses until some other form of pressure support enables a new equilibrium configuration to be attained.

The possible equilibrium configurations which can exist when the star collapses are *white dwarfs, neutron stars* and *black holes*. Although we have great confidence that these are the correct end-points of stellar evolution, we have much less understanding of the processes by which they are formed or of the progenitor stars from which they are formed. The problem can be understood by considering the ranges of masses which these 'dead stars' can have. In white dwarfs and neutron stars, the pressure which holds up the star is the quantum mechanical pressure associated with the fact that electrons, protons and neutrons are fermions, that is, only one particle is allowed to occupy any one quantum mechanical state. The white dwarfs are held up by *electron degeneracy pressure*, and can have any mass less than about $1.5M_\odot$. Neutron stars, in which *neutron degeneracy pressure* is responsible for the pressure support can have masses up to about $1.5M_\odot$, possibly slightly higher if the neutron star is rapidly rotating. Thus, according to current understanding, dead stars more massive than about $2M_\odot$ must be black holes. We will discuss the origin of these limits in Section 15.3.1.

Whilst this knowledge is gratifying, it does not help us decide which types of stars become white dwarfs, neutron stars or black holes. For example, low-mass stars, $M < 2M_\odot$, can, in principle, end up in any of the three forms. Even stars with masses much greater than $2M_\odot$ can form white dwarfs or neutron stars if they lose mass sufficiently rapidly. Theoretical calculations have shown that even $10M_\odot$ stars may lose mass very effectively towards the ends of their lifetimes and produce non-black hole remnants.

Some clues are provided by the statistics of stars of different types. Although the statistics are not very well known, the following general picture emerges. The lifetime of a star of mass equal to that of the Sun is about 10^{10} years, roughly the age of the Galaxy. The number of white dwarfs observed locally is roughly consistent with the total number of stars of mass between about 1 and $4M_\odot$ which have completed their evolution since the Galaxy formed. The corresponding rate of formation of white dwarfs is also similar to the present rate of formation of planetary nebulae, suggesting that the latter are important progenitors of white dwarfs. For neutron stars, the statistics are poorer. The total number of neutron stars in the Galaxy derived from the statistics of radio pulsars is roughly the same as the number of stars in the mass range $4–10M_\odot$, assuming a typical lifetime for a pulsar. This formation rate of pulsars is also similar to the supernova rate in our Galaxy, which is probably about one every 50 years.

These statistics are broadly consistent with a general picture in which low-mass stars with $1 \leq M \leq 4M_\odot$ evolve through the planetary nebula phase into white dwarfs. More massive stars, $4 \leq M \leq 10M_\odot$, are the progenitors of neutron stars, and many of these form in supernova explosions. It is normally assumed that stellar-mass black holes are formed in supernova explosions associated with more massive stars, $M \geq 10M_\odot$.

15.2 Supernovae

15.2.1 *General considerations*

The likely sequence of events which leads to the formation of white dwarfs was described in Section 13.3, and corresponds to what we might call the 'peaceful' demise of stars. In contrast, the formation of neutron stars and black holes must be associated with the rapid liberation of huge amounts of energy, the gravitational binding energy of a $1M_\odot$ neutron star being $\sim 10^{46}$ J. These events can be naturally associated with the violent events known as *supernovae*, in which the whole star explodes and its envelope is ejected at high velocity, giving rise to *supernova remnants*. Supernovae are extremely violent and luminous stellar explosions in which the optical luminosity of the star at maximum light

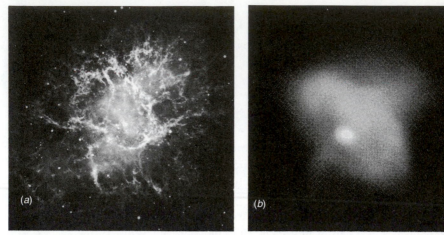

Figure 15.1. (*a*) The Crab Nebula (also known as M1 and NGC 1952), photographed in red light. (Photograph from the Hale Observatories.) (*b*) An X-ray image of the Crab Nebula taken by the ROSAT X-ray observatory. The bright central X-ray source is the Crab pulsar, which has pulse period 33.2 ms and which is the energy source for the nebula. (Courtesy of the Max Planck Institut für Extraterrestrische Physik, Garching, Germany.)

can be as great as that of a small galaxy. Four such supernovae have been observed in our own Galaxy in the last 1000 years: the supernova of 1006; the supernova of 1054, which gave rise to the Crab Nebula (Fig. 15.1); Tycho's supernova of 1572; and Kepler's supernova of 1604 (see Murdin and Murdin 1985). In each of these cases, when the star exploded it became the brightest star in the sky. The supernova of 1006 must have been particularly spectacular, probably having reached magnitude −9, almost 10000 times brighter than the brightest stars. These four supernovae are all relatively nearby, and the natural interpretation is that more distant supernovae which exploded in the plane of our Galaxy would be much more difficult to observe because of the obscuring effect of interstellar dust. For example, only about 250 years ago, the supernova which gave rise to the supernova remnant Cassiopaeia A must have exploded, but it was not recorded by astronomers, presumably because it was too faint to be observed, although its distance is believed to be only about 2.8 kpc. Most recently, the supernova SN1987A exploded in the Large Magellanic Cloud, the dwarf companion galaxy of our own Galaxy, and it is of outstanding importance for understanding supernovae and the late stages of stellar evolution.

Supernovae are frequently observed in other galaxies as part of 'supernova patrols', and it is from these that the properties of supernovae have been determined. Normally, after the initial outburst, which lasts only a few days, the light decays exponentially, with a half-life of about 77 days. Supernovae come in two broad categories. In the case of Type I supernovae, their properties are remarkably uniform, with essentially identical light curves and intrinsic luminosities. In contrast, Type II supernovae exhibit a much more diverse range of properties.

The Type I supernovae pose an intriguing problem since we have to devise a physical scenario by which essentially identical explosions can occur. One appealing idea is that the explosion may result from the collapse of a white dwarf which is accreting matter from a nearby companion in a binary system. When the mass of accreted matter brings the total mass of the star over the critical mass for stability as a white dwarf, the star collapses, liberating about 10^{46} J as a neutron star of mass about $1M_\odot$ is formed. Since the progenitor stars are all of the same type and mass, this can account rather elegantly for the fact that their observed properties are so uniform.

Type II supernovae result from the collapse of more massive stars, probably those with masses greater than about $8M_\odot$. As the collapse of the core of such a star proceeds, the densities and temperatures become sufficiently great for the inverse β-decay process to become important. In this process, energetic electrons interact with protons to form neutrons through the reaction

$$p + e^- \rightarrow n + \nu_e \qquad (15.1)$$

This results in the production of large fluxes of neutrons and neutrinos. The densities are, however, so great that even the neutrinos cannot escape from the collapsing core, the dominant source of opacity involving the weak neutral currents, which gives rise to scattering of neutrinos from nucleons. This is an important example of the impact of discoveries in particle physics upon astrophysics. Thus, the neutrinos and the energy released in the collapse are trapped. How the energy is ultimately released in the form of a supernova outburst is not entirely clear, but one possible picture is that the very central regions collapse to form a neutron star, which slightly overshoots and reverberates, sending a shock wave out through the infalling material. The outer layers of the pre-supernova star are ejected at velocities exceeding its escape velocity. The light curves of Type II supernovae in their initial outburst phases can be satisfactorily modelled by the impulsive liberation of 10^{46} J of energy into the red giant outer envelope of a massive evolved star (Chevalier 1992).

Almost certainly the final stages of collapse of the central regions of a collapsing star are complex. We have not taken account of any rotation or magnetic fields present in the collapsing star, nor the effects of any asymmetry in the collapse. The latter is of the greatest interest because any asymmetric collapse which results in a net quadrupole moment is a strong source of *gravitational waves*. This process is one of the most powerful mechanisms for generating gravitational waves with intensities which should be detectable by the next generation of gravitational wave detectors.

Supernova explosions are probably the origin of most of the heavy elements found in nature. When the core of a massive star collapses, the central temperatures become very high indeed, and, in the ensuing explosion, non-stationary nucleosynthesis can take place. This process is in contrast to nucleosynthesis in the interiors of stars which takes place over long timescales and results in more or less equilibrium abundances of the elements. In non-stationary nucleosynthesis encountered in exploding stars, a much more diverse distribution of elements is found. This process, known as *explosive nucleosynthesis*, is particularly effective in producing many of the heavier elements in the periodic table. Examples of the results of numerical computations of the explosive nucleosynthesis of shells of carbon, oxygen and silicon are shown in Fig. 15.2, indicating how this process can account quantitatively for the observed abundances of many of the heavy elements.

Woosley (1986) provides an excellent up-to-date survey of modern calculations of the products of nuclear burning, both in the case of steady state hydrostatic burning inside stars and in the case of explosive nucleosynthesis. He presents the results of theoretical calculations of the burning of the shells of different chemical compositions expected during the late stages of stellar evolution, as illustrated in Fig. 14.6. The computations have been performed for different astrophysical circumstances, for example, in low- and high-mass stars, in novae and in Types I and II supernovae. Of particular interest is his summary of the likely processes of formation of many of the isotopes in the periodic table, which is reproduced in Table 15.1. The code at the bottom of the Table lists the various processes of nucleosynthesis, the key distinction being between those with the prefix 'E', meaning 'explosive nucleosynthesis', and those without an 'E', indicating hydrostatic nucleosynthesis. It can be seen that many of the most abundant elements are synthesised by steady state hydrostatic processes, for example, the CNO cycle synthesising ^{12}C and ^{16}O, carbon burning producing ^{20}Ne, and oxygen burning producing ^{28}Si and ^{32}S. On the other hand, the processes responsible for creating many of the other isotopes involve explosive nucleosynthesis, for example, most of the heavy elements between sulphur (S) and iron (Fe). Important radioactive species such as ^{26}Al are attributed to explosive nuclear burning.

It would be important to find independent evidence for the process of explosive nucleosynthesis. One of the major puzzles has been to explain why the light curves of supernovae follow closely an exponential decline with a half-life of about 77 days. There is now excellent evidence that this is associated with the radioactive decay of ^{56}Ni into ^{56}Co and then to ^{56}Fe, a decay chain which has exactly the correct half-life to explain the light curve of supernovae (see the reaction chain 15.2). It is expected that large amounts of iron peak elements are synthesised in explosive nucleosynthesis, in particular $0.08 M_{\odot}$ of ^{56}Ni is expected to be one of the principal products. These expectations have been spectacularly confirmed by observations of the supernova SN1987A in the Large Magellanic Cloud (see Section 15.2.2). Another important aspect of this process of explosive nucleosynthesis is that the explosion itself provides an effective mechanism for dispersing the products of nucleosynthesis throughout the interstellar medium.

Two other aspects of supernovae are of special importance in the context of high energy astrophysics. The first is that the kinetic energy of the matter ejected in the explosion is a powerful source of heating for the ambient interstellar gas. The shells of supernova remnants are observable until they are about 100000 years old (Fig. 15.3). At most stages they are observable as intense X-ray sources, in the early stages through the radiation of hot gas originating in the explosion

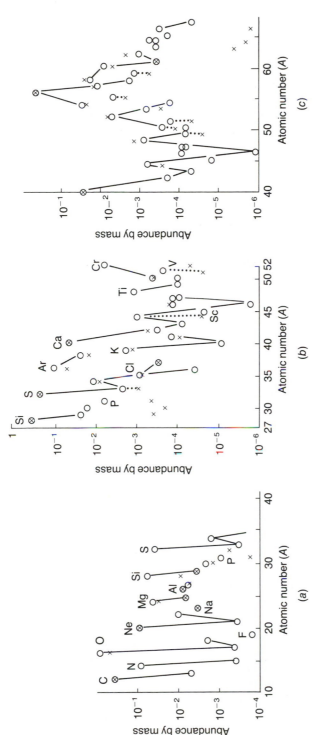

Figure 15.2. Examples of the products of explosive nucleosynthesis. In these computer simulations, shells of carbon, oxygen and silicon are rapidly heated to a very high temperature, as in a supernova explosion, and the nucleosynthesis of heavier elements takes place in an expanding cooling shell. The peak temperatures reached were 2×10^9 K in the case of carbon burning (a), 3.6×10^9 K in the case of oxygen burning (b), and $(4.7 - 5.5) \times 10^9$ K in the case of silicon burning (c). The circles represent the observed solar abundances of the elements, and the crosses the products of explosive nucleosynthesis. It can be seen that there is encouraging agreement between the predicted and observed abundances. (From W.D. Arnett and D.D. Clayton (1970). *Nature*, **227**, 780; see also S.E. Woosley, W.D. Arnett and D.D. Clayton (1973). *Astrophys. J. Suppl.*, **26**, 231.)

Table 15.1. *Important processes in the synthesis of various isotopes[a,b] (from Woosley (1986))*

^{12}C	He	^{32}S	O, EO	^{49}Ti	ESi^c, EHe^c
^{13}C	H, EH	^{33}S	EO	^{50}Ti	nnse
^{14}N	H	^{34}S	O, EO	^{50}V	ENe, nnse
^{15}N	EH^c	^{36}S	EC, Ne, ENe	^{51}V	ESi^c
^{16}O	He	^{35}Cl	EO, EHe, ENe	^{50}Cr	EO, ESi
^{17}O	EH, H	^{37}Cl	EO, C, He	^{52}Cr	ESi^c
^{18}O	H, EH, He	^{36}Ar	EO, ESi	^{53}Cr	ESi^c
^{19}F	EH, He(?)	^{38}Ar	O, EO	^{54}Cr	nnse
^{20}Ne	C	^{40}Ar	?, Ne, C	^{55}Mn	ESi^c, nse^c
^{21}Ne	C, ENe	^{39}K	EO, EHe	^{54}Fe	ESi, EO
^{22}Ne	He	^{40}K	He, EHe, Ne, ENe	^{56}Fe	ESi^c, nse, αnse^c
^{22}Na	EH, ENe	^{41}K	EO^c	^{57}Fe	nse^c, ESi^c, αnse^c
^{23}Na	C, Ne, ENe	^{40}Ca	EO, ESi	^{58}Fe	He, nnse, C, ENe
^{24}Mg	Ne, ENe	^{42}Ca	EO, O	^{59}Co	αnse^c, C
^{25}Mg	Ne, ENe, C	^{43}Ca	EHe, C	^{58}Ni	αnse, ESi
^{26}Mg	Ne, ENe, C	^{44}Ca	EHe	^{60}Ni	αnse^c
^{26}Al	ENe, EH	^{46}Ca	EC, C, Ne, ENe	^{61}Ni	αnse^c, ENe, C, EHe^c
^{27}Al	Ne, ENe	^{48}Ca	nnse	^{62}Ni	αnse^c, ENe, O
^{28}Si	O, EO	^{45}Sc	EHe, Ne, ENe	^{64}Ni	ENe
^{29}Si	Ne, ENe, EC	^{46}Ti	EO	^{63}Cu	ENe, C
^{30}Si	Ne, ENe, EO	^{47}Ti	EHe^c	^{65}Cu	ENe
^{31}P	Ne, ENe	^{48}Ti	ESi^c	^{64}Zn	EHe^c, αnse^c

[a]The most important process is listed first, and additional (secondary) contributions follow.

[b] The coding of the different nuclear reactions is as follows:

H = hydrogen burning; EH = explosive hydrogen burning, novae

He = hydrostatic helium burning; EHe = explosive helium burning (especially Type I supernovae)

C = hydrostatic carbon burning; EC = explosive carbon burning

Ne = hydrostatic neon burning; ENe = explosive neon burning

O = hydrostatic oxygen burning; EO = explosive oxygen burning

Si = hydrostatic silicon burning; ESi = explosive silicon burning

nse = nuclear statistical equilibrium (NSE)

αnse = α-rich freeze out of NSE

nnse = neutron-rich NSE.

[c] Radioactive progenitor.

itself (Fig. 7.14), and in the later stages through the heating of the ambient gas to a high temperature as the shock wave advances ahead of the shell of expelled gas (Fig. 15.3(*b*)). In both cases, the emission mechanism is the bremsstrahlung, or free–free emission, of hot ionised gas. Thus, the kinetic energy of the expanding supernova remnant is a powerful heating source for the interstellar gas, regions up to about 50 pc about the site of the explosion being heated to temperatures of

Figure 15.3 (*a*). The Cygnus Loop (NGC 6960-92) observed in red light by the Palomar
48-inch Schmidt Telescope (photograph from the Hale Observatories). It is an old
supernova remnant, probably about 50000 years old.

10^6 K or greater. The overlapping of these expanding spheres of hot gas can result
in a substantial fraction of the interstellar gas being heated to this temperature.
This picture is consistent with observations of the soft X-ray flux observed from
the interstellar medium, and of highly excited ions, such as five-times ionised
oxygen (OVI) and triply ionised carbon (CIV) in absorption in the interstellar
medium (see Section 17.2).

The second important aspect is that supernovae are sources of very high energy
particles. Direct evidence for this comes from the synchrotron radio emission of
supernova remnants. This topic is central to the study of high energy processes
in astrophysics, and we deal with it and its many ramifications in Section 19.5.

15.2.2 *The supernova SN1987A*

One of the most exciting and important astronomical events of recent
years has been the explosion of a supernova in one of the dwarf companion
galaxies of our own Galaxy, the Large Magellanic Cloud. This supernova, known
as SN1987A, was first observed on 24 February 1987, and reached about third
visual magnitude by mid-May 1987 (Fig. 15.4). It was therefore the brightest
supernova since Kepler's supernova of 1604, and the first bright supernova to be

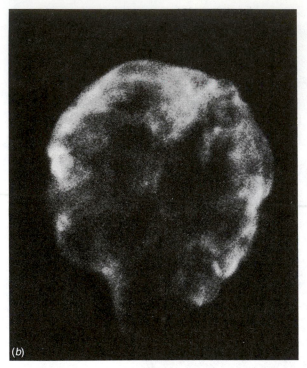

Figure 15.3 (*b*). The Cygnus Loop observed by the ROSAT X-ray Observatory. (Courtesy of the Max Planck Institut für Extraterrestrische Physik, Garching, Germany.)

studied in detail with all the power of modern instrumentation. It presented a unique opportunity for studying the physics of supernovae, and many astronomers have regarded it as 'the astronomical event of the century'. Ironically, it appears that it was a peculiar Type II supernova in that the light curve showed a much more gradual increase to maximum light than is typical of Type II supernovae (Fig. 15.5(*a*)) – it took 80 days to reach maximum light, and its bolometric luminosity then remained roughly constant at magnitude 4 for about two months after the explosion, despite the fact that there was a rapid decline in its surface temperature. It was also subluminous as compared with the typical Type II supernova. Chevalier (1992) provides an excellent survey of what is known about the supernova from the first five years of observation.

The supernova coincided precisely with the position of the bright blue supergiant star Sanduleak -69 202, which disappeared following the supernova explosion as revealed by observations with the International Ultraviolet Explorer (IUE). This observation indicated that the progenitor of the supernova was a massive early-type B3 star. The fact that the progenitor was a highly luminous blue star was a surprise, because it was expected that, following the considerations of Section 13.3, it should have been a red supergiant star. A clue to the evolution of the progenitor was provided by the observation of dense gas shells close to the supernova. It is now believed that the progenitor did evolve to become a red giant,

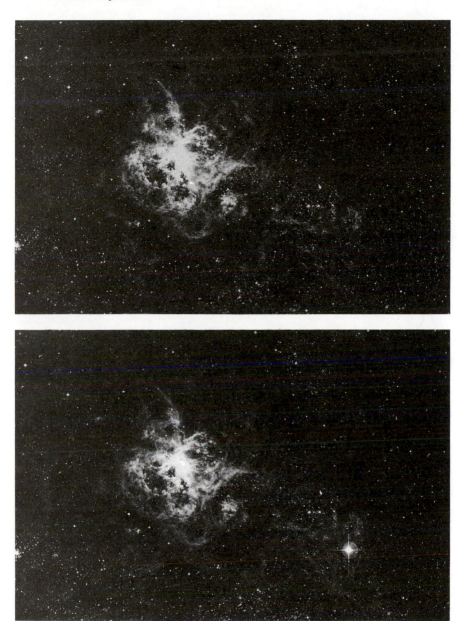

Figure 15.4. The field of SN1987A before and after the supernova explosion, which was first observed on 24 February 1987. (Courtesy of Photolabs, Royal Observatory, Edinburgh. Copyright © 1987 Royal Observatory, Edinburgh.)

but that strong mass loss occurred which blew off the outer layers and so the star became a blue rather than a red supergiant, as suggested by the evolutionary tracks with strong mass loss shown in Fig. 14.11. This phase of mass loss must have ended about 10^4 years ago, before the star exploded as a blue supergiant.

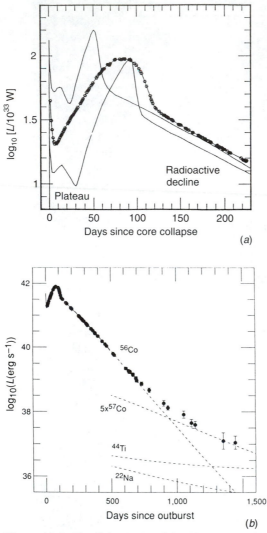

Figure 15.5. The light curve of SN1987A. (*a*) The light curve for the first 200 days, showing the prolonged initial outburst and the subsequent exponential decay with a half-life of 77 days. (*AAO Newsletter*, December 1989.) (*b*) The bolometric luminosity of the supernova in the ultraviolet, optical and infrared wavebands during the first five years. The energy deposited by radioactive nuclides (dashed lines) is based upon the following initial masses: $0.075 M_\odot$ of ^{56}Ni (and subsequently ^{56}Co), $10^{-4} M_\odot$ of ^{44}Ti, $2 \times 10^{-6} M_\odot$ of ^{22}Na, and $0.009 M_\odot$ of ^{57}Co, the last being five times the value expected from the solar ratio of ^{56}Fe/^{57}Fe. (From R.A. Chevalier (1992). *Nature*, **355**, 691.)

One of the pieces of great good fortune was that, at the time of the explosion, neutrino detectors were operational at the Kamiokande experiment in Japan and at the Irvine–Michigan–Brookhaven (IMB) experiment located in an Ohio salt-mine in the USA (see Section 7.5). Both experiments were designed for an entirely different purpose, which was to search for evidence of proton decay, but the signature of the arrival of a burst of neutrinos was convincingly demonstrated in both experiments. Only 20 neutrinos with energies in the range 6–39 MeV were detected (12 at Kamiokande and 8 at IMB), but they arrived almost simultaneously at the two detectors, the duration of this pulse being about 12 s (Bahcall (1989)). This signal was far above the background noise in the detectors. The timescale of 12 s is consistent with what would be expected when allowance is made for neutrino trapping in the core of the collapsing star. This observation, coupled with the measured energies of the neutrinos, enable limits to be set to the rest mass of the neutrino. If the electron neutrino had a finite rest mass, the more energetic neutrinos would be expected to arrive before the less energetic ones, since they would have velocities closer to that of light. It is a useful exercise in special relativity, which is left to the reader, to determine the upper limit to ⸤he mass of the electron neutrino from these data, given that the distance of the Large Magellanic Cloud is about 55 Mpc. The limits derived from these considerations correspond to $m_{\nu_e} \leq 20$ eV, a limit intriguingly close to the value needed to close the Universe by primordial electron neutrinos with finite rest mass. This is another example of astronomy providing information of importance for fundamental physics. What makes this identification of the neutrino pulse with the supernova so convincing is the fact that the supernova was only observed optically some hours after the neutrino pulse. The neutrinos escape more or less directly from the centre of the collapse of the progenitor, whereas the optical light has to diffuse out through the supernova envelope.

The observation of the neutrino flux from the supernova is uniquely important for the theory of stellar evolution. The supernova was ideally placed from the astronomical point of view in that the distance to the Large Magellanic Cloud is accurately known, and hence the neutrino luminosity can be found. Adopting standard cross-sections for the neutrino interactions and the size of the detectors, it turns out that the neutrino luminosity of the supernova was of the same order as that expected from the formation of a neutron star ($E \approx 10^{46}$ J). In so far as the neutrino energy spectrum can be determined from the small number of neutrinos detected, it is consistent with what would be expected when a neutron star forms. Notice, also, that these observations provide strong support for the essential correctness of our understanding of the late stages of stellar evolution.

The light curve of the supernova has now been followed for almost five years after the initial explosion (Fig. 15.5(*b*)). After the initial outburst, which lasted much longer than is usual in Type II supernovae, the luminosity decayed exponentially with the characteristic half-life of 77 days until roughly 800 days after the explosion, after which time the rate of decline decreased. The early phases of development of the light curve have suggested that the progenitor star must have been massive, $M \approx 20 M_{\odot}$, which is consistent with the mass of the B-star Sanduleak -69 202. Stellar evolution models, which begin with this mass, have been developed in which the progenitor first becomes a red giant and then, because of strong mass loss, moves to the blue region of the H–R diagram for

10^4 years, before exploding as a supernova. This picture is consistent with the best models of the early light curve, which require the star to have a smaller envelope than is usual for the B-star and a lower abundance of heavy elements than the standard cosmic abundances, roughly one-third of the solar value. This last result is consistent with the general trend of the heavy element abundances of stars in the Large Magellanic Cloud.

One of the most intriguing observations has been the search for the products of the radioactive decay chains in the optical-infrared spectrum of the supernova. The radioactive chain responsible for the exponential decrease in the luminosity of the supernova is as follows:

$$^{56}\text{Ni} \xrightarrow{\beta^+} {}^{56}\text{Co} \xrightarrow{\beta^+} {}^{56}\text{Fe} \qquad (15.2)$$

The first β-decay has a half-life of only 6.1 days, whereas the second β-decay, which has half-life 77.1 days, is inferred to be the source of energy for the exponential decay of the luminosity of the supernova. 3.5 MeV of energy is liberated in the form of γ-rays in the decay of each ^{56}Co nucleus. It is therefore possible to work out the amount of ^{56}Ni produced in the supernova explosion directly from the bolometric luminosity of the supernova. It is inferred that about $0.07 M_\odot$ of ^{56}Ni was deposited in the supernova envelope, and this figure agrees very well with the theoretical expectations of explosive nucleosynthesis. It is therefore expected that the ratio of abundances of ^{56}Ni and ^{56}Co to iron should decrease as the radioactive elements decay. As illustrated in Figs. 5.8 and 5.9, evidence was found for the γ-ray lines of ^{56}Co within six months of the explosion, as well as fine structure lines of cobalt and nickel in the infrared spectrum of the supernova once the exponential decrease in luminosity began. It can be seen that a substantial amount of cobalt is present in the supernova, a key diagnostic of the reaction chain (15.2).

More detailed studies have revealed an enormous amount about the nature of supernova explosions. The γ-rays released in the radioactive decay of ^{56}Co have energies of the order 1 MeV, and this energy is communicated to the gas by Compton scattering until, at an energy of about 20 keV, photoelectric absorption becomes the dominant loss process (see Fig. 4.16, but modified for the cosmic abundances of the elements). If all the ^{56}Co were created at the centre of the explosion, it was expected that, after about one year, the envelope would become optically thin and reveal a Comptonised X-ray spectrum with a low energy cut-off at 20 keV due to photoelectric absorption. In fact, this X-ray spectrum, as well as the γ-ray lines, were observed after only six months. This strongly suggests that, at an early stage, the radioactive products were mixed into the material of the envelope of the star. The likely causes of this mixing are hydrodynamic instabilities related to the classical Kelvin–Helmholtz instability. Fig. 15.6 shows the results of numerical simulations of the distribution of the oxygen abundance at an age of 3.5 hours, according to the computations of W.D. Arnett and his colleagues.

There are other important features of the light curve shown in Fig. 15.5. The optical light curve alone shows a break at about 500–600 days, but, at that same

Figure 15.6. The oxygen abundance structure found in a two-dimensional simulation for SN1987A at an age of 3.5 hours. The complex structure results from hydrodynamic instabilities at the composition interface. (From R. A. Chevalier (1992). *Nature*, **355**, 691, after computations by W. D. Arnett and his colleagues.)

time, the far infrared flux increased so that the total luminosity continued to decrease exponentially, as seen in Fig. 15.5(*b*). In addition, observations of the near infrared lines of iron showed that less than $0.075 M_\odot$ of iron was present. At about the same time, the emission lines showed absorption of the redshifted gas, indicating that absorption was taking place within the supernova. All these observations are consistent with the formation of dust within the supernova ejecta after about 500 days.

After about 900 days, the rate of decline of the total luminosity of the supernova decreased. The natural interpretation of this phenomenon is that another longer lived radioactive nuclide had taken over from ^{56}Co, and the expected candidate is ^{57}Co. In Fig. 15.5(*b*), the expected light curve is shown, assuming that ^{57}Co is five times more abundant than it is locally; also shown are the expected contributions of the longer lived radionuclides ^{44}Ti and ^{22}Na. The totality of these observations provides direct confirmation for the radioactive theory of the origin of the supernova light curve and the formation of iron peak elements in supernova explosions.

In addition to information about supernovae, SN1987A has provided a vast amount of information about the surrounding interstellar medium. In a word, the supernova outburst has illuminated the material ejected in previous mass-loss events, particularly from the period of strong mass loss during the red giant phase before the progenitor became a blue supergiant. There is evidence for bipolar emission structures, as well as for 'light echoes' from dust 'sheets' lying along the line of sight between the supernova and the observer (Fig. 15.7(*a*)). Perhaps the most intriguing phenomenon is the discovery of a ring of emission in the forbidden line of doubly ionised oxygen [OIII] about the supernova by the Hubble Space Telescope (Fig. 15.7(*b*)). This ring was excited by the initial outburst of ultraviolet radiation from the supernova. The ring is in the form of an almost perfect ellipse, and so, from the ratio the lengths of its major and minor axes, the angle of inclination can be found, assuming the ring is actually circular.

The ultraviolet spectrum of the supernova has been regularly monitored by the International Ultraviolet Observatory (IUE), and, after a certain time t_0, forbidden ultraviolet emission lines were observed, which increased to a maximum intensity after a time t_{max}. Panagia *et al.* (1991) make the natural interpretation that the forbidden ultraviolet emission lines also originate from the ring. Now, if the ring were circular and lying in the plane of the sky, it would all have been illuminated at the same time according to the observer on Earth, but, because it is inclined and the speed of light is finite, the distant observer observes the emission from the nearside of the ring first at the time t_0 after the explosion. Only after some later time t_{max} is the total radiation from the ring observed, when the light from the far side of the ring reaches the observer. If we measure time from the point at which the supernova was observed to explode, it is a simple calculation to show that $t_0 = (R_r/c)(1 - \sin i)$ and $t_{max} = (R_r/c)(1 + \sin i)$, where R_r is the radius of the circular ring and i is the angle of inclination, the angle between the plane of the ring and the plane of the sky. Thus, the ratio of t_0 to t_{max} gives an independent estimate of the angle i, which turns out to be in excellent agreement with that determined from the ellipticity of the ring. From these relations, the radius of the ring, R_r, can be found simply from light travel time arguments. In their more detailed analysis, Panagia and his colleagues find the diameter of the ring to be $(1.27 \pm 0.07) \times 10^{16}$ m, and, combining this with the observed angular diameter of the ring, 1.66 ± 0.03 arcsec, a distance of 51.2 ± 3 kpc is obtained. This is a remarkably accurate distance for the Large Magellanic Cloud, and is in excellent agreement with independent estimates. In particular, the distance has also been estimated using the Baade–Wesselink technique applied to the expanding photosphere of the supernova (see Appendix A). B.P. Schmidt and R.P. Kirshner (1993, personal communication) find a distance of 49 ± 3 kpc using this technique.

This is a lovely example of how quite unexpected bonuses can be found when remarkable phenomena such as the supernova SN1987A occur and a facility such as the Hubble Space Telescope is available. The ring of gas must have been created during the mass-loss phase of the progenitor star. If the outflow during the red to blue supergiant transition was indeed in the form of a bipolar outflow,

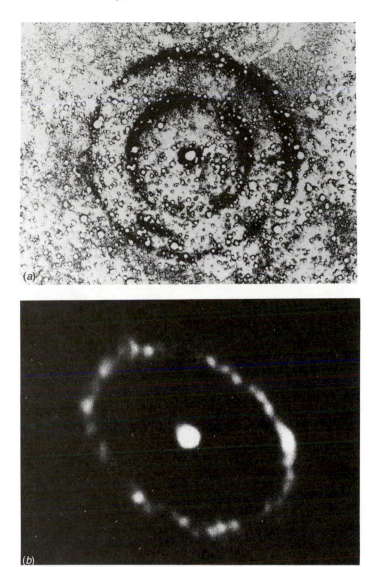

Figure 15.7. (*a*) 'Light echo' rings about the supernova SN1987A due to clouds or 'sheets' of interstellar dust lying along the line of sight between the supernova and the observer. (From AAO Annual Report 1990.) (*b*) The ring of ionised gas about the supernova excited by the ultraviolet radiation emitted in the initial outburst. This image, taken in the forbidden line of doubly ionised oxygen [OIII], was made with the Faint Object Camera of the Hubble Space Telescope. (From N. Panagia, R. Gilmozzi, F. Macchetto, H-M. Adorf and R. P. Kirshner (1991). *Astrophys. J.*, **380**, L23.)

the circular ring may well have formed in the equatorial plane of the outflow, similar to what is believed to occur in the bipolar outflows about protostars and young stars (see Section 17.5.1). Incidentally, the reasoning used to determine the physical diameter of the ring is exactly the same as that used to evaluate the dynamics of superluminal radio sources (see Volume 3).

15.3   The structures of white dwarfs and neutron stars

15.3.1   *White dwarfs and neutron stars as degenerate stars*

In the cases of both white dwarfs and neutron stars, there is no internal heat source – the stars are held up by degeneracy pressure. The role of degeneracy pressure comes about very naturally because, in the centres of stars at an advanced stage in their evolution, the densities become high and the use of the pressure formulae for a classical gas is inappropriate. The combination of Heisenberg's uncertainty principle and Fermi's exclusion principle for fermions ensures that, at very high densities, when the interparticle spacing becomes small, the particles of the gas must possess large momenta and cannot occupy the same quantum state. According to the uncertainty relation, $\Delta p \Delta x \approx \hbar$, and it is these large quantum mechanical momenta which provide the pressure of the degenerate gas. The results described in Section 15.1 concerning the upper mass limits for white dwarfs and neutron stars are a consequence of the equation of state for degenerate matter, and it is such an important result that it is worthwhile showing how it comes about.

First of all, we work out the physical conditions under which degeneracy pressure is important. If the electron–proton plasma is in thermal equilibrium at temperature T, the root mean square velocity of the particles is given by $\frac{1}{2}m\langle v^2 \rangle = \frac{3}{2}kT$, and hence the typical momentum of the particles is $p = mv \approx (3mkT)^{1/2}$. According to the Heisenberg uncertainty principle, the interparticle spacing at which quantum mechanical effects become important is $\Delta x \approx \hbar/\Delta p$, and hence, setting $\Delta p = p$, the density of the plasma, which is mostly contributed by the protons, is

$$\rho \approx m_{\mathrm{p}}/(\Delta x)^3 \approx m_{\mathrm{p}} \left(\frac{3mkT}{\hbar} \right)^{3/2} \tag{15.3}$$

where m is the mass of the particle which provides the degeneracy pressure. Because the electrons are much lighter than the protons and neutrons, they become degenerate at much larger interparticle spacings, and hence at lower densities, than the protons and neutrons. Thus, the density at which degeneracy occurs in the non-relativistic limit is proportional to $T^{3/2}$.

A better calculation for this critical density is performed by Kippenhahn and Weigert (1990), who give the complete expression for the pressure of a denerate gas applicable for any chemical composition of the stellar material. What complicates the problem is that, in general, the material can be in any state of ionisation. Following their conventions, the density of material ρ can be written in terms of the atomic mass unit m_{u} in three ways:

$$\rho = (n + n_{\mathrm{e}})\mu m_{\mathrm{u}} = n\mu_0 m_{\mathrm{u}} = n_{\mathrm{e}}\mu_{\mathrm{e}} m_{\mathrm{u}} \tag{15.4}$$

n_{e} is the number density of electrons and n is the number density of nuclei in the plasma; μm_{u}, $\mu_0 m_{\mathrm{u}}$ and $\mu_{\mathrm{e}} m_{\mathrm{u}}$ are the average particle masses per free particle (μ), per nucleus (μ_0) and per electron (μ_{e}), respectively. Thus, for example, for a fully ionised hydrogen plasma, $\mu = 0.5$, $\mu_0 = 1$ and $\mu_e = 1$; for fully ionised helium, $\mu = 1.33$, $\mu_0 = 4$ and $\mu_e = 2$; for fully ionised iron, $\mu = 56/29 \approx 2$, $\mu_0 = 28$ and $\mu_e = 2$. Obviously, for mixtures and partially ionised gases, the values of the μ's differ from these simple cases.

Equating the pressure of a degenerate electron gas in the non-relativistic limit (equation 15.9 below) to the pressure of a classical gas $p = \rho k T / \mu m_u$, we find that the critical density is given by

$$\frac{T}{\rho^{2/3}} = \frac{(3\pi^2)^{2/3}\hbar^2}{5 m_e m_u^{2/3} k} \frac{\mu}{\mu_e^{5/3}}$$

or

$$\rho = 2.38 \times 10^{-5} \left(\frac{T}{\mu}\right)^{3/2} \mu_e^{5/2} \text{ kg m}^{-3} \tag{15.5}$$

where T is the temperature in kelvin. Fig. 15.8 is a plot of temperature against density, showing the regions in which the different forms of equation of state apply. Also shown on this diagram is a heavily dashed line, showing the conditions of temperature and density from the centre to the surface of the Sun. It can be seen that, in stars like the Sun, the equation of state can always be taken to be that of a classical gas. When the star moves off the main sequence, however, the central regions contract, and, although there is a modest increase in temperature, the matter in the core can become degenerate, and this plays a crucial role in the evolution of stars on the giant branch. Ultimately, in the white dwarfs, the densities are typically about 10^9 kg m^{-3}, and so they are degenerate stars.

The next important consideration is whether or not the electrons are relativistic. To order of magnitude, we can find the condition for the electrons to become relativistic by setting $\Delta p \approx m_e c$ in Heisenberg's uncertainty relation, and then, by the same arguments as above, we find that the density is

$$\rho \sim \frac{m_p}{(\Delta x)^3} \sim m_p \left(\frac{m_e c}{\hbar}\right)^3 \sim 3 \times 10^{10} \text{ kg m}^{-3} \tag{15.6}$$

A better calculation, with exactly the same physics but expressed in a slightly different way, is to require the Fermi momentum of a degenerate Fermi gas in the zero temperature limit to be $m_e c$ (see Kippenhahn and Weigert 1990). In this case, the density at which the electrons become relativistic is

$$\rho = \frac{m_u}{3\pi^2} \left(\frac{m_e c}{\hbar}\right)^3 \mu_e = 9.74 \times 10^8 \mu_e \text{ kg m}^{-3}$$

This limit is indicated on Fig. 15.8. In the centres of the most massive white dwarfs, the densities attain these values, and so the equation of state for a relativistic degenerate electron gas has to be used. We will find that it is this feature which determines the upper mass limit for white dwarfs and neutron stars.

Next, we can work out by these rough methods the equations of state for degenerate matter in the non-relativistic and relativistic regimes. In general, the relation between pressure and energy density can be written $p = (\gamma - 1)\epsilon$, where p is the pressure, ϵ is the energy density of the matter or radiation which provides the pressure and γ is the ratio of specific heats. In the non-relativistic regime, the

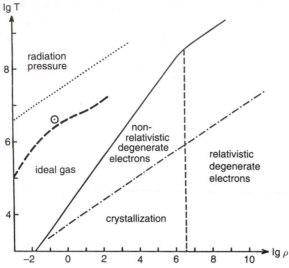

Figure 15.8. A sketch of the temperature–density plane showing the regions in which different types of equation of state are applicable. In addition to the regions discussed in the text, the diagram also shows the regions of the plane in which radiation pressure exceeds the gas pressure and also the region in which the degenerate gas is expected to become a solid, that is, it represents the melting temperature of the stellar material. The heavy dashed line shows the location of the Sun on this diagram from its core to envelope. (After R. Kippenhahn and A. Weigert (1990). *Stellar structure and evolution*, p. 130. Berlin: Springer-Verlag.)

energy of an electron in the degenerate limit is $E = \frac{1}{2}m_e v^2 = p^2/2m_e \approx \hbar^2/2m_e a^2$, where $a \approx \Delta x$ is the interelectron spacing. Therefore, to order of magnitude, the energy density of the material is $\epsilon \approx E/a^3 = \hbar^2/2m_e a^5$. Since the density of matter is $\rho \sim m_p/a^3$, it follows that $p \propto \rho^{5/3}$, and hence the ratio of specific heats $\gamma = \frac{5}{3}$. The pressure of the gas is therefore roughly

$$p \approx \frac{\hbar^2}{3m_e a^5} \approx \frac{\hbar^2}{3m_e}\left(\frac{\rho}{m_p}\right)^{5/3} \tag{15.7}$$

We can repeat this calculation for a relativistic electron gas, in which case $E \approx pc \approx \hbar c/a$, and hence $\epsilon \approx E/a^3 \approx \hbar c/a^4$. Since $\rho \sim m_p/a^3$, $p \propto \rho^{4/3}$ and $\gamma = \frac{4}{3}$. The pressure of the gas is roughly

$$p \approx \frac{\hbar c}{3a^4} \approx \frac{\hbar c}{3}\left(\frac{\rho}{m_p}\right)^{4/3} \tag{15.8}$$

The exact results, which come out of the application of statistical mechanics to a Fermi–Dirac distribution in the ground state, are as follows:

$$\text{non-relativistic} \qquad p = \frac{(3\pi^2)^{2/3}}{5}\frac{\hbar^2}{m_e}\left(\frac{\rho}{\mu_e m_u}\right)^{5/3} \tag{15.9}$$

$$\text{relativistic} \qquad p = \frac{(3\pi^2)^{1/3}\hbar c}{4}\left(\frac{\rho}{\mu_e m_u}\right)^{4/3} \tag{15.10}$$

We obtain the corresponding results for degenerate neutrons if we substitute neutrons for the electrons in the above expresions and set $\mu_e = 1$. Then, the expressions for the pressure of the neutron gas in the two limits are

$$\text{non-relativistic} \qquad p = \frac{(3\pi^2)^{2/3}}{5} \frac{\hbar^2}{m_n} \left(\frac{\rho}{m_n} \right)^{5/3} \tag{15.11}$$

$$\text{relativistic} \qquad p = \frac{(3\pi^2)^{1/3} \hbar c}{4} \left(\frac{\rho}{m_n} \right)^{4/3} \tag{15.12}$$

What makes these results so important is that, in all these cases, the pressure is independent of the temperature. As a consequence, it is remarkably straightforward to find solutions for the internal pressure and density structure inside these stars. Because of its central importance for so much of high energy astrophysics, we outline this calculation here.

15.3.2 *The Chandrasekhar limit for white dwarfs and neutron stars*

We need only two of the equations of stellar structure to carry out the analysis. The first is the *equation of hydrostatic equilibrium*, which states that the inward force of gravity acting on an element of the star should be balanced by the outward pressure gradient. For the case of a spherically symmetric star, all properties depend only upon the radius, r, and so we can write the equations in term of total derivatives. Applying Gauss's law to a unit volume of material of the star at radius r (Fig. 15.9), we find that

$$\frac{dp}{dr} = -\frac{GM\rho}{r^2} \tag{15.13}$$

In this equation, M is the mass within radius r inside the star. The second equation is the *law of conservation of mass*, which tells us that the mass within a shell of thickness dr at distance r from the centre is

$$\frac{dM}{dr} = 4\pi r^2 \rho \tag{15.14}$$

We now reorganise equation (15.13) and differentiate to find a second order differential equation relating p and ρ:

$$\frac{d}{dr} \left(\frac{r^2}{\rho} \frac{dp}{dr} \right) + 4\pi G \rho r^2 = 0 \tag{15.15}$$

Now, we are interested in the cases in which the pressure p depends only upon the density ρ through an equation of state of the form $p = \kappa \rho^\gamma$ with $\gamma = \frac{5}{3}$ and $\frac{4}{3}$. The solutions of this type are known as *polytropes* and are written in terms of the *polytropic index* n such that $\gamma = 1 + \frac{1}{n}$. Thus, if $\gamma = \frac{5}{3}$, $n = \frac{3}{2}$ and if $\gamma = \frac{4}{3}$, $n = 3$. The next step is to change variables so that equation (15.15) is reduced to a more manageable form. First, we write the density at any point in the star in terms of the central density ρ_c in the following way: $\rho(r) = \rho_c w^n$. Then, we write the distance r from the centre in terms of the dimensionless distance z

$$r = az, \qquad \text{where} \qquad a = \left[\frac{(n+1)\kappa \rho_c^{(1/n)-1}}{4\pi G} \right]^{1/2} \tag{15.16}$$

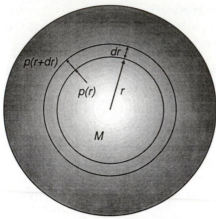

Figure 15.9. Illustrating how to derive the equation of hydrostatic equilibrium. Consider the forces acting on either side of a shell of thickness dr at radius r from the centre of a spherical star. On the inner face of the shell, the outward force is $4\pi r^2 p(r)$, and at radius $r + dr$ it is $4\pi(r + dr)^2 p(r + dr)$. Performing a Taylor expansion on the latter force for small dr and subtracting the first force, we find the net outward force to be $4\pi r^2 (dp/dr)dr$. This outward force is balanced by the inward force of gravity acting on the shell, that is, $-GM(4\pi r^2 \rho dr)/r^2$, where M is the mass contained within radius r. Equation (15.13) follows immediately.

With a little bit of algebra, equation (15.15) becomes

$$\frac{1}{z^2}\left[\frac{\mathrm{d}}{\mathrm{d}z}\left(z^2\frac{\mathrm{d}w}{\mathrm{d}z}\right)\right] + w^n = 0 \tag{15.17}$$

This equation is known as the *Lane–Emden equation*.

Kippenhahn and Weigert (1990) give a splendid account of how this equation can be used to obtain insights into many different phases of stellar evolution. In addition, they provide a very accessible account of its solutions. Analytic solutions exist only for $n = 0, 1$ and 5. For all values of n less than 5, the density goes to zero at some finite radius z_n, which corresponds to the surface of the star at radius $R = az_n$. The solutions of the Lane–Emden equation for $\gamma = \frac{5}{3}$ ($n = \frac{3}{2}$) and $\gamma = \frac{4}{3}$ ($n = 3$) are displayed in Fig. 15.10. The exact values of z at which w goes to zero are $z_{3/2} = 3.654$ and $z_3 = 6.897$ for $n = \frac{3}{2}$ and 3, respectively. From the definition of a, we now find the relation between the central density of the star, ρ_c, and its radius, R, since the latter lies at a fixed value of z for a given value of n. It is obvious from the expression (15.16) that

$$\rho_c \propto R^{2n/(1-n)} \tag{15.18}$$

Thus, for $n = \frac{3}{2}$, $\rho_c \propto R^{-6}$, so that the central density increases as the radius decreases but, notice, much faster than R^{-3}.

Next, we can find the mass–radius relation by integrating the density distribu-

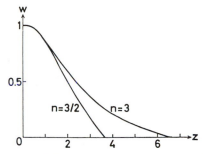

Figure 15.10. Solutions of the Lane–Emden equation for values of the polytropic index $n = \frac{3}{2}$ and 3, corresponding to ratios of specific heat $\gamma = \frac{5}{3}$ and $\frac{4}{3}$, respectively. In both cases, the density falls to zero at a finite value of z. (From R. Kippenhahn and A. Weigert (1990). *Stellar structure and evolution*, p. 177. Berlin: Springer-Verlag.)

tions shown in Fig. 15.7 from $r = 0$ to R:

$$M = \int_0^R 4\pi\rho r^2 \, dr = 4\pi\rho_c \int_0^R w^n r^2 \, dr$$

$$= 4\pi\rho_c a^3 \int_0^{z_n} w^n z^2 dz = 4\pi\rho_c \left(\frac{r}{z}\right)^3 \int_0^{z_n} w^n z^2 dz$$

But from the expression (15.17), we know that

$$\int_0^{z_n} z^2 w^n dz = -\left(z^2 \frac{dw}{dz}\right)_R$$

Therefore,

$$M = 4\pi\rho_c \left(\frac{R}{z_c}\right)^3 \left[-z^2\left(\frac{dw}{dz}\right)\right]_R \tag{15.19}$$

Now, for any polytrope, the expression in square brackets in equation (15.19) is just a constant for a fixed value of n. From the figures quoted by Kippenhahn and Weigert, we find that $[-z^2(dw/dt)]_R$ is 2.71406 if $n = \frac{3}{2}$ and 2.01824 if $n = 3$. Therefore,

$$M \propto \rho_c R^3 \propto R^{(3-n)/(1-n)} \tag{15.20}$$

Thus, if $n = \frac{3}{2}$, $M \propto R^{-3}$. This is an important result – the greater the mass of the star, the smaller its radius and the greater the central density. Consequently, the central density increases rapidly with increasing mass until a critical density is reached, at which the relativistic equation of state with $n = 3$ has to be used instead of $n = \frac{3}{2}$. We can see immediately from equation (15.20) that the consequence of this is that the mass of the star is independent of its radius!

Let us therefore work out the mass of the star in the extreme relativistic case for which $n = 3$. Substituting for $n = 3$ into equation (15.19), we find

$$M = 4\pi \left(\frac{\kappa}{\pi G}\right)^{3/2} \times 2.01824$$

$$= \frac{(3\pi)^{3/2}}{2} \left(\frac{\hbar c}{G}\right)^{3/2} \times \frac{2.01824}{(\mu_e \mu_u)^2}$$

$$= \frac{5.836}{\mu_e^2} M_\odot \tag{15.21}$$

In white dwarf stars, the chemical abundances have evolved through to helium, carbon or oxygen, and therefore we expect the limiting mass for the white dwarfs to correspond to $\mu_e = 2$, and so

$$M_{Ch} = 1.46 M_\odot \tag{15.22}$$

This is the famous *Chandrasekhar mass*. According to the biography by Wali (1991), Chandrasekhar first found the upper limit to the mass of white dwarfs during the sea voyage from India when he first came to England in August 1930. The derivation of this upper mass limit, and the consequence that there is no stable state for a more massive star, led to the famous controversy with Eddington.

The same analysis can be carried out for neutron stars, for which we set $m_u = m_n$ and $\mu_e = 1$. The formal result found from the expression (15.21) is that the upper limit is $M_{ns} \leq 5.73 M_\odot$. As discussed by Shapiro and Teukolsky in Chapter 9 of their text (1983), this is a significant overestimate because a proper general relativistic treatment is needed as well as a realistic equation of state. The relativity parameter $2GM/Rc^2$ for neutron stars of mass $1 M_\odot$ and radius $R = 10$ km is 0.15, and so, for objects of greater mass, a general relativistic treatment is required. The effect of general relativity is to make the effective force of gravity stronger, since the gravitational potential energy contributes to the total mass (see Section 15.6). The various considerations which Shapiro and Teukolsky give in their treatment of this problem suggest that the upper limit for neutron stars must be less than about $3 M_\odot$.

Expressions (15.21) and (15.22) are such important results that it is worthwhile giving a much more physical approach to the problem. We gave a crude derivation of the equations of state for non-relativistic and relativistic degenerate gases, which resulted in expressions (15.7) and (15.8). Using the same approximate methods, we can work out the total internal energy of the star in the ultrarelativistic limit

$$U = V\epsilon = 3Vp \approx V\hbar c \left(\frac{\rho}{m_p}\right)^{4/3}$$

One of the standard results of stellar structure is the *virial theorem*, according to which, for a stable star, the total internal energy, U, is one-half of the total gravitational potential energy, Ω_g, that is,

$$2U = |\Omega_g| \qquad 2V\hbar c \left(\frac{\rho}{m_p}\right)^{4/3} = \frac{1}{2}\frac{GM^2}{R} \tag{15.23}$$

Now, $V \approx R^3$ and $\rho V = M$. Therefore, we find that the left-hand side of equation (15.23) becomes

$$\frac{2\hbar c}{R} \left(\frac{M}{m_p} \right)^{4/3} \tag{15.24}$$

Note the key point that, because we have used the relativistic equation of state, the left-hand side of equation (15.23) depends upon radius as R^{-1}, exactly the same dependence as the gravitational potential energy. Just as in the analysis proceeding for the Lane–Emden equation, the mass of the star does not depend upon its radius. From equation (15.23), we find

$$M \approx \frac{1}{m_p^2} \left(\frac{\hbar c}{G} \right)^{3/2} \approx 2 M_\odot \tag{15.25}$$

dropping constants of order unity. This expression is of exactly the same form as the expression (15.21). Furthermore, this is an upper limit to the mass of the star because inspection of equations (15.23) and (15.24) shows that $|\Omega_g| \propto M^2$, whereas $U \propto M^{4/3}$. Therefore, with increasing mass, the gravitational energy always exceeds twice the internal energy of the star since both energies depend upon the radius R in the same way – consequently, there is no equilibrium state. For lower mass stars, the question of whether or not the star is stable depends upon how close n is to 3, since stable degenerate stars are found for $n < 3$.

Notice that the Chandrasekhar mass depends only upon fundamental constants. One of the more intriguing ways of rewriting the expression (15.25) is in terms of a 'gravitational fine structure constant', α_G. The usual fine structure constant is $\alpha = e^2/4\pi\epsilon_0\hbar c$. The equivalent formula for gravitational forces can be found by replacing $e^2/4\pi\epsilon_0$ in the inverse square law of electrostatics, $F = e^2/4\pi\epsilon_0 r^2$, by GM^2 in Newton's law of gravity, $F = Gm_p^2/r^2$, where m_p is the mass of the proton. Thus, $\alpha_G = Gm_p^2/\hbar c$. Putting in the values, we find $\alpha^{-1} = 137.04$ and $\alpha_G = 5.6 \times 10^{-39}$, the ratio of these constants, $\alpha_G/\alpha = 2.32 \times 10^{40}$, reflecting the differing strengths of the electrostatic and gravitational forces. Therefore, the Chandrasekhar mass is roughly

$$M \approx m_p \alpha_G^{-3/2}$$

In other words, from basic constants of physics we find that the stars typically have about 10^{60} protons. Notice also that the calculation applies equally to white dwarfs and neutron stars, the only difference being that the neutron stars are very much denser than the white dwarfs.

15.3.3   *The internal structures of white dwarfs and neutron stars*

The determination of the internal structures of the white dwarfs and neutron stars depends upon detailed knowledge of the equation of state of the degenerate electron and neutron gases, and this has been the subject of much study (see Shapiro and Teukolsky 1983).

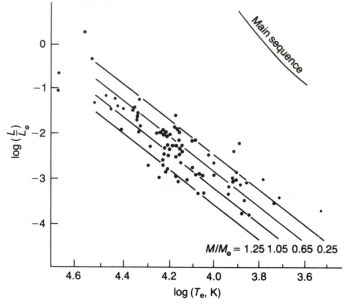

Figure 15.11. Comparison of the theoretical Hertzsprung–Russell diagram for white dwarfs with their observed properties. The location of the cooling curve on the H–R diagram depends upon the mass of the white dwarf. (From S.L. Shapiro and S.A. Teukolsky (1983). *Black holes, white dwarfs and neutron stars: the physics of compact objects*, p. 70, New York: Wiley-Interscience.)

The white dwarfs The case of white dwarfs is the more straightforward. At the typical densities found in white dwarfs, $\rho \sim 10^9$ kg m^{-3}, the equation of state is well understood, the main uncertainty being the chemical composition of the star. Spectroscopic observations of their surface properties show that most of the white dwarfs have lost their hydrogen envelopes. For stars with masses roughly that of the Sun, it is unlikely that nuclear burning proceeds beyond carbon. On the other hand, some massive stars which form iron in their cores may be able to lose mass and angular momentum effectively, and thus form white dwarfs which are composed of iron nuclei. The energy of the star is derived from the internal energy with which the star was endowed when it was formed. The cooling times for the white dwarfs are about $10^9 - 10^{10}$ years, much longer than the thermal cooling timescale for a star like the Sun because their surface areas are much smaller than those of main sequence stars. For the few white dwarf stars which have been found in star clusters, the ages of the clusters are of the same order as the cooling lifetimes of the white dwarfs.

As expected, when converted into an H–R diagram, the white dwarf sequence lies below the main sequence and can give a good account of the observed properties of white dwarfs (Fig. 15.11). In fact, the solid lines are no more than the cooling curves for black bodies, the luminosity L being proportional to T^4.

The neutron stars With increasing density, the degenerate electron gas becomes relativistic, and, when the total energy of the electron exceeds the mass difference between the neutron and the proton, $E = \gamma m_e c^2 \geq (m_n - m_p)c^2 = 1.29$ MeV, the inverse β-decay process (15.1) can convert protons into neutrons. In a non-degenerate electron gas, the neutrons would decay into protons and electrons after about 11 minutes, but this is not possible if the electron gas is highly degenerate and there are no available states for the emitted electron to occupy. The condition that this stabilisation takes place is that the Fermi energy of the degenerate electron gas is greater than the kinetic energy of the emitted electron. In other words, the presence of the Fermi sea of degenerate electrons stabilises the neutrons. For the case of a hydrogen plasma, the critical density can be found as follows. The total energy of the electron must be

$$E_{tot} = (m_n - m_p)c^2 = 1.29 \text{ MeV}$$

We can immediately find the corresponding critical Fermi momentum, p_F, from the standard relation between total energy and momentum:

$$p_F = \gamma m_e v = \left(\frac{E_{tot}^2}{c^2} - m_e^2 c^2 \right)^{1/2}$$

The number density of a degenerate electron gas is given by the standard formula $n_e = (8\pi/3h^3)p_F^3$, from which we find the total density, $\rho = n_e m_u \mu_e$. Taking $\mu_e = 1$ for a hydrogen plasma, we find $\rho = 1.2 \times 10^{10}$ kg m^{-3}. This process is often referred to as *neutronisation*.

For the case of heavier nuclei, which are expected to form the bulk of the matter in white dwarfs and neutron stars, the situation is more complicated. At densities $\sim 10^{10}$ kg m^{-3}, the nuclei form a non-degenerate Coulomb lattice, and the nuclei are the conventional stable elements such as carbon, oxygen and iron. As the density increases, the inverse β-decay reaction (15.1) favours the formation of neutron-rich nuclei. However, the energies needed to achieve this transition are greater than in the case of protons because the neutrons are degenerate within the nuclei, and therefore the electrons must be sufficiently energetic to exceed the Fermi energy within the nucleus. If the nuclei become too neutron rich, however, they begin to break up, and an equilibrium state is set up consisting of neutron-rich nuclei, a free neutron gas and a degenerate relativistic electron gas. This process of releasing neutrons from the neutron-rich nuclei is referred to as *neutron drip*, and sets in at a density of about 4×10^{14} kg m^{-3}.

These processes result in profound changes in the equation of state such that stable stars cannot form until much higher central densities are attained, $\sim 10^{17}$ kg m^{-3}, at which the neutron-drip process has resulted in the conversion of almost all of the matter into neutrons. It is the degeneracy pressure of this neutron gas which prevents collapse under gravity and results in the formation of a *neutron star*. Exactly the same physics described above for the white dwarfs is responsible for providing pressure support for the neutron star, the only difference being that the neutrons are about 2000 times more massive than the electrons, and consequently degeneracy sets in at a correspondingly higher density (expression

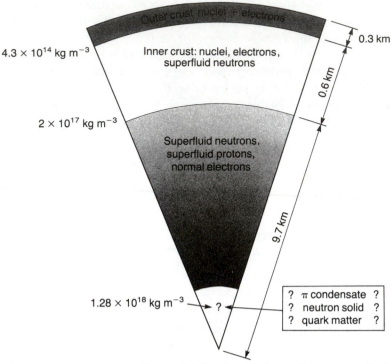

4.3×10^{14} kg m^{-3}

Outer crust: nuclei + electrons

Inner crust: nuclei, electrons, superfluid neutrons

0.3 km

0.6 km

2×10^{17} kg m^{-3}

Superfluid neutrons, superfluid protons, normal electrons

9.7 km

1.28×10^{18} kg m^{-3}

?

? π condensate ?
? neutron solid ?
? quark matter ?

Figure 15.12. A representative model showing the internal structure of a $1.4M_\odot$ neutron star. (After S.L. Shapiro and S.A. Teukolsky (1983). *Black holes, white dwarfs and neutron stars: the physics of compact objects*, p. 251, New York: Wiley-Interscience.)

(15.3)). In addition, as indicated above, a general relativistic treatment is needed to determine the structures of the most massive neutron stars

The internal structures of neutron stars are less well determined because of uncertainties in the equation of state of degenerate nuclear matter. The problems involved in determining the equation of state are elegantly presented by Shapiro and Teukolsky (1983), and Fig. 15.12 shows a representative example of the internal structure of a neutron star. Following Shapiro and Teukolsky, the various zones in the models may be described as follows:

1 The *surface layers* are taken to be the regions with densities less than about 10^9 kg m^{-3}. At these large densities, the matter consists of atomic polymers of ^{56}Fe in the form of a close packed solid. In the presence of strong surface magnetic fields, the atoms become cylindrical. The matter behaves like a one-dimensional solid with high conductivity parallel to the magnetic field and with essentially zero conductivity across it.

2 The *outer crust* is taken to be the region with density ρ in the range $10^9 \le \rho \le 4.3 \times 10^{14}$ kg m^{-3}, and consists of a solid region composed of matter similar to that found in white dwarfs, that is, heavy nuclei forming a Coulomb lattice embedded in a relativistic degenerate gas of electrons.

When the energies of the electrons are great enough, inverse β-decay increases the numbers of neutron-rich nuclei, which would be unstable on Earth; for example, ^{62}Ni forms at a density of 3×10^{11} kg m^{-3}, ^{80}Zn at 5×10^{13} kg m^{-3}, ^{118}Kr at 4×10^{14} kg m^{-3} and so on.

3 The *inner crust* has densities between about 4.3×10^{14} and about 2×10^{17} kg m^{-3} and consists of a lattice of neutron-rich nuclei together with free degenerate neutrons and a degenerate relativistic electron gas. As the density increases, more and more of the nuclei begin to dissolve, and the neutron fluid provides most of the pressure.

4 The *neutron liquid* phase occurs at densities greater than about 2×10^{17} kg m^{-3} and consists mainly of neutrons with a small concentration of protons and electrons.

5 In the very centre of the neutron star, a *core region* of very high density ($\rho \geq 3 \times 10^{18}$ kg m^{-3}) may or may not exist. The existence of this phase depends upon the behaviour of matter in bulk at very high energies and densities. For example, is there a phase transition to a neutron solid, or to quark matter, or to some other phase of matter quite distinct from the neutron liquid at extremely high densities? Many of the models of stable neutron stars do not possess this core region, but it is certainly not excluded that quite exotic forms of matter could exist in the centres of some neutron stars.

In many ways, a neutron star may be thought of as one huge nucleus containing about 10^{60} nucleons. One of the remarkable consequences of the fact that neutron stars consist of matter in bulk at nuclear densities is that, in the inner regions, the neutron stars are likely to be superfluid and the protons superconducting. It is interesting to contrast the physical conditions in neutron stars with laboratory superfluids and superconductors. At low temperatures, liquid helium undergoes a phase transition to the superfluid state. Below the critical temperature, which is about 2.14 K in the case of ^{4}He, the viscosity of the fluid is reduced to zero, and this results in many remarkable phenomena. The reason is that, at these very low temperatures, helium forms a quantum liquid in which most of the atoms are in the same ground state, and the liquid as a whole is subject to the rules of quantum mechanics. Being bosons, the ^{4}He atoms have a great preference for remaining in the ground state, forming what is known as a Bose condensation. ^{3}He also becomes superfluid at a low enough temperature, but now the atoms are fermions and so, in order to create a Bose condensation, ^{3}He atoms pair up with opposite spins so that the pairs obey Bose–Einstein statistics. The physical cause of the phenomenon of superfluidity in this case is long-range attractive forces between the ^{3}He atoms, which result in the fluid being in a lower energy state if pairs of helium atoms remain correlated. If the energy difference, Δ, between the 'paired-state' and the 'unpaired-state' is greater than kT, where T is the temperature of the fluid, the system remains in the lower energy state, the particles forming long-range pairs.

Pairing processes are also responsible for the phenomenon of superconductivity in metals. At low temperatures, almost all the electronic states up to the Fermi level of the metal are filled, and the electrical conductivity is due to a very small fraction of them which are close to the Fermi level. At low enough temperatures, the conduction electrons close to the Fermi level can form pairs with opposite spins because of long-range attractive forces between them, which are due to inter-actions between the electrons and the lattice vibrations. If the energy gap, Δ, asso-ciated with the energy difference between the paired and unpaired states is greater than kT, the lower energy state with the electrons forming *Cooper pairs* is pre-ferred, and, since the pairs of electrons are bosons, they prefer to occupy the same state. As Weisskopf expresses it, the Cooper pairs form a superconducting 'frozen crust' on top of the Fermi distribution (see Weisskopf (1981) for a simple physical explanation of the origin of superconductivity and the formation of Cooper pairs).

There is no attractive force between free neutrons, but there is a net attractive force between the neutrons within an atomic nucleus which is mediated by the bulk nuclear forces in the nucleus. When we consider nuclear matter in bulk, as in the central regions of a neutron star, there are long-range attractive forces between pairs of neutrons, the interaction energy being about 3 MeV. This energy is much greater than that corresponding to the typical internal temperature of a neutron star, which is probably of the order of $kT \sim 1 - 10$ keV (see below). Therefore, it is likely that the neutrons in the central regions of neutron stars form pairs and are superfluid. Specifically, according to Shapiro and Teukolsky (1983), the free neutrons can form a superfluid in the inner crust among the neutron–rich nuclei (region 3). Likewise, in region 4, the liquid neutron phase, in which the nuclei have dissolved into neutrons and protons, the neutron fluid is expected to be superfluid. Finally, the protons in the quantum liquid phase (region 4) are expected to be superconducting. In all these phases, the electrons remain 'normal' in the sense that the interactions between them are not significant enough to produce superconductivity at these temperatures. These phenomena do not have an important influence upon the overall internal structure of the neutron star, but they have a profound impact upon the way in which the interior of the neutron star rotates and upon the way in which the magnetic field behaves in the superconducting fluid.

To this catalogue of exotic phenomena we now add a magnetic field. The observation of polarised radio emission from radio pulsars, and, in particular, the observed rotation of the plane of polarisation of the radiation within the pulses, are powerful evidence for the presence of a magnetic field in pulsars. Field strengths in the range 10^6–10^9 T are suggested from the observed rate of deceleration of pulsars (see Section 15.4). Further evidence for such strong magnetic fields is provided by the observation of a cyclotron radiation feature in the X-ray spectra of the X-ray pulsar Hercules X-1 and others, which indicates that field strengths of the order of 10^8 T are present in the source regions (Fig. 18.2). Although these seem very strong fields, there is no problem in accounting for them astrophysically. This is because, in the collapse of a star, the magnetic field is very strongly tied to the ionised plasma by the process of *magnetic flux freezing* (Section 10.5). When a star collapses spherically, the magnetic field strength increases as $B \propto r^{-2}$ because of conservation of magnetic flux, and so, if a star like the Sun possesses a magnetic field of strength 10^{-2} T, there is no problem in accounting for a field strength of 10^8 T if the star collapses to 10^{-5} times its initial radius. It might be thought that the magnetic field would be expelled from the central regions of the neutron star because of the superconducting proton fluid inferred to be present in the neutron fluid phase. The presence of the normal relativistic degenerate electron gas, however, ensures that the magnetic field can exist within the central regions.

To complete the picture, we must add the rotation of neutron stars. It is the rotation of neutron stars which is responsible for the observation of their pulsed emission at radio and X-ray wavelengths, the pulses being attributed to the swinging of a beam of radiation from the poles of the neutron star past the observer (see Fig. 15.14). It is interesting to compare the observed rotation periods of the neutron stars with the maximum that they could possess. A rough estimate of this may be made by assuming that the neutron star would break up due to centrifugal forces if its rotational kinetic energy is greater than half its gravitational potential energy, that is, the star no longer satisfies the virial theorem. For a $1M_\odot$ neutron star, the break-up rotational period is about 0.5 ms. This is shorter than the observed rotation periods of all pulsars, although pulsars with periods in the range 1–10 ms, the *millisecond pulsars*, are well known objects, the shortest period being only 1.5 ms, which is within a factor of about three of the break-up rotational period.

There are now over 500 *radio pulsars* known. Most of them are isolated objects, but 28 of them are now known to be members of binary systems. Among these, the close binary systems with large orbital eccentricities are of special importance for testing general relativity (Section 15.6.2) and for estimating the masses of neutron stars (Section 15.5). Among the more remarkable objects discovered in the long term monitoring of the arrival times of the pulses at the Earth, has been the system PSR B1257+12. It consists of a pulsar about which two planet-sized objects are orbiting with orbital periods of 66.540 and 98.203 days. It is intriguing that these periods are close to a 3:2 resonance. These binary and multiple systems must have formed by variants upon the evolutionary scenarios outlined in Section 14.6 (Fig. 14.13).

The *X-ray binary stars* are members of close binary star systems, and thus are subject to a wide variety of other phenomena which can influence the properties and evolution of the neutron star. Specific features to be added include the facts that the neutron star, with its rotation and magnetic field, is located in a frame of reference rotating about the centre of mass of the binary system, and that the neutron star accretes matter from the primary star, the accreted matter bringing with it its own angular momentum and magnetic fields.

On general grounds it is expected that neutron stars should be very hot when they form because they have to get rid of their gravitational binding energy, and the cores of the collapsing stars are heated to extremely high temperatures during collapse. The neutron star then cools by thermal radiation from its surface and by neutrino emission from the interior of the star. Below temperatures of about 10^9 K, the neutron star is transparent to neutrinos, and so they provide a very efficient means of getting rid of the thermal energy of the star. The neutrino emission processes are the dominant cooling mechanisms at high temperatures; they involve a variety of neutrino emission processes, including β-decay modified by various neutrino interaction processes which occur in the neutron sea inside the neutron star. The predicted *surface* temperatures of the neutron stars are about 2×10^6 K after about 300 years, and they remain in the range about $(0.5–1.5) \times 10^6$ K for at least 10^4 years. Only the radio pulsars can be used to test this prediction, because the surfaces of those which are members of X-ray binaries are heated by accretion to much greater temperatures. The youngest pulsars known, such as the pulsars in the Crab and Vela supernova remnants, and the pulsar in the Magellanic Clouds, PSR 0540-693, lie in a suitable age range. Despite deep observations with the Einstein X-ray Observatory, only upper limits to the temperatures of the surfaces of these neutron stars of about 2×10^6 K are available. There is thus no inconsistency with theory at the moment, but it will not require a large increase in sensitivity to provide a critical test of the theory of the cooling of neutron stars, and consequently for the physics of condensed matter at nuclear densities.

The discovery of pulsars and X-ray binaries has opened up vast new fields of study for physicists and astrophysicists – the challenge to the theorist and the observer is to devise ways in which, through careful study of the observed properties of these objects, further insight may be obtained into the behaviour of matter under extreme conditions which are inaccessible in terrestrial laboratories.

15.4   Isolated neutron stars – radio pulsars

Radio pulsars were more or less a complete surprise when they were discovered in 1967 by Hewish and Bell (see Hewish *et al.* 1968). They had been predicted as long ago as the 1930s by Baade and Zwicky (1934), soon after the discovery of the neutron, but models of neutron stars suggested that, as compact stars, the only detectable emission should be the thermal radiation from their surfaces. In a prescient paper of 1967, before the announcement of the discovery of pulsars, Pacini predicted that they might be observable at long radio wavelengths if they were magnetised and were oblique rotators. An excellent up-to-date survey of the history, observation and astrophysics of pulsars is given in the monograph *Pulsar astronomy* by Lyne and Graham-Smith (1990).

In 1964, Hewish had established that the fluctuating radio signals observed at low radio frequencies from compact radio sources were due to electron density fluctuations in the interplanetary medium. This provided a new method for finding compact radio sources, many of which were quasars, and also of studying the properties of the interplanetary medium. Hewish designed a large array to undertake these studies, and was awarded a grant of £17286 by the Department of

Scientific and Industrial Research to build it, as well as outstations for measuring the Solar Wind. To obtain adequate sensitivity at the low observing frequency of 81.5 MHz (3.7 m wavelength), the array had to be very large, $4\frac{1}{2}$ acres (1.8 hectares) in area, in order to record the rapidly fluctuating intensities of bright radio sources on a timescale of one-tenth of a second. This was the key technological development which led to the discovery of pulsars, since normally radio astronomical observations require long integration periods to detect faint sources. The first sky surveys began in July 1967, and Jocelyn Bell, Hewish's research student, discovered a strange source which seemed to consist entirely of scintillating radio signals (Fig. 15.13(*a*)). The problem was that the source was not always present, and its nature remained a mystery. In November 1967, the source was observed using a receiver with a much shorter time-constant, and it was found to consist entirely of a series of pulses, with a pulse period of about 1.33 s (Fig. 15.13(*b*)). This pulsar, PSR1919+21, was the first to be identified, and over the next few months three further sources were discovered with pulse periods in the range 0.25 to almost 3 s. This remarkable story has been described by Bell-Burnell (1983) and Hewish (1986).

It was not long after the discovery of pulsars that they were convincingly identified with isolated, rotating, magnetised neutron stars following the proposals by Gold and Pacini. The key observations were the very stable, short periods of the pulses and the observation of polarised radio emission. To account for the observation of radio pulses, the magnetic axis of the star and its rotation axis must be misaligned. The pulses are assumed to originate from beams of radio emission emitted along the magnetic axis, as illustrated in Fig. 15.14. The discoveries of the pulsars in the Crab Nebula and the Vela supernova remnant were of special importance because they are both young pulsars with ages consistent with the ages of these supernova remnants. The very short period of the Crab pulsar, 33 ms, could only be consistent with the presence of a neutron star, and consequently these observations provided convincing evidence that pulsars are rotating neutron stars which form in supernova explosions.

The pulse periods of pulsars can be measured with very high accuracy indeed, and one of the most important parameters is the rate at which the pulse period changes with time. For most pulsars, the rate at which the pulse period increases can be measured, and this can be used to derive an age estimate. The slowing down can be described by a *braking index*, n, which is defined by $\dot{\Omega} = -K\Omega^n$, Ω being the angular frequency of rotation. The braking index provides information about the energy loss mechanism, which is slowing down the rotation of the neutron star. Among the most important of these is magnetic braking. In order to produce pulsed radiation from the magnetic poles of the neutron star, the magnetic dipole must be oriented at an angle with respect to the rotation axis, and then the magnetic dipole displays a varying dipole moment at a large distance (Fig. 15.14). This results in the radiation of electromagnetic energy from the dipole, which is extracted from the rotational energy of the neutron star. It is a useful exercise to show that, by exact analogy with the radiation of an electric

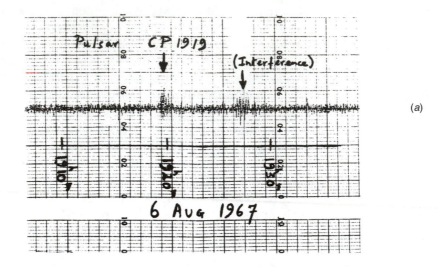

(*a*)

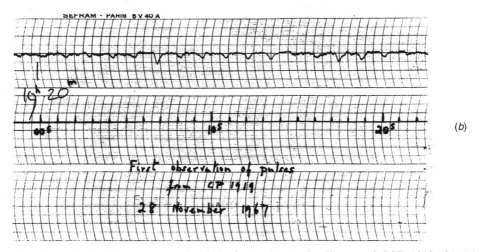

(*b*)

Figure 15.13. The discovery records of the first pulsar to be discovered, PSR 1919+21. (*a*) The first record of the strange scintillating source, labelled CP 1919. Note the subtle differences between the signal from the source and the neighbouring signal due to terrestrial interference. (*b*) The signals from PSR 1919+21 observed with a shorter time-constant than the discovery record, showing that the signal consists entirely of regularly spaced pulses with period 1.33 s. (From A.G. Lyne and F. Graham-Smith (1990). *Pulsar astronomy*, p. 5. Cambridge: Cambridge University Press.)

dipole (Section 3.3), a magnetic dipole, of magnetic dipole moment p_m, radiates electromagnetic radiation at a rate

$$-\frac{\mathrm{d}E}{\mathrm{d}t} = \frac{\mu_0|\ddot{\mathbf{p}}_m|^2}{6\pi c^3}$$

This expression can be simply derived by replacing the electrostatic constants

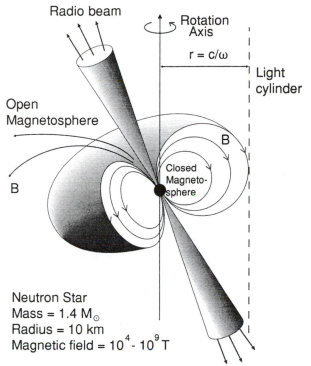

Figure 15.14. A schematic model of a pulsar as a magnetised rotating neutron star, in which the magnetic and rotation axes are misaligned. The radio pulses are assumed to be due to beams of radio emission from the poles of the magnetic field distribution, and are associated with the passage of the beam across the line of sight to the observer. Typical parameters of the neutron star are indicated on the diagram.

$|\ddot{\mathbf{p}}|^2/4\pi\epsilon_0$ in the expression (3.9) by the corresponding magnetostatic constants $\mu_0|\ddot{\mathbf{p}}_{\mathrm{m}}|^2/4\pi$, where $\mathbf{p}_{\mathrm{m}}$ is the magnetic dipole moment of the neutron star. In the case of a rotating magnetic dipole, we can write $\mathbf{p}_m = \mathbf{p}_{\mathrm{m0}}\sin\Omega t$, where Ω is the angular velocity of the neutron star and $\mathbf{p}_{\mathrm{m0}}$ is the component of the magnetic dipole perpendicular to the rotation axis. Consequently, we find that

$$-\left(\frac{\mathrm{d}E}{\mathrm{d}t}\right) = \frac{\mu_0\Omega^4\mathbf{p}_{\mathrm{m0}}^2}{6\pi c^3}$$

Now, the magnetic dipole radiation extracts rotational energy from the neutron star. If I is the moment of inertia of the neutron star,

$$-\frac{\mathrm{d}(\frac{1}{2}I\Omega^2)}{\mathrm{d}t} = -I\Omega\frac{\mathrm{d}\Omega}{\mathrm{d}t} = \frac{\mu_0\Omega^4\mathbf{p}_{\mathrm{m0}}^2}{6\pi c^3} \tag{15.26}$$

Consequently, $\mathrm{d}\Omega/\mathrm{d}t \propto -\Omega^3$, so that the braking index for magnetic dipole radiation is $n = 3$. Direct measurements of the braking index n can be made if the second derivative of the pulsar angular frequency $\ddot{\Omega}$ can be measured. If $\dot{\Omega} = -K\Omega^n$, then $\ddot{\Omega} = -nK\dot{\Omega}^{(n-1)}$. Dividing the latter by the former, we find

$n = \Omega\ddot{\Omega}/\dot{\Omega}^2$. This can also be written

$$n = \frac{\Omega\ddot{\Omega}}{\dot{\Omega}^2} = \frac{v\ddot{v}}{\dot{v}^2} = 2 - \frac{P\ddot{P}}{\dot{P}^2}$$

where $P = 2\pi/\Omega$ is the period of the pulsar.

Braking indices n have been measured for a number of pulsars. For example, for the Crab pulsar, $n = 2.515 \pm 0.005$; for PSR 1509-58, $n = 2.8 \pm 0.2$; for PSR 0540-69, $n = 2.01 \pm 0.02$ (data from Lyne and Graham-Smith (1990)). It is interesting that, in the case of the Crab pulsar, it has also been possible to measure the third derivative of the angular frequency with respect to time, $d^3\Omega/dt^3$, and it is also consistent with the value $n = 2.515$. Thus, although the magnetic breaking may be the cause of the deceleration in some cases, it cannot be the whole story. Part of the deceleration may also be associated with torques exerted on the neutron star by the outflow of particles, which also remove angular momentum from the star. It is interesting that, if the neutron star possessed a significant quadrupole moment in the early stages of formation and evolution, it would radiate gravitational radiation for which the braking index is 5, plainly inconsistent with the observed value for the Crab pulsar.

The age of the pulsar can be estimated if it is assumed that its deceleration can be described by a constant braking index throughout its lifetime. Integrating $\dot{\Omega} = -K\Omega^n$, we find

$$\frac{1}{(n-1)}\left[\Omega^{-(n-1)} - \Omega_0^{-(n-1)}\right] = K\tau$$

where τ is the age of the pulsar and Ω_0 is its initial angular velocity. If $n > 1$ and $\Omega_0 \gg \Omega$, the age of the pulsar is found to be

$$\tau = \frac{\Omega^{-(n-1)}}{K(n-1)} = -\frac{\Omega}{(n-1)\dot{\Omega}} = \frac{P}{(n-1)\dot{P}}$$

It is conventional to set $n = 3$ to derive the age of pulsars, and so $\tau = P/(2\dot{P})$. A plot of $\dot{P}$ against P is shown in Fig. 15.15, from which it can be seen that the typical lifetime for the majority of pulsars is about 10^7 years. The millisecond pulsars have much smaller values of $\dot{P}$ than those with periods greater than about 0.1 s, and consequently they have much greater ages, in the most extreme cases of the order of the age of the Universe, $\sim 10^{10}$ years. It is also important that many of the millisecond pulsars are members of binary systems. The Crab Nebula pulsar has the largest spin-down rate of all pulsars, and, using the above formula, $\tau = P/(2\dot{P})$, we find that $\tau = 1400$ years, which is of the same order of magnitude as the age of the Crab Nebula, which was observed to explode in 1054.

It can be seen from equation (15.26) that the rate of loss of rotational energy from the neutron star can be determined directly from the slow-down rate of the pulsar. A particularly interesting result for the Crab pulsar is that the rate at which it loses rotational energy, $dE/dt \sim 6.4 \times 10^{31}$ W, is similar to the energy requirements of the surrounding supernova remnant in non-thermal radiation and bulk kinetic energy of expansion, $dE/dt \sim 5 \times 10^{31}$ W. The origin of the continuous supply of high energy particles to the Nebula had been a major

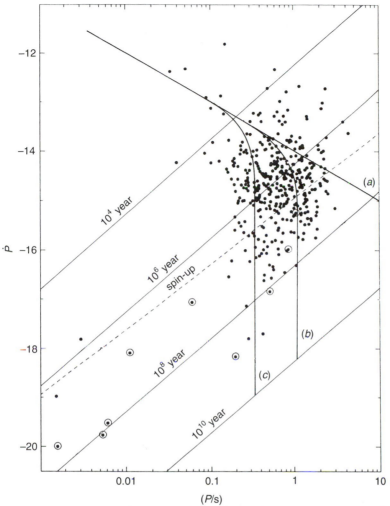

Figure 15.15. A plot of $\dot{P}$ versus P for pulsars. The dots enclosed in circles represent pulsars which are members of binary systems. Lines of constant age according to the formula $\tau = P/2\dot{P}$ are shown. Schematic evolutionary tracks of pulsars on this diagram are shown assuming (a) that there is no decay in the strength of the magnetic field and that the timescale for decay of the magnetic fields is (b) 10^7 years and (c) 10^6 years. The upper limit to the spin-up periods for dead pulsars according to the models of van den Heuvel (1987) is also shown (see the discussion leading to expression (16.48)). (After A.G. Lyne and F. Graham-Smith (1990). *Pulsar astronomy*, p. 129. Cambridge: Cambridge University Press.)

mystery prior to the discovery of the Crab pulsar, because the radiation lifetimes of the particles emitting X-ray and optical synchrotron radiation in the Nebula are much less than the age of the supernova remnant (see Sections 18.1 and 18.4). The continuous injection of energy into the Nebula from the pulsar solves this problem.

If the magnetic braking mechanism is responsible for the slow-down of the neutron star, estimates can be made of the magnetic field strengths at the surface of the neutron star. Approximating the magnetic field at the surface of the neutron star by a dipole field, we can use the classical result that the magnetic field strength at its surface is

$$\mathbf{B} = \frac{\mu_0 p_{m0}}{4\pi r^3}[2\cos\theta\mathbf{i}_r + \sin\theta\mathbf{i}_\theta]$$

Thus, at $r = R$, the surface magnetic field strength is $B_s \approx \mu_0 p_{m0}/4\pi R^3$. Substituting into equation (15.26), we find

$$-\frac{d\Omega}{dt} = \frac{\mu_0\Omega^3 p_{m0}^2}{6\pi c^3 I} = \frac{\mu_0\Omega^3}{6\pi c^3 I}\left(\frac{4\pi R^3 B_s}{\mu_0}\right)^2 = \frac{8\pi\Omega^3 R^6 B_s^2}{3\mu_0 c^3 I} \qquad (15.27)$$

For a uniform sphere rotating about its axis, $I = 2MR^2/5$, and so we find

$$B_s = -\left(\frac{3\mu_0 c^3 M\dot{\Omega}}{20\pi\Omega^3 R^4}\right)^{1/2} = \left(\frac{3\mu_0 c^3 M}{80\pi^3 R^4}\right)^{1/2}(P\dot{P})^{1/2} \approx 3 \times 10^{15}(P\dot{P})^{1/2} \quad \text{T} \quad (15.28)$$

The magnetic field strengths typically lie in the range 2×10^6–2×10^9 T for most pulsars, although the millisecond pulsars have much weaker magnetic fields.

Pulsar periods are remarkably stable once account is taken of their steady decelerations. There are, however, discontinuous changes in the slow-down rates, and two types of behaviour have been identified. One type is 'noisy', what Lyne and Graham-Smith describe as 'fairly continuous erratic behaviour'. The second is much more dramatic, and consists of large discontinuous changes in the pulsar's rotation speed – these are known as *glitches*. These phenomena occur about once every few years in the Crab and Vela pulsars, and are of special interest because they enable unique insights to be gained into the internal structures of neutron stars and the behaviour of the superfluid components in the inner regions.

The nature of the discontinuity is shown in Fig. 15.16, in which it can be seen that the pulsar eventually settles down to a steady slow-down following the abrupt glitch. These phenomena can be attributed to changes in the moment of inertia of the neutron star as it slows down. An attractive picture to explain the general features of Fig. 15.16 is provided by a two-component model for the interior of the neutron star, in which the superfluid neutron component is only weakly coupled to the other components, namely the normal component, the crust and the charged particles. Let us call the moments of inertia of these components I_s and I_n, respectively. After a glitch has taken place, it is assumed that the angular frequency of the normal component decreases discontinuously. Following Shapiro and Teukolsky (1983), the rate at which the superfluid component is spun up is determined by weak coupling between the superfluid and normal components, the coupling being described by a single parameter, τ_c, the relaxation time for frictional dissipation, which is also the timescale for exchange of angular momentum between the two components (see also Section 16.3.3). The change of angular frequency with time, $\dot{\Omega}$, is then governed by two linear differential

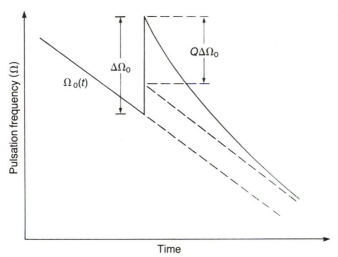

Figure 15.16. Illustrating the phenomenon of *glitches* observed in the Crab pulsar and a few other pulsars, including that in the Vela supernova remnant. The pulse period increases smoothly as the the rotation rate of the neutron star decreases, but there are sudden discontinuities in the pulse period, after which the steady increase in period continues. The variation of the pulse period with time during the glitches provides information about the internal structure of the neutron star. (From S. L. Shapiro and S. A. Teukolsky (1983). *Black holes, white dwarfs and neutron stars: the physics of compact objects*, p. 296, New York: Wiley-Interscience.)

equations:

$$I_n \dot{\Omega} = -\alpha - \frac{I_n (\Omega - \Omega_s)}{\tau_c}$$
$$I_s \dot{\Omega}_s = \frac{I_n (\Omega - \Omega_s)}{\tau_c} \tag{15.29}$$

where α describes the loss of rotational energy due to external torques, for example, by magnetic dipole radiation. These equations can be solved to find the rate of change of Ω with time (see, for example, Shapiro and Teukolsky 1983)

$$\Omega(t) = \Omega_0(t) + \Delta \Omega_0 \left[Q \exp(-t/\tau) + 1 - Q \right] \tag{15.30}$$

where Q is a 'healing parameter' which describes the degree to which the angular frequency returns to its extrapolated value, $\Omega_0(t) = \Omega_0 - \alpha t / I$, the pulsar angular frequency in the absence of the glitch, where Ω_0 is a constant. The significance of these quantities is illustrated in Fig. 15.16. The expression (15.30) is called the *glitch function*, and can give a good description of the behaviour of the angular frequency of the neutron star following a glitch. The values of τ found in this analysis are of the greatest interest because they are related to the physical processes of coupling between the superfluid and normal components of the neutron star. For the Vela pulsar, τ is of the order of months, whereas for the Crab pulsar, it is of the order of weeks. These are very long timescales, and indicate that a considerable fraction of the neutron fluid must be in the superfluid state. The two-component model can provide a good explanation of the glitches observed in the Crab and Vela pulsars. Of particular interest is the fact that the values of τ_c and Q for different glitches in the same pulsar seem to be more or less the same, as is required by the model.

One possible mechanism by which the moment of inertia of the neutron star can change is the result of what is called a 'starquake', by analogy with the deformations of the Earth's crust which take place during an earthquake. The crust takes up an equilibrium configuration in which the gravitational, centrifugal and the solid state forces in the crust are in balance. As the pulsar slows down, the centrifugal forces weaken, and the crust attempts to establish a new equilibrium figure with a lower moment of inertia. In a *starquake*, the crust establishes its new shape by cracking the surface. Since the moment of inertia decreases, this results in a speed up of the normal component, that is, the crust, the normal component, the charged particles and the magnetic field. As these components are weakly coupled to the neutron superfluid, the latter is spun up through frictional forces over a timescale τ_c, resulting in a slow-down of the crust and the associated magnetic field structure.

The starquake model provides a good explanation for the glitches observed in the Crab pulsar, but it cannot account for those observed in the Vela pulsar because these glitches occur too frequently. The shape of the crust of the neutron star turns out to be similar to that of a rotating liquid mass, the ellipticity being given by $\epsilon \approx E_{\rm rot}/E_{\rm grav}$, where $E_{\rm rot}$ is the rotational energy of the neutron star and $E_{\rm grav}$ is its gravitational potential energy. Estimates of ϵ for the Crab and Vela pulsars suggest that $\epsilon \approx 10^{-3}$ and 10^{-4}, respectively. The changes in ellipticity can also be evaluated for these pulsars as well, $\Delta\epsilon = \Delta I/I = -\Delta\Omega/\Omega$, and the values for the Crab and Vela pulsars are $\Delta\epsilon \sim 10^{-8}$ and 10^{-6}, respectively. It is amusing to note that these pronounced effects can be attributed to the shrinkage of the neutron star by only a fraction of a millimetre. As noted above, glitches are observed about once every few years, and so, although this model may be applicable to the Crab pulsar, the glitches in the Vela pulsar occur too frequently since the age of the pulsar is about 10^4 years.

Other models for the phenomenon of glitches involve investigating the properties of the rotating neutron superfluid. Superfluid liquids display many remarkable properties, in particular, on a macroscopic scale, the fluid must rotate irrotationally, so that, within the superfluid, $\nabla \times \mathbf{v} = 0$. In a superfluid, the angular velocity is quantised, so that in the lowest energy state $\oint \mathbf{v} \cdot \mathrm{d}\mathbf{l} = h/2m_{\rm n}$, where $\mathbf{v}$ is the velocity of the fluid and $2m_{\rm n}$ is the mass of a neutron pair. The resolution of these conflicting requirements is that the rotation of the neutron fluid is the sum of a discrete array of vortices which are parallel to the rotation axis. The finite vorticity of the fluid is confined to the very core of each vortex tube which consists of normal fluid (see Feynman (1972) for an explanation of how this comes about). In the case of the Crab pulsar, the number of vortex lines per unit area is about 2×10^9 m^{-2}, that is, their spacing is about 10^{-4} m. The relevance of the vortices to the origin of pulsar glitches concerns how they interact with the crustal material. In some models the vortices are 'pinned' to nuclei in the crust, and in others they thread the spaces in between them. As the star slows down, angular momentum is transferred outwards by the migration of the vortices. If the vortices are pinned, this process is jerky, and may lead to small glitches. In the case of the giant glitches, there may be a catastrophic unpinning of the vortices, leading to a large change in rotation speed.

The immediate environment of the pulsar is referred to as its *magnetosphere*, by analogy with the magnetically dominated regions around the Earth (see Section 10.7). To an excellent approximation, the pulsar may be considered as a non-aligned rotating magnet, but the magnetic field strengths at the surface are very strong indeed. Much of the electrodynamics is best appreciated from the simpler case of an aligned rotating magnet, which was first analysed by Goldreich and Julian (1969). The physics is exactly the same as that of a *unipolar inductor*. Just outside the surface of the neutron star, the Lorentz forces on the charged particles are very great indeed and far exceed the force of gravitational attraction. In quantitative terms, the ratio of these forces is $e(\mathbf{v} \times \mathbf{B})/(GMm/r^2) \sim e\Omega r^3 B / GMm_e \approx 10^{12}$, for the case of the Crab pulsar. Thus, the structure of the magnetosphere of the neutron star is completely dominated by electromagnetic forces. Furthermore, the induced electric fields at the surface of the neutron star are so strong that the force on an electron in the surface exceeds the work function of the surface material, and consequently there must be a plasma surrounding the neutron star. As a result, it is inevitable that there is a fully conducting plasma surrounding the neutron star, and electric currents can flow in the magnetosphere.

The result is a complex distribution of magnetic and electric fields in the magnetosphere of the neutron star. The induced electric field, $\mathbf{E}_i = (\mathbf{v} \times \mathbf{B})$, is neutralised by the flow of charges in the plasma, so that the net field is reduced to zero, that is, $\mathbf{E} + (\mathbf{v} \times \mathbf{B}) = 0$, and the space charge distribution can be found from Maxwell's equation div $\mathbf{E} = \rho_e/\epsilon_0$, where $\rho_e = e(n_+ - n_-)$ is the electric charge density. The typical picture for the magnetic field and charge distribution about a rotating magnetised neutron star is shown in Fig. 15.17. There is a certain radius, called the *light cylinder* or *corotation radius*, at which the velocity of rotation of material corotating with the neutron star is equal to the velocity of light. Within this radius, the charged particles are tied to the magnetic field lines, and, just as in the case of the Solar Wind, the magnetic field takes up a spiral configuration when viewed from above (see Fig. 10.7). Within the light cylinder, some of the field lines are closed, but those which extend beyond the light cylinder are open, and particles dragged off the poles of the neutron star can escape beyond the light cylinder. Because of the strong electric fields which must be present within the magnetosphere, particle acceleration can take place, and we take up this topic in Chapter 21.

As is often the case in remarkable objects like pulsars, the most difficult part to understand is the signature which led to their discovery, that is, the physical mechanism by which the *radio pulses* themselves are generated. The one clear requirement of all models of the radio emission mechanism is that the radiation cannot

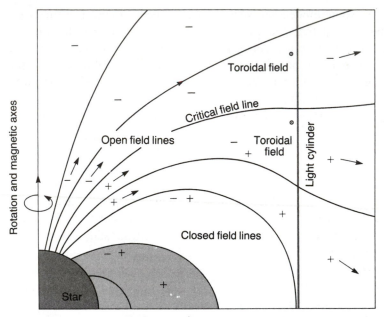

Figure 15.17. A diagram illustrating the magnetic field and charge distribution about a rotating magnetised neutron star, according to the analysis of Goldreich and Julian (P. Goldreich and W.H. Julian (1969). *Astrophys. J.*, **157**, 839.) It is assumed that the magnetic axis is parallel to the rotation axis of the neutron star. The star behaves like a unipolar inductor, and the charge distribution in the vicinity of the neutron star is shown. Electrons are removed from the surface of the star by the induced electric field in the surface layers, so that there must be charge and current distributions within the magnetosphere of the neutron star. Particles attached to closed magnetic field lines corotate with the star and form a corotating magnetosphere. The *light cylinder* is defined as that radius at which the rotational velocity of the corotating particles is equal to the velocity of light. The magnetic field lines which pass through the light cylinder are open and are swept back to form a toroidal field component. Charged particles stream out along these open field lines. The critical field line is at the same potential as the exterior interstellar medium, and divides regions of positive and negative current flows from the star. The plus and minus signs indicate the sign of the electric charge in different regions about the neutron star. (From S.L. Shapiro and S.A. Teukolsky (1983). *Black holes, white dwarfs and neutron stars: the physics of compact objects*, p. 285, New York: Wiley-Interscience. After R.N. Manchester and J.H. Taylor (1977). *Pulsars*, p. 179. San Francisco: W.H. Freeman and Co.)

be incoherent radiation. An effective radiation temperature, or *brightness temperature*, $T_b = (\lambda^2/2k)(S_v/\Omega)$, can be estimated from the known distances of the pulsars, the duration of the pulses and their observed flux densities, S_v. Typically, brightness temperatures in the range 10^{23}–10^{26} K are found. This far exceeds the temperatures of thermal material which radiates in the radio waveband, and the way around this problem is to assume that the radiation is some form of coherent radiation; in other words, the particles radiate in bunches rather than singly. In order to radiate coherently, the bunches, consisting, say, of N charges, must have dimension less than the wavelength of the emitted radiation, and then, because the intensity of the radiation depends upon the square of the total oscillating charge, the intensity of the radiation can be N^2 times that of an individual charge.

The problem is to understand how coherent emission can occur in the vicinity of the magnetic poles of pulsars. One attractive scenario is associated with the enormous induced electric fields which are present in the vicinity of the neutron star. The potential difference between the pole and equator of a magnetised neutron star with surface magnetic field 10^8 T rotating once per second is 10^{16} V. As a result, electrons are dragged off the surface of the neutron star and are rapidly accelerated to very high energies, more than 10^6 times their rest mass energies. As they stream out along the curved magnetic field lines, they radiate *curvature radiation*, which is associated with the acceleration of the particles as they move in curved trajectories (see Section 18.1.3). In the strong magnetic fields close to the pulsar, these high energy photons interact with the transverse component of the magnetic field, or with ambient low energy photons, to produce electron–positron pairs. In turn, these electrons and positrons radiate high energy photons, which generate electron–positron pairs, and so on. The generation of these electron–photon cascades results in bunches of particles which can radiate coherently in sheets. This may be the process responsible for producing the very high brightness temperatures observed in pulsars. We will return to this problem in Section 18.1.3 when we deal with curvature radiation in the context of synchrotron radiation. It is natural to asssociate these processes with the open field lines which emanate from the poles of the magnetic field structure, and this can account for the narrowness of the pulses.

An attractive feature of this model, which is due to Ruderman and Sutherland (1975), is that it may explain why most pulsars have periods about 1 s and magnetic field strength about 10^8 T – only if the magnetic field is strong enough, and if the rotation period is sufficiently short, will electron–photon cascades take place. According to this theory, there is a 'death-line' in the magnetic field–period relation for pulsars such that $BP^2 \geq 10^7$ T s^{-1}, where P is the period of the pulsar in seconds (see Fig. 15.15).

In the case of the Crab pulsar, infrared, optical, X- and γ-ray pulses are also observed with similar pulse profiles to that observed at radio wavelengths. The important distinction between the radio pulses and those observed at higher energies is that the brightness temperatures of the radiation at optical and shorter wavelengths are consistent with the incoherent emission of high energy electrons. For example, the brightness temperature of the X-ray emission is about 10^{11} K, which corresponds to electron energies of only about 10 MeV. There is therefore no problem in assuming that the emission at the higher energies is the incoherent radiation of the electrons accelerated in the pulsar's magnetosphere. There are problems in identifying exactly where these processes take place within the magnetosphere. Lyne and Graham-Smith (1990) provide a detailed survey of the various constraints and possible models for the radio and higher energy emissions.

The example of the Crab pulsar shows convincingly how magnetic torques exerted by the pulsar's magnetic field on its surroundings can account for its deceleration. More than simply the effect of magnetic braking is, however, necessary to account for the observed distribution of pulsar ages. If the magnetic dipole moment of the pulsar remained constant, then, according to equation (15.26), $d\Omega/dt \propto -\Omega^3$, that is, $dT/dt \propto T^{-1}$. This relation is shown schematically in Fig. 15.15. It is evident that there is an absence of pulsars with very long periods which would be expected to predominate in this diagram. The preferred interpretation of Fig. 15.15 is that the magnetic dipole moment of the pulsar decays as it ages. The other lines on Fig. 15.15 show the expected evolutionary tracks if the characteristic decay times of the magnetic dipole moment are 10^6 and 10^7 years. According to Lyne and Graham-Smith, a characteristic age of 9×10^6 years can also account for the observed distribution of radio luminosities of pulsars. The mechanism responsible for the decay of the magnetic dipole moment is not understood.

One interesting result derived from the period distribution of pulsars is that there are remarkably few short period pulsars such as that present in the Crab Nebula. Although more than 500 pulsars are now known, and more than 150 supernova remnants observed in our Galaxy, there are only eight good associations of supernova remnants with pulsars, and these are all with the youngest known pulsars. Another surprise is how rare Crab Nebula-type supernova remnants are. This result suggests that only very rapidly rotating young pulsars form Crab Nebula-type remnants. Searches have been made for other young pulsars in our Galaxy, but none have been found. Because of the absence of young pulsars, it is inferred that most pulsars must be created with periods about 1 s and not close to the break-up rotation speeds, which correspond to periods $T < 1$ ms. Evidently, the progenitors of pulsars are normally able to get rid of the angular momenta of their central cores in the process of collapse to form neutron stars.

The millisecond pulsars, of which several are now known, appear to be somewhat different objects from the standard pulsars. All of them have very stable periods, from which it is inferred that they must have very weak magnetic fields. This makes good sense because, if the magnetic field is weak, there is only weak coupling to the external medium and little deceleration. An attractive picture for their formation is that they were once members of binary systems. If mass transfer takes place from the companion star onto a dead pulsar, it will be spun up. In fact, a weak magnetic field is a great advantage for effective spin-up because the magnetic pressure determines the accretion radius about the star, and, if the field is weak, the angular momentum transfer can occur close to the surface of the neutron star, resulting in a large spin-up. According to the models of van den Heuvel (1987), there is a maximum spin-up rate which is limited by the value of the surface magnetic field strength of the pulsar. This limit can be written $P = 1.9B_g^{6/7}$ ms, where B_g is the surface magnetic field strength measured in units of 10^5 T (see Section 16.4.2, expression (16.48)). This relation has been plotted on Fig. 15.15. It can be seen that virtually all the millisecond pulsars lie below this limiting spin-up line. It is noteworthy that many of the millisecond pulsars are members of binary systems. On the other hand, if the companion star explodes, disruption of the system may occur, resulting in isolated millisecond pulsars. Notice that, in this picture, the dead pulsar becomes alive again because its period is spun up and it recrosses the 'death-line'. It will also be noted that this picture involves the dead pulsars having magnetic fields $\sim 10^5$ T. These ideas have important implications for the role of neutron stars in the theory of stellar evolution, particularly the theory of the evolution of binary stars. Radhakrishnan (1986) provides an excellent survey of the full implications of the properties of pulsars for the evolution of a wide variety of binary stellar systems.

15.5   Neutron stars in binary systems – X-ray binaries

There is no reason to suppose that the neutron stars which are the optically invisible components of binary X-ray sources are different in nature from the isolated variety which are observed as radio pulsars. The distinctive phenomena which they exhibit are associated with the fact that they are members of binary systems. This is an enormous bonus astrophysically because it enables the properties of the neutron stars, such as their masses, distances and physical environments, to be determined.

The discovery of binary X-ray sources was made by the UHURU X-ray observatory (Section 7.3.1). The discovery records of the X-ray pulsations in the source Hercules X-1 is shown in Fig. 7.11, and in Fig. 15.18 the X-ray eclipses and sinusoidal variation of the arrival times of the X-ray pulses are shown. This is convincing evidence that the X-ray sources are members of binary star systems. Subsequent X-ray satellites such as SAS-III, Ariel V, HEAO-1, Hakucho, Tenma and EXOSAT have elucidated many of their properties. The distribution of bright X-ray sources as observed by HEAO-1 is shown in Fig. 1.8(e). It can be seen that they are concentrated towards the plane of the Galaxy, with many of the brightest sources lying in the general direction of the centre of the Galaxy. Convincing evidence has been found for the binary nature of more than half of the Galactic X-ray sources, and, because of selection effects which can prevent sources being identified as binary systems, particularly if the plane of the binary orbit is at a large angle to the line of sight, it is likely that a large fraction of the point-like X-ray sources of the Galactic population are binary in nature. The other X-ray sources are associated with supernova remnants, isolated hot white dwarfs and normal stars with very hot coronae. Some of the X-ray sources are associated with globular clusters, and a number of these are certainly binary systems.

The binary X-ray sources come in two principal varieties. In all cases, the neutron star associated with the X-ray binary is invisible optically, and hence the optical observations refer only to the primary companion star. Many of the binary systems contain *late O or early B type stars*, these being among the most luminous and massive stars known with rather short main sequence lifetimes, $\sim 10^7$ years. These are very rare classes of star, and belong to the youngest of the stellar populations known. The second class is associated with cooler low-

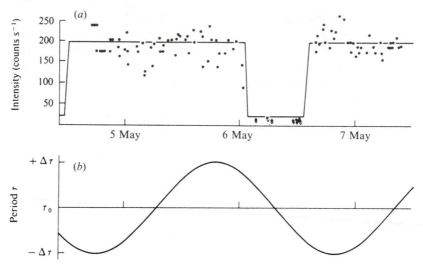

Figure 15.18. (*a*) The rate of arrival of X-ray photons from the X-ray source Hercules X-1 (Her X-1). The source is observed for about 34 hours and then is eclipsed for 6 hours. (*b*) Variations in the arrival time of pulses from Her X-1. The sinusoidal variation of the pulse arrival time is naturally attributed to orbital motion of the X-ray source in a binary system.

mass main sequence stars, having masses, luminosities and temperatures similar to those of the Sun. Members of this class are often referred to as *Galactic bulge sources* because they lie in the general direction of the Galactic Centre, and a number of them have been identified as belonging to globular clusters; the latter are members of the Galactic bulge population and are amongst the oldest stellar systems in our Galaxy. These low-mass X-ray binaries have the distinctive property that they exhibit the phenomenon of X-ray bursts.

Since the X-ray sources are members of binary systems, the masses of the neutron stars can be estimated by the classical methods of dynamical astronomy. In the best cases, the velocity curves of both the primary and secondary stars are measured. In the case of the high-mass binaries, some of the O and B stars are sufficiently bright for it to be possible to measure accurately the variation of radial velocity with phase in the binary orbit. Although the X-ray pulsar is not visible optically, its velocity can be found from the Doppler shifts of its X-ray pulse period. The X-ray pulsars have periods which range from a fraction of a second to about 15 minutes, the lower end of this range being similar to the periods found in the radio pulsars. Thus, the radial velocities of both components of massive binary systems can be measured. From the amplitude of the velocity excursions about the mean value for the members of the binary, the ratio of masses of the two stars, M_1/M_2, can be measured. Absolute values of the masses cannot be determined, however, but only the quantity $(M_1 + M_2)\sin^3 i$, where i is the angle of inclination of the orbit to the plane of the sky. It is therefore necessary to make some estimate of the angle i to make progress. Compared to the size of the primary star, the X-ray source can be considered as a point object, and so the X-ray source may be occulted by the primary if the plane of the orbit lies close to the line of sight from the Earth. In a number of cases, such X-ray occultations are observed with periods equal to those of the binary orbits. In addition, the X-ray source itself may influence the surface properties of the primary star, either by distorting the figure of the surface into an ovoid shape, because of the gravitational influence of the neutron star (see Fig. 14.12), or possibly by heating up the face of the primary star closest to the X-ray source, thus causing that face of the primary star to be more luminous optically when pointing towards the Earth. In the first case, the optical luminosity of the primary is expected to vary at half the period of the binary, whereas, in the second, the optical luminosity varies with the same period as the binary period. There is convincing evidence for both of these phenomena among the binary X-ray sources. Most of the uncertainties in the masses are associated with the problem of determining the inclinations of the binary orbits to the line of sight.

These procedures have been used to estimate the masses of the neutron stars in X-ray binaries, and a compilation of mass estimates is shown in Fig. 15.19. Also included in this diagram are the masses of the components of the binary radio pulsar systems for which masses can be found from very accurate pulsar timing (see Section 15.6.2). The derived masses of the neutron stars are entirely consistent with the theoretical expectation that their masses should lie close to $1M_\odot$. Taylor (1992) finds that all the masses lie in the range $1.35 \pm 0.27 M_\odot$. This is a remarkably beautiful example of the symbiosis between the traditional techniques of optical astronomy and modern procedures in radio and X-ray astronomy.

For the low-mass systems, it is much more difficult to estimate masses because normally it is not possible to measure the velocity curve for the primary star, which is faint and the light of which can be overwhelmed by the light from the accretion disc. The velocity curve can still be measured for the X-ray pulsar, but this is equivalent to studying what are known classically as *single-line spectroscopic binaries*, in which high-resolution optical spectroscopy can provide the radial velocity of only one star as a function of the phase. In this case, observations of the velocity curve of the X-ray source determine what is known as the *mass function* of the binary system:

$$f(M_X, M_0, i) = \frac{M_0^3 \sin^3 i}{(M_X + M_0)^2}$$

where M_X is the mass of the X-ray pulsar and M_0 is the mass of the primary star. Thus, more assumptions need to be made to derive the masses of these stars. It turns out that the best black hole candidates are single-line spectroscopic binaries of this type in which the velocity curve of the primary can be measured.

The case for associating the X-ray emission of binary X-ray sources with *accretion* of matter from the primary onto the neutron star is wholly convincing, as can be appreciated from the following simple arguments, which will be developed in more detail in Chapter 16. First of all, if matter falls from the primary onto the

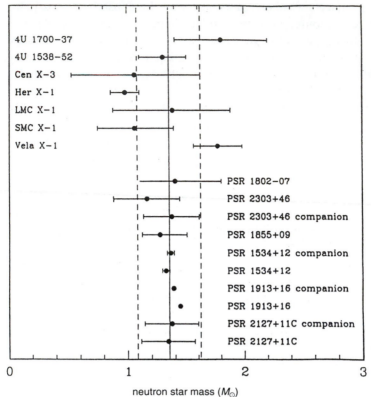

Figure 15.19. Mass estimates for the neutron stars in X-ray binary systems and binary radio pulsars, for which good mass determinations are available from their velocity curves and other information. (Courtesy of Professor J. Taylor. See also S. L. Shapiro and S. A. Teukolsky (1983). *Black holes, white dwarfs and neutron stars: the physics of compact objects*, p. 256, New York: Wiley-Interscience.)

secondary star, the kinetic energy which it gains is converted into thermal energy when the infalling matter hits the surface of the neutron star. The thermal energy released is just the gravitational binding energy of matter on the surface of the neutron star, so that about 5% of the rest mass energy of the infalling matter is available for emission as radiation. This represents a very efficient source of energy, at least five times the efficiency available from nuclear energy sources.

The second simple calculation is to ask what the typical temperature of the radiating matter would have to be to account for the observed X-ray luminosities of binary X-ray sources. If we assume the luminosity of a typical luminous X-ray binary to be 10^{30} W, and assume that this is emitted as black-body radiation from the surface of a neutron star, then, equating the radiation flux of a black body at temperature T to this luminosity, the lower limit to the temperature of the emitting region is about 10^7 K. Thus, it is entirely natural that the radiation should be emitted in the X-ray waveband (Fig. 1.7(*a*)).

The third simple argument concerns the steady state X-ray luminosity of accreting compact objects. If the luminosity of the source were too great, the radiation pressure acting on the infalling gas would be sufficient to prevent the matter falling onto the surface of the compact object. In the simplest calculation, which we carry out in Chapter 16, it is assumed that the radiation pressure acting on the infalling matter is due to Thomson scattering of the emergent radiation by infalling electrons. If other sources of opacity are also important, these increase the radiation pressure and result in a lower value for the critical luminosity, above which accretion is suppressed. The resulting critical luminosity is known as the *Eddington luminosity*, and depends only upon the mass of the gravitating body

$$L_{\text{Edd}} = 1.3 \times 10^{31} (M/M_\odot) \text{ W} \tag{15.31}$$

It is remarkable that the luminosities of most of the binary X-ray sources in the Galaxy and the Magellanic Clouds are more or less consistent with this upper limit. Their luminosities extend up to about 10^{31} W, and the X-ray luminosity function cuts off sharply above this value. The calculation which leads to the Eddington limit is a very general one and does not depend strongly upon the details of the accretion process or the means by which the X-ray emission is produced.

These three arguments show how naturally accretion can account for the properties of these X-ray sources, and also demonstrate the importance of accretion as a source of energy in astrophysics. The ramifications of these ideas are profound, and they have been extended to the cases of accretion onto black holes, both those of stellar mass, which may be present in some of the X-ray binaries, and the supermassive varieties, which are likely to be present in active galactic nuclei. We take up the story of accretion in Chapter 16.

15.6   Black holes

15.6.1   *Introduction to general relativity and the Schwarzschild metric*

Neutron stars are the last known forms of stable star. For more compact objects, the attractive force of gravity becomes so strong that no physical force can prevent collapse to a physical singularity and electromagnetic radiation cannot escape from it. In a prescient paper of 1783, the Reverend John Michell noted that, if a star is sufficiently compact and massive, the escape velocity from its surface exceeds the speed of light, and therefore light cannot escape from it. Performing a wholly classical calculation, the escape velocity from the surface of a star of mass M and radius r is $v = (2GM/r)^{1/2}$, and therefore, naively setting v equal to the speed of light, c, the radius of such a star would be $r = 2GM/c^2$. As we will find in a moment, this is just the expression for the *Schwarzschild radius* of a black hole of mass M, and the spherical surface of radius r plays the role of the 'surface' of the black hole.

This is not the place to develop the details of the *General Theory of Relativity*, but my aim is to derive some of the important results directly from the Schwarzschild metric and to make other results plausible. Within months of the Einstein's definitive formulation of the General Theory in 1915, Schwarzschild discovered the solution for the metric of space-time about a point mass M:

$$ds^2 = \left(1 - \frac{2GM}{rc^2}\right) dt^2 - \frac{1}{c^2} \left[\frac{dr^2}{\left(1 - \frac{2GM}{rc^2}\right)} + r^2(d\theta^2 + \sin^2\theta \, d\phi^2) \right] \qquad (15.32)$$

This metric, known as the *Schwarzschild metric*, has the same meaning as the interval ds^2 in special relativity, which is known as the *Minkowski metric*:

$$ds^2 = dt^2 - \frac{1}{c^2}[dr^2 + r^2(d\theta^2 + \sin^2\theta \, d\phi^2)] \qquad (15.33)$$

In both cases, the metrics have been written in spherical polar coordinates. The key point is that the coordinates r and t have different meanings in the metrics (15.32) and (15.33).

In the limit of large distances from the point mass, $r \to \infty$, the two metrics become the same. They are, however, very different at small values of r, reflecting the influence of the mass M upon the geometry of space-time. These differences are explored in more detail in Chapter 14 of my book, *Theoretical concepts in physics* (TCP; Longair 1992). Let us consider the simplest case of two events in space-time separated by coordinates dr, dt, with $d\theta = d\phi = 0$, that is, the spatial increment is purely in the radial direction. In special relativity, dr and dt are the differences in the distance and time coordinates in the inertial frame of reference, S; the *proper time* interval, $d\tau$, between events is

$$d\tau = ds = \left(dt^2 - \frac{dr^2}{c^2} \right)^{1/2}$$

In general relativity, the distance coordinate r appearing in the metric (15.32) gives the correct result for proper distances measured perpendicular to r, as can be seen from inspection of the metric – the proper distance dl is just $dl = r \, d\theta$, and hence r is often referred to as an *angular diameter distance*. The increment of proper distance in the radial direction in the reference frame in which the point mass is at rest is *not* dr, but is given by the appropriate quantity in the metric $dx = dr/(1 - 2GM/rc^2)^{1/2}$; dx is known as the increment of *geodesic distance* in the radial direction in the case $d\theta = d\phi = 0$. In general relativity, geodesics replace straight lines as the shortest distance between two points. Thus, the r's appearing in metrics (15.32) and (15.33) have quite different meanings – the reasons for this are intimately related to the fact that the geometry of space-time is not flat.

In general relativity, there are three different time intervals. The reason for this is that, unlike the case of special relativity, the rate at which time passes depends upon the gravitational potential, a result which follows directly from the Principle of Equivalence (see TCP, Section 14.3.3). In Newtonian terms, the gravitational potential at the point r is $-GM/r$. In the case of the Schwarzschild metric, the time interval between events according to an observer who is stationary in S and located at the point r is $dt' = dt(1 - 2GM/rc^2)^{1/2}$. This notation makes it clear how the time interval dt' depends upon the gravitational potential in which the observer is located. It can be seen that the time interval dt' only reduces to dt in the limit of very large distances from the origin, $r \to \infty$, at which the gravitational potential goes to zero. The time interval dt is a very convenient *coordinate time*, since every observer can relate the time dt' to dt because it only depends upon the gravitational potential in which the observer is located.

In general relativity, the increment of *proper time* is

$$ds = d\tau = dt \left[\left(1 - \frac{2GM}{rc^2} \right) - \frac{\left(\dfrac{dr}{dt} \right)^2 / c^2}{\left(1 - \dfrac{2GM}{rc^2} \right)} \right]^{1/2} = dt' \left(1 - \frac{1}{c^2} \frac{dx^2}{dt'^2} \right)^{1/2} \quad (15.34)$$

again for the case $d\theta = d\phi = 0$. Let us examine what this means. Since $dx' = 0$ in the proper frame of reference S', the observer in S' must be moving at a velocity $v = dx/dt'$ through S. In other words, the proper time interval is just

$$d\tau = dt' \left(1 - \frac{v^2}{c^2} \right)^{1/2} = \frac{dt'}{\gamma} \quad (15.35)$$

where γ is the Lorentz factor, $\gamma = (1 - v^2/c^2)^{-1/2}$. The expression (15.35) is just the standard formula for time dilation between the frame S and the proper frame S' of an observer travelling in the radial direction.

The metric (15.32) enables us to derive the formula for the redshift of electromagnetic waves emitted from the point r from the origin as observed at infinity. If we let the time interval $\Delta t'$ correspond to the period of the waves emitted at the point r, the observed period of the wave at infinity Δt is given by

$$\Delta t' = \left(1 - \frac{2GM}{rc^2} \right)^{1/2} \Delta t \quad (15.36)$$

Therefore, the emitted and observed frequencies, ν_e and ν_∞, respectively, are related by

$$\nu_\infty = \nu_e \left(1 - \frac{2GM}{rc^2} \right)^{1/2} \quad (15.37)$$

This expression is the general relativistic result corresponding to Michell's profound insight. If the radiation is emitted from radial coordinate $r = 2GM/c^2$, the frequency of any wave is redshifted to zero frequency. This means that no information can reach infinity from radii $r \le r_g = 2GM/c^2$. This radius, r_g, is known as the *Schwarzschild radius*. Since according to general relativity, no radiation can escape from within this radius, the surface $r = r_g$ is 'black'. Notice that, in the general relativistic treatment, the r coordinate is the angular diameter distance.

The other key aspect of the Schwarzschild metric is the dynamics of test masses in the field of the point mass. It is simplest to begin with the Newtonian expression for the dynamics of a test particle in terms of the conservation of energy in a gravitational field. The velocity of the particle, v, can be written in spherical polar coordinates as $v^2 = \dot{r}^2 + (r\dot{\theta})^2$, and therefore, writing the gravitational potential energy of the particle as $-GmM/r$, conservation of energy results in the relation

$$\frac{1}{2}mv^2 - \frac{GMm}{r} = \frac{1}{2}mv_\infty^2$$

$$\dot{r}^2 + (r\dot{\theta})^2 - \frac{2GM}{r} = v_\infty^2 \qquad (15.38)$$

where v_∞ is the velocity of the particle at infinity. Now, conservation of angular momentum requires $m\dot{\theta}r^2 = $ constant. For a test particle, it is convenient to write the conservation of angular momentum in terms of the *specific angular momentum* of the particle, $h = \dot{\theta}r^2$, that is, its angular momentum per unit mass. Then, equation (15.38) can be written as

$$\dot{r}^2 + \frac{h^2}{r^2} - \frac{2GM}{r} = \dot{r}_\infty^2 \qquad (15.39)$$

where $\dot{r}_\infty$ is the radial velocity of the particle at infinity. Notice that, in this formulation, r and t have their usual Newtonian meanings in Euclidean geometry. Another important observation is that, according to equation (15.39), so long as h is non-zero, the particle cannot reach $r = 0$ because the term h^2/r^2, the energy term associated with the centrifugal force, becomes greater than the gravitational potential energy $2GM/r$ for small enough values of r.

The corresponding equation which is obtained from a general relativistic treatment is as follows:

$$\dot{r}^2 + \frac{h^2}{r^2} - \frac{2GM}{r} - \frac{2GMh^2}{r^3c^2} = (A^2 - 1)c^2 \qquad (15.40)$$

where h and A are constants of the motion (see TCP, pp. 295–301). This equation looks formally similar to equation (15.39), but there are two important differences. First, and most obviously, there is an extra term $(-2GMh^2/r^3c^2)$ on the left-hand side, and, secondly, it will be recalled that the constants and the variables have different meanings in general relativity. Thus, r is angular diameter distance, $\dot{r}$ means dr/ds, where s is proper time, and the constants h and (A^2-1) are constants equivalent to h and $\dot{r}_\infty^2$, but which come out of the formal analysis of dynamics according to the Schwarzschild metric. The crucial difference between equations (15.39) and (15.40) is the additional term $(-2GMh^2/r^3c^2)$, which has the effect of enhancing the attractive force of gravity, even when the particle has a finite specific angular momentum h. The greater the value of h, the greater the enhancement. One way of thinking about this result is to recall that the kinetic energy associated with the rotational motion about the point mass contributes to the inertial mass of a test particle and thus enhances the gravitational force upon it.

A number of elegant results can be found from the analysis of equation (15.40). First of all, it can be seen by inspection that, for sufficiently small values of r, the general relativistic term $(-2GMh^2/r^3c^2)$ becomes greater than the centrifugal potential term, meaning that this purely general relativistic term 'increases the strength of gravity' close enough to $r = 0$. It is a useful exercise to show that, if the specific angular momentum of the particle $h \leq 2r_gc$, it will inevitably fall in to $r = 0$, where r_g is the Schwarzschild radius, $r_g = 2GM/c^2$. Another important related result is that there is a *last stable circular orbit* about the point mass. This

orbit has radius $r = 3r_g$. There do not exist circular orbits with radii less than
this value, the particles spiralling rapidly in to $r = 0$. This is why the black hole
is called a 'hole' – matter inevitably collapses in to $r = 0$ if it comes too close to
the point mass.

The Schwarzschild radius, r_g, plays a special role in the theory of black holes.
Putting in the values of the constants, we find,

$$r_g = 3 \left(\frac{M}{M_\odot} \right) \quad \text{km} \tag{15.41}$$

Thus, for our own Sun, we find $r_g = 3$ km, which is negligible compared with
the solar radius of 695 980 km. A neutron star which has $M = M_\odot$ has radius
$r_g \approx 10$ km, and then general relativity is important in determining the stability
of these stars.

From the metric (15.32), it appears that there is a singularity at the Schwarzschild radius $r_g = 2GM/c^2$. It can be shown, however, that this is not a physical
singularity, perhaps the simplest demonstration of this being the fact that a
particle can fall through r_g and into $r = 0$ in a finite proper time. The apparent
singularity in the metric results from the particular choice of coordinates in
which the metric (15.32) is written – if Kruskal or Finkelstein coordinates are
used, the singularity disappears (see, for example, Section 8.6 of Rindler (1977)
or Section 12.6 of Shapiro and Teukolsky (1983)).

At $r = 0$, there is, however, a real *physical singularity*, and, according to the
classical theory of general relativity, the infalling matter collapses to a singular
point. This is a very unappealing situation, but it is perhaps of some comfort
that these Schwarzschild singularities are unobservable because no information
can arrive to the external observer from within the Schwarzschild radius r_g. For
all practical purposes, the black hole may be considered to have a black spherical
surface at r_g. From the classical point of view, physics breaks down at $r = 0$.
There has been a great deal of study of the nature of these singularities, in
particular, the general conditions under which black holes inevitably form. The
singularity theorems developed by Penrose and Hawking give a precise answer
to this question (see, for example, Hawking and Ellis (1973)). If, for example,
matter had a negative energy equation of state, it might be possible to avoid
the singularity. It should also be noted that the singularity theorems are based
upon classical theories of gravity, such as general relativity, or theories in which
the force of gravity is attractive. As yet, there is no quantum theory of gravity,
but intuitively one feels that taking proper account of quantum phenomena in
collapse to a black hole must provide a deeper understanding of the nature of
singularities present in the classical theory of black holes.

15.6.2   *Experimental tests of general relativity*

Before going any further, we should assess the status of general relativity
on the basis of the various experimental tests of the theory which have been
carried out. The problem is that, generally, the magnitudes of the effects of the
curvature of space-time are very small, typically being of the order $2GM/rc^2$,

where M is the mass of the object and r is the radial distance from it. Thus, at the surface of the Earth, we find $2GM_E/r_Ec^2 = 1.4 \times 10^{-8}$; at the surface of the Sun, $2GM_\odot/r_\odot c^2 = 4 \times 10^{-6}$. Thus, within the Solar System, the effects of general relativity are very small.

Traditionally, there have been four tests of the theory (see, for example, Will (1989)). The first is the measurement of the *gravitational redshift* of electromagnetic waves in a gravitational field. This was first carried out using the *Mössbauer effect* to measure very precisely the change in frequency of photons falling in a gravitational field from the top to the bottom of a high tower. The more recent versions of the test have involved placing hydrogen masers in rocket payloads and measuring very precisely the change in frequency with altitude. These experiments have demonstrated directly the gravitational redshift of light, and it may be thought of in two ways. One approach is to regard it as no more than the conservation of energy in a gravitational field. A better way of thinking about it is as direct evidence for Einstein's Principle of Equivalence, according to which a gravitational field is exactly equivalent to an accelerated frame of reference. In the rocket experiments, the principle was tested to a precision of about 5 parts in 10^5.

The oldest test, and the first great triumph of general relativity, was the explanation of the perihelion shift of the orbit of the planet Mercury. In 1859, Le Verrier showed that, once account is taken of the influence of all the other planets in the Solar System, there remained a small but significant precession of the orbit of Mercury which amounted to about 43 arcsec per century. The origin of this precession remained a mystery, possible explanations including the presence of a hitherto unknown planet, which was named Vulcan, close to the Sun, oblateness of the solar interior, deviations from the inverse square law of gravity near the Sun and so on. In fact, Einstein's theory of general relativity predicts that, because of the curvature of space-time in the gravitational field of the Sun, the orbit of Mercury should precess and, for an elliptical orbit of eccentricity e, the angular precession of the perihelion per orbit amounts to

$$\mathrm{d}\phi = \frac{3\pi}{2} \left(\frac{c}{v}\right)^2 \left[\frac{r_g(M_\odot)}{r}\right]^2 \frac{1}{(1-e^2)} \quad \mathrm{rad}$$

Now, for Mercury, the mean radius of its orbit $r = 5.8 \times 10^{10}$ m, the period of the orbit about the Sun is 88 days and the Schwarzschild radius of the Sun $r_g(M_\odot) = 3$ km. Substituting these values, it is found that the predicted precession is 43 arcsec per century, in remarkable agreement with the observed value. There has been some discussion as to whether or not the agreement really is as good as this comparison suggests because there might be a contribution to the precession if the core of the Sun were rapidly rotating. The results of helioseismology suggest that the core of the Sun is not rotating sufficiently rapidly to perturb the excellent agreement between the predictions of general relativity and the observed precession.

The next test historically was the measurement of the deflection of light by

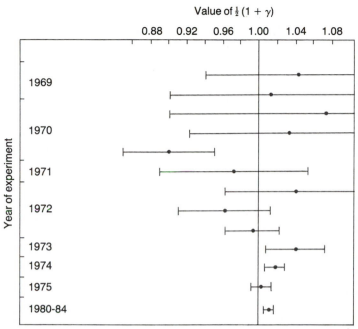

Figure 15.20. The results of experiments to measure the deflection of radio waves by the Sun using the angular separation between compact radio sources measured by VLBI. The early experiments used specific groups of quasars to measure the deflections, whereas the experiments from 1980–4 used quasars and compact radio sources distributed over the celestial sphere. The results are plotted in terms of the parameter $\frac{1}{2}(\gamma + 1)$, which takes the value unity in general relativity. The parameter γ measures how much space curvature is produced by unit mass and takes the value unity in general relativity. (From C.M. Will (1989). In *The new physics*, ed. P.C.W. Davies, p. 13. Cambridge: Cambridge University Press.)

the Sun. The bending of space-time in the vicinity of the Sun means that light, travelling along null-cones in bent space-time, suffers a small deflection when the light from a distant object grazes the Sun's limb. According to general relativity, the deflection amounts to 1.75 arcsec. The measurement of this small deflection was the objective of the famous eclipse expeditions of 1919 which were undertaken by Eddington and Crommelin. Although the precision was not great, their observations agreed with the predictions of Einstein's theory, which is twice the deflection expected according to a Newtonian calculation. The modern version of the test involves measuring very precisely the angular separations between compact radio sources as they are observed close to the Sun. By means of very long baseline interferometry (VLBI), angular deviations of less than a milliarcsecond are quite feasible. Fig. 15.20 shows how the precision of the test has improved over the years. The most recent observations are in good agreement with general relativity.

The last of the four traditional tests is closely related to the deflection of light by the Sun, and concerns the time delay expected when an electromagnetic

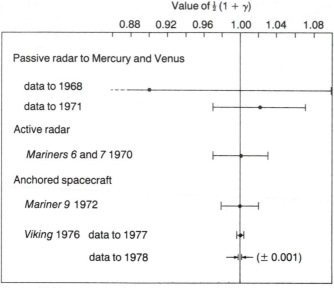

Figure 15.21. A comparison of the results of the Shapiro time-delay experiments with the expectations of theory. The notation is the same as in Fig. 15.20. (From C.M. Will (1989). In *The new physics*, ed. P.C.W. Davies, p. 15. Cambridge: Cambridge University Press.)

wave propagates through a varying gravitational potential. It was only in 1964 that Irwin Shapiro realised that the gravitational redshift of signals passing close to the Sun causes a small time delay, which can be measured by very precise timing of signals which are reflected from planets or space vehicles as they are about to be occulted by the Sun. Originally, the radio signals were simply reflected from the surface of the planets, but later experiments used transponders on space vehicles which passed behind the Sun. The most accurate results were obtained using 'anchored' transponders on space vehicles which landed on Mars, and these are displayed in Fig. 15.21, which has the same format as Fig. 15.20. It can be seen that the agreement with general relativity is excellent.

The most spectacular recent results have come, however, from radio observations of short-period pulsars. Observations by Taylor and his colleagues using the Arecibo radio telescope have demonstrated that these are the most stable clocks over long timescales that we know of in the Universe. They have been able to detect variations in the time-keeping of even the most accurate laboratory clocks, as has been demonstrated by comparing the times measured by two pulsars against a standard clock.

The pulsars have enabled a wide variety of very sensitive tests to be carried out of general relativity and the possible existence of a background flux of gravitational radiation, but by far the most intriguing systems are those pulsars which are members of binary systems. Twenty-eight of these are now known, the most important being those in which the other member of the binary system is

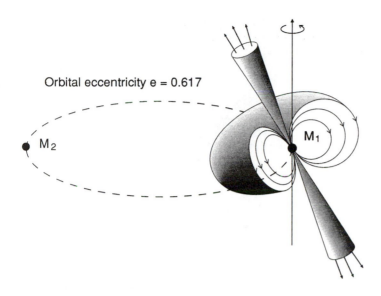

Orbital eccentricity e = 0.617

M_2 M_1

Binary period = 7.751939337 hours
Pulsar period = 59 milliseconds
Neutron star mass M_1 = 1.4411(7) $M_\odot$
Neutron star mass M_2 = 1.3874(7) $M_\odot$

Figure 15.22. A schematic diagram showing the binary pulsar PSR 1913+16. As a result of the ability to measure precisely many parameters of the binary orbit from ultra-precise pulsar timing, the masses of the two neutron stars have been measured with very high precision. (Data courtesy of Professor J. Taylor.)

also a neutron star and in which the neutron stars form a close binary system. The first of these to be discovered was the binary pulsar PSR 1913+16, which is illustrated schematically in Fig. 15.22. The system has a binary period of only 7.75 hours and the orbital eccentricity is large, $e = 0.617$. This system is a pure gift for the relativist. To test general relativity, we require a perfect clock in a rotating frame of reference, and systems such as PSR 1913+16 are ideal for this purpose. The neutron stars are so inert and compact that the binary system is very 'clean', and so can be used for some of the most sensitive tests of general relativity yet devised. To give just a few examples of the precision which can be obtained, I reproduce some recent results with the kind permission of Professor J. Taylor and his colleagues.

In Fig. 15.23, the determination of the masses of the two neutron stars in the binary system PSR 1913+16 is shown, assuming that general relativity is the correct theory of gravity. Various parameters of the binary orbit can be measured very precisely, and these provide different estimates of functions involving the masses of the two neutron stars, M_1 and M_2. In Fig. 15.23, the various parameters of the binary orbit are shown, those which have been measured with very good accuracy being indicated by an asterisk. It can be observed that the different loci

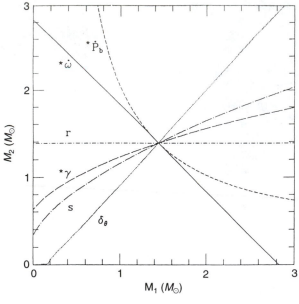

Figure 15.23. The measurement of the masses of the neutron stars in the binary system PSR 1913+16, resulting from very precise timing of the arrival times of the pulses at the Earth. The different parameters of the neutron star's orbit depend upon different combinations of the masses M_1 and M_2 of the neutron stars. It can be seen that the lines intersect very precisely at a single point in the M_1/M_2 plane. (Courtesy of Professor J. Taylor.)

intersect at a single point in the M_1/M_2 plane. Some measure of the precision with which the theory is known to be correct can be obtained from the accuracy with which the masses of the neutron stars have been estimated, as indicated in Fig. 15.22.

A second remarkable measurement has been the rate of loss of orbital rotational energy by the emission of gravitational waves. The binary system loses energy by the emission of gravitational radiation, and the rate at which energy is lost can be precisely predicted once the masses of the neutron stars and the parameters of the binary orbit are known. The rate of change of the angular frequency, Ω, of the orbit due to gravitational radiation energy loss is precisely known, $-\mathrm{d}\Omega/\mathrm{d}t \propto \Omega^5$. The change in orbital phase due to the emission of gravitational waves has been observed over a period of 17 years, and the observed changes over that period agree precisely with the predictions of general relativity (Fig. 15.24). Thus, although the gravitational waves themselves have not been detected, exactly the correct energy loss rate from the system has been measured – it is generally assumed that this is convincing evidence for the existence of gravitational waves, and this observation acts as a spur to their direct detection by future generations of gravitational wave detectors. This is a very important result for the theory of

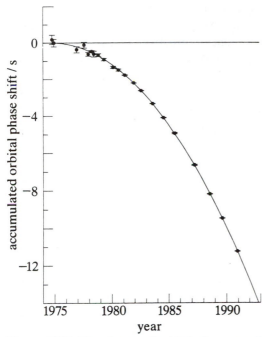

Figure 15.24. The change of orbital phase as a function of time for the binary neutron star system PSR 1913+16 compared with the expected changes due to gravitational radiation energy loss by the binary system. (Courtesy of Professor J. Taylor.)

gravitation, since this result alone enables a wide range of alternative theories of gravity to be eliminated. For example, since general relativity predicts only quadrupolar emission of gravitational radiation, any theory which, say, involved the dipole emission of gravitational waves can be eliminated.

Thus, general relativity has passed all of the most stringent tests attempted so far, and we can have much greater confidence than in the past that it is an excellent description of the relativistic theory of gravity. The same techniques of accurate pulsar timing can also be used to determine whether or not there is any evidence for the gravitational constant G changing with time. These tests are slightly dependent upon the equation of state used to describe the interior of the neutron stars, but, for the complete range of possible equations of state, the limits of $\dot{G}/G$ are less than about 10^{-11} year^{-1}. Thus, there can have been little change in the value of the gravitational constant over typical cosmological timescales, which are about $(1-2) \times 10^{10}$ years. Continued observations of certain of the binary pulsars should enable this limit to be improved by an order of magnitude.

15.6.3   *The general case of black holes in general relativity*

The discussion of Section 15.6.1 gives an impression of how the simplest forms of black hole occur in general relativity. In 1916, the next spherically symmetric solution of the Einstein field equations was discovered and concerned the case of point mass M with electric charge Q; the external solutions for the space-time geometry are given by the Reissner–Nordstrøm metric, which we will not reproduce here (see Misner, Thorne and Wheeler (1973), pp. 877 *et seq.* for more details). Then, in 1962, Kerr (1963) discovered the general solution for a black hole with angular momentum J. It has been shown that isolated black holes can only possess these three properties: mass M, charge Q and angular momentum J – all other quantum numbers and properties are radiated away during the formation of the black hole. Thus, black holes are the simplest macroscopic objects in nature.

The rotating black holes, which are sometimes referred to as *Kerr black holes*, are of the greatest interest because they possess a number of properties which are of relevance to many aspects of high energy astrophysics, particularly in the study of X-ray sources and of active galactic nuclei. The mathematical analysis of the properties of black holes is of quite daunting complexity (see, for example, Chandrasekhar (1985), especially his footnote on p. 530). We can only scrape the surface of the subject and indicate some the the properties of rotating black holes which may have application in understanding astrophysical phenomena. To give a flavour of the subject, we can write down the *Kerr metric* in what are known as Boyer–Lindquist coordinates:

$$ds^2 = \left(1 - \frac{2GMr}{\rho c^2}\right) dt^2 - \frac{1}{c^2}\left[\frac{4GMra\sin^2\theta}{\rho c}\, dt d\phi\right.$$
$$\left. + \frac{\rho}{\Delta} dr^2 + \rho\, d\theta^2 + \left(r^2 + a^2 + \frac{2GMra^2\sin^2\theta}{\rho c^2}\right)\sin^2\theta\, d\phi^2\right] \qquad (15.42)$$

Unlike the above texts, I have preserved the dimensional constants G and c and adopted the same sign conventions as in the metrics (15.32) and (15.33). In this formalism, the black hole rotates in the positive ϕ direction, and the symbols have the following meanings: $a = (J/Mc)$ is the angular momentum of the black hole per unit mass and has the dimensions of distance, $\Delta = r^2 - (2GMr/c^2) + a^2$ and $\rho = r^2 + a^2\cos^2\theta$. If the black hole is non-rotating, $J = a = 0$, and the Kerr metric reduces to the standard Schwarzschild metric (15.32). The properties of this metric are described by Misner, Thorne and Wheeler (1973, pp. 878–87) and by Shapiro and Teukolsky (1983, pp. 357–64). Let us summarise some of these properties.

(1) It can be seen that the metric coefficients are independent of t and ϕ and so the metric is stationary and axisymmetric about the ϕ-axis.

(2) By inspection, it can be seen that, just as in the case of the Schwarzschild metric, the metric coefficient of dr^2 becomes singular at a certain radial distance, in the case of the Kerr metric when $\Delta = 0$. This radius corresponds to the surface of infinite redshift or to the *horizon* of the rotating black hole. The radius at

which this occurs is the solution of

$$\Delta = r^2 - (2GMr/c^2) + a^2 = 0$$

Taking the larger of the two roots of this quadratic equation, we find that the horizon occurs at radius r_+, given by

$$r_+ = \frac{GM}{c^2} + \left[\left(\frac{GM}{c^2}\right)^2 - \left(\frac{J}{Mc}\right)^2\right]^{1/2} \tag{15.43}$$

This spherical surface has exactly the same properties as the Schwarzschild radius in the case of non-rotating black holes. Particles and photons can fall in through this radius, but they cannot emerge outwards through it according to the classical theory of general relativity.

(3) It can be seen from the expression (15.43) that, if the system has too much angular momentum J, no black hole will be formed. Inspection of expression (15.43) shows that this maximum angular momentum corresponds to $J = GM^2/c$. This result is of special interest because the matter which falls into a black hole takes with it angular momentum as well as mass. For a maximally rotating black hole, the horizon radius is $r_+ = GM/c^2$, which is just half the result for the case of Schwarzschild black holes, $r_g = 2GM/c^2$.

(4) The effect of the rotation of the black hole is that, as one approaches the black hole, the phenomenon of the 'dragging of inertial frames' becomes more and more important. As expressed by Thorne, Price and MacDonald (1986), 'The hole's rotation drags *all* physical objects near it into orbital motion in the same direction as the hole rotates; nothing can resist. The nearer one is to the horizon, the stronger the dragging effect. Therefore, it is inevitable that our fiducial observers have some radius-dependent finite angular velocity as seen by a distant static (undragged) observer.' In practice, this means that any gyroscope will be observed to precess relative to the distant stars. Before the horizon equivalent to the Schwarzschild radius is reached, the dragging of the inertial frames becomes so strong that there is a limiting radius, known as the *static limit*, within which no observer can remain at rest relative to the background stars. Thus, all observers within this limit must rotate in the same direction as the black hole. No matter how hard an observer tries to remain static, for example by firing rocket engines in the direction opposite to the rotation, rotation relative to the background stars cannot be prevented. Within the static radius, light cones point in the direction of rotation. The static radius about the black hole, r_{stat}, is

$$r_{\text{stat}} = \frac{GM}{c^2} + \left[\left(\frac{GM}{c^2}\right)^2 - \left(\frac{J}{Mc}\right)^2 \cos^2\theta\right]^{1/2} \tag{15.44}$$

where θ is the polar angle measured relative to the rotation axis of the black hole. It will be noted that, in the polar directions $\theta = 0$ and π, r_+ and r_{stat} coincide, but that they are distinct in all other directions. Only in the case of no rotation, $J = 0$, do the two radii coincide for all values of θ. A diagram showing the comparative structures of rotating and non-rotating black holes is presented in Fig. 15.25.

Non-rotating, Schwarzschild black hole

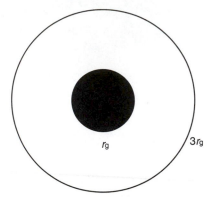

r_g $3r_g$

Maximally rotating Kerr black hole

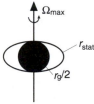

Ω_{max}

r_{stat}

$r_g/2$

Figure 15.25. A schematic diagram illustrating the various regions about a maximally rotating black hole compared with the corresponding dimensions for a non-rotating black hole of the same mass.

(5) The shrinking of the horizon for rotating black holes described in (3) above is an important result because it suggests that, in the case of the maximally rotating black hole, matter can move in stable circular orbits with much smaller radii about the black hole as compared with the Schwarzschild case, before it falls into $r = 0$. The corollary of this is that we expect to be able to extract more of the rest mass energy of infalling matter in the case of rotating black holes as compared with the non-rotating case.

Shapiro and Teukolsky analyse the orbits of particles moving in the equatorial plane of a rotating black hole, and we simply quote their results. It will be recalled that, in the case of a non-rotating black hole, the last stable orbit lies at $r = 3r_g = 6GM/c^2$. In the case of rotating black holes, the corresponding results depend upon whether the particle is moving in the direction of rotation of the black hole (corotation) or in the opposite direction (counter-rotation). The most interesting cases are those of maximally rotating black holes for which $J = GM^2/c$. The last stable circular orbits lie at $r = r_+ = GM/c^2$ for corotating test particles and at $r_+ = 9r_+ = 9GM/c^2$ for counter-rotating particles. Notice that the latter orbit lies outside the static limit, which comes as no surprise. Correspondingly, the maximum binding energies of these orbits can be found, that is, the amount of energy which has to be lost in order that the material attains a bound stable circular orbit with radius r. For the corotating case, the maximum binding energy is a fraction $(1 - 1/\sqrt{3})$ of the rest mass energy of the orbiting material and $(1 - \sqrt{(25/27)})$ for counter-rotating orbits. The corotating case is of the greatest interest because this result means that up to 42.3% of the rest mass energy of the material can be released as it spirals into the black hole through a sequence of almost circular equatorial orbits. This is the process by which energy is liberated through the accretion of matter onto black holes, and is likely to be the source of energy in some of the most extreme astrophysical objects. Note that the energy available is very much greater than that attainable from nuclear fusion processes, which at most can release about 1% of the rest mass energy of the matter.

(6) The region of space-time between the radii r_+ and r_{stat} is referred to as the *ergosphere*, and it has a number of important properties which may be of importance astrophysically. Although the matter has to rotate in the same direction as the black hole within the ergosphere, particles can escape from it by a process known as the *Penrose process*. This depends upon the fact that there exist negative energy orbits within the region defined by the static limit. Penrose showed that, if a particle enters one of these negative energy orbits and splits into two pieces, one of them can fall down the hole while the other escapes to infinity with greater energy than the original particle had when it fell in. The source of the extra energy is the rotational energy of the black hole. The reason that the volume between the horizon at r_+ and the static limit at r_{stat} is called the *ergosphere* is that this is the region from which energy can, in principle, be extracted by the Penrose process.

(7) An interesting point concerns whether or not we can tap the rotational energy of the black hole as a source of energy for the exterior Universe. It turns out that the fraction of the rest mass energy of the black hole which can be made available to the exterior Universe is

$$1 - 2^{-1/2}[1 + [1 - (J/J_{max})^2]^{1/2}]^{1/2}$$

For a maximally rotating black hole, this percentage amounts to 29%. Once again, we observe the large efficiencies available, in principle, from black holes for powering astrophysical phenomena.

(8) Another interesting phenomenon concerns the electrodynamics of black holes. We have already stated that an isolated black hole cannot possess any properties other than mass, electric charge and angular momentum. However, there is no reason why a magnetic field cannot be associated with the black hole, provided it is 'tied onto' the surrounding medium. This is illustrated in Fig. 15.26, which shows the magnetic field lines in the vicinity of a black hole with a thick accretion disc (from Thorne, Price and Macdonald (1986)). The electrodynamics of such configurations may well be important in understanding the origin of high energy particles in active galactic nuclei. In a remarkable book, *Black holes: the membrane paradigm*, Thorne, Price and Macdonald show that the

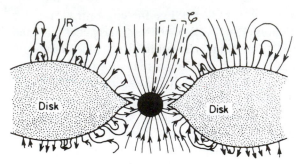

Figure 15.26. A possible configuration for the magnetic field distribution about a black hole which is surrounded by an accretion disc. It can be seen that the magnetic field lines are linked to both the accretion disc and the surrounding medium. (From K.S. Thorne, R.H. Price and D.A. Macdonald (1986). *Black holes: the membrane paradigm*, p. 135. New Haven: Yale University Press.)

electrodynamics of all types of black holes can be understood if the horizon is replaced by a two-dimensional membrane with the resistivity of free space.

(9) Finally, we should note that, in addition to a resistivity, other properties can be attributed to the surface of a black hole, in particular a temperature, defined by

$$T = \frac{\hbar c^3}{8\pi GMk} \approx \frac{10^{-7}}{(M/M_\odot)} \quad \text{K} \tag{15.45}$$

According to classical general relativity, nothing can escape from a black hole, and therefore it cannot emit radiation. In a brilliant analysis, Hawking showed, however, that, when quantum electrodynamical processes are considered at the horizon of a black hole, there is a finite probability of particles escaping to infinity (Hawking (1975a,b)). The simple picture is that of virtual photons being created from the vacuum at the horizon, and, before they can annihilate, one of them falls down the hole while the other escapes to infinity. Hawking predicted that the emitted particles should have a thermal spectrum at the temperature given by expression (15.45). Evidently, the rate of evaporation of solar-mass black holes by this process is very low indeed. However, since the temperature is inversely proportional to mass, if very-low-mass black holes were formed in the very early Universe, all those with mass less than about 10^{12} kg would have completely evaporated by now. Black holes with roughly this mass should be ending their existence by now, and might be the sources of a strong pulse of electromagnetic radiation. Inserting $M = 10^{12}$ kg into expression (15.45), the temperature of the thermal emission should be about 10^{11} K, so that a pulse of γ-rays would be expected.

We will find that these rather exotic phenomena, and many others, will find important applications, particularly in the astrophysics of active galactic nuclei.

15.6.4   *Observational evidence for black holes*

It may come as a surprise that astrophysicists nowadays write so confidently about objects which are nothing less than singularities in space-time. What is the evidence that black holes exist? There are two theoretical arguments which I find compelling. First of all, as mentioned in Section 15.6.1, it has been proved that black-hole type singularities are a very general property of all theories in which the force of gravity is attractive. It is the combination of Einstein's mass–energy relation $E = mc^2$ and the attractive nature of gravity which are the root causes of the black hole phenomenon. The second argument is more intuitive. The fact that neutron stars have been discovered, and that their masses match so well the values predicted by theory, is for me a key result. It will be recalled that the typical radius of a $1M_\odot$ neutron star is about 10 km, just greater then three times the radius of a black hole of the same mass. There must be other collapsing remnants of higher mass or perhaps remnants which collapsed more vigorously and so were unable to form stable neutron stars. There is also the strong possibility that neutron stars in binary systems accrete mass from the primary star, and so their masses eventually exceed the upper mass limit for stability – this is the analogue for neutron stars of the process believed to be responsible for the formation of Type I supernovae through accretion of mass onto white dwarfs (Section 15.2.1).

The best way of finding stellar-mass black holes is to search for binary X-ray sources with invisible companions which are inferred to have masses greater than $3M_\odot$. A great deal of effort has been devoted to discovering whether or not any of the X-ray binary sources contain black holes. McClintock (1992) identified four high luminosity X-ray binaries in which there is the clearest evidence for the presence of black holes. These are the galactic sources Cygnus X-1 and A0620-00 and the sources LMC X-1 and LMC X-3 in the Large Magellanic Cloud. Cowley (1992) comes to similar conclusions, but has some doubt about the case of LMC X-1. The characteristics of these systems are that they are high luminosity X-ray sources which do not contain X-ray pulsars and which must be radiating at close to the Eddington limit for stellar mass objects, $L_X \sim 10^{31}$ W. The primary companions are single-lined spectroscopic binaries. They exhibit rapid variability or 'flickering' in their X-ray intensities, and this would be consistent with the X-rays originating close to a black hole, since the minimum timescale for such variability should be $\tau \sim r_g/c \sim 10\,(M/M_\odot)$ μs. Evidence of flickering on the timescales of milliseconds has been claimed in a number of sources, including Cygnus X-1, but, in a number of cases, it has turned out that the flickering sources contain X-ray pulsars. This indicates that sources containing X-ray binaries can exhibit very short timescale fluctuations, and that this is not necessarily a definite signature of black holes in X-ray binaries.

The best evidence is provided by the classical astrometric and astrophysical techniques of the analysis of single-lined spectroscopic binaries. According to McClintock (1992), it is likely that the masses of the unseen companions in the four above binary X-ray sources are greater than $3M_\odot$. The estimation of the actual masses of the binary components is model dependent, in particular upon knowledge of the inclination of the binary orbit to the plane of the sky and the mass of the visible companion. Specific models for the four X-ray binaries are

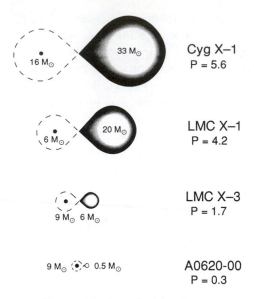

Cyg X–1
P = 5.6

LMC X–1
P = 4.2

LMC X–3
P = 1.7

A0620–00
P = 0.3

P = orbital period in days

Figure 15.27. Schematic sketches, to scale, of plausible models for the four stellar mass black hole candidates discussed by McClintock. The visible companion is shown as filling its critical Roche lobe (shaded regions) in each case. (After J.E. McClintock (1992). *Proc. Texas-ESO-CERN symposium on relativistic astrophysics, cosmology and fundamental particles*, eds J.D. Barrow, L. Mestel and P.A. Thomas, p. 495. New York: New York Academy of Science Publications.)

shown in Fig. 15.27, but there is some uncertainty about the actual values of the masses shown.

Arguments similar to those given in Section 15.5 can be used to demonstrate that accretion is the likely source of energy for X-ray binaries which contain black holes, but there is an important distinction, in that there is no longer a solid surface onto which the matter is accreted. As we will show in the next chapter, the accretion process continues to be a very effective source of energy down to the last stable orbit about the black hole, and so many of the considerations developed concerning accretion onto neutron stars in binary systems are also applicable to accretion onto black holes.

Other observational evidence for black holes comes from the study of active galactic nuclei, which is discussed in Section 16.2.3. In these cases, the black holes must be 'supermassive', in the sense that masses in the range $10^5 \leq M/M_\odot \leq 10^{10}$ are required to account for the enormous luminosities of objects such as quasars and BL-Lac objects, as well as the short timescales of their variability.

16

Accretion power in astrophysics

16.1 Introduction

By accretion, we mean the accumulation of diffuse gas or matter onto some object under the influence of gravity. Until the discovery of X-ray binary systems in the early 1970s, the process of accretion had received relatively limited attention in astrophysics. The discovery of intense X-ray sources in close binary systems which contained neutron stars, for example, the source Hercules X-1 (see Section 7.3.2), initiated a new epoch in high energy astrophysics, when it was realised just how efficient the process of accretion onto compact objects could be. The process was soon applied to binary systems containing white dwarfs, to account for cataclysmic variables, and to active galactic nuclei, where it can naturally account for even the most extreme luminosities observed. Let us begin our analysis by deriving some of the simple relations which show how naturally accretion can account, in principle, for many of the essential features of galactic X-ray sources and active galactic nuclei.

16.2 Accretion – general considerations

16.2.1 The efficiency of the accretion process

Consider the accretion of matter onto a star of mass M and radius R. If the matter falls onto the star in free-fall from infinity, it acquires kinetic energy as its gravitational potential energy becomes more negative. Considering a proton falling in from infinity, we can write

$$\frac{1}{2}m_{\mathrm{p}}v_{\mathrm{ff}}^2 = \frac{GMm_{\mathrm{p}}}{r} \tag{16.1}$$

When the matter reaches the surface of the star at $r = R$, it is rapidly decelerated, and, assuming all the matter accumulates on the surface of the star, the kinetic energy of free-fall has to be radiated away as heat which is available to power the X-ray source. If the rate at which mass is accreted onto the star is $\dot{m}$, the rate

133

at which kinetic energy is dissipated at the surface of the star is $\frac{1}{2}\dot{m}v_{\rm ff}^2$, and hence the luminosity of the source is

$$L = \frac{1}{2}\dot{m}v_{\rm ff}^2 = \frac{GM\dot{m}}{R} \qquad (16.2)$$

It is convenient to introduce the Schwarzschild radius of a star of mass M, $r_{\rm g} = 2GM/c^2$. Inserting this radius into the expression (16.2), we obtain

$$L = \frac{1}{2}\dot{m}c^2 \left(\frac{r_{\rm g}}{R}\right) \qquad (16.3)$$

We can therefore write the luminosity as $L = \xi\dot{m}c^2$. This is a remarkable formula. It can be seen that, written in this form, ξ is the efficiency of conversion of the rest mass energy of the accreted matter into heat. According to the above calculation, the efficiency is roughly $\xi = \frac{1}{2}(r_{\rm g}/R)$. Thus, the efficiency of energy conversion simply depends upon how compact the star is. For a white dwarf, for which $M = M_\odot$ and $R \approx 5 \times 10^6$ m, $\xi \approx 3 \times 10^{-4}$. For a neutron star with mass $M = M_\odot$ and $R = 15$ km, $\xi \sim 0.1$. Thus, accretion onto neutron stars is a remarkably powerful source of energy. This efficiency of energy conversion can be compared with nuclear energy generation. The largest release of nuclear binding energy occurs in the conversion of hydrogen into helium in the p–p chain for which $\xi \approx 7 \times 10^{-3}$. Thus, accretion onto neutron stars is an order of magnitude more efficient as an energy source as compared with nuclear energy generation.

A simple interpretation of the expression (16.3) suggests that we would do even better if the matter were accreted onto a black hole. The problem is that there is no solid surface onto which the matter can be accreted as there is in the case of a neutron star. The properties of black holes were described in Section 15.6 where it was shown that, for many purposes, the Schwarzschild radius plays the role of the surface of the black hole. If, however, the matter simply fell directly into the black hole, there would be no heat released. In practice, what happens is that the infalling matter must have some angular momentum, and conservation of angular momentum prevents matter from falling directly into the black hole.

Consider a small mass element m falling from a large distance onto a black hole. Because of small fluctuations in the gravitational potential, the element is bound to acquire some small amount of angular momentum with respect to the black hole, $\mathscr{L} = I\Omega = mv_\perp r$. Because of the conservation of angular momentum, $mv_\perp r = $ constant, and hence the rotational energy of the element increases as it approaches the hole: $E_{\rm rot} = \frac{1}{2}I\Omega^2 = \frac{1}{2}\mathscr{L}^2/I \propto r^{-2}$, since $I = mr^2$. Thus, as the mass element approaches the hole, its rotational energy increases more rapidly than its gravitational potential energy, which increases only as r^{-1}. This is a familiar result in the analysis of orbits according to Newtonian dynamics – no matter how small the angular velocity of a mass element at a very large distance, this is sufficient to prevent collapse to $r = 0$. As discussed in Section 15.6.1, this result is modified in the general relativistic treatment because of the presence of an 'attractive' term in a general relativistic version of the law of conservation of energy. In the case of a spherically symmetric black hole, this term is $-(2GMh^2/r^3c^2)$ (see equation (15.40)). The net result is that particles with specific angular momenta $h \leq 2r_{\rm g}c$

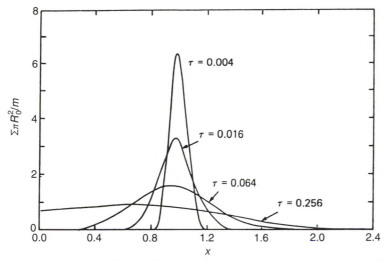

Figure 16.1. The evolution of a ring of matter of mass m which is placed initially at radius $r = R_0$. The ring spreads out under the action of viscous forces in the ring. The surface density, Σ, is shown as a function of radius $x = r/R_0$ and the dimensionless time variable $\tau = 12vtR_0^{-2}$. v is the kinematic viscosity of the matter of the disc. (From J. Frank, A. King and D. Raine (1992). *Accretion power in astrophysics*, p. 71. Cambridge: Cambridge University Press. After J.E. Pringle (1981). *Ann. Rev. Astron. Astrophys.*, **19**, 140.)

can fall directly in to $r = 0$. This is, however, a tiny specific angular momentum relative to the typical values found in, say, the mass transfer in close binary systems. Thus, the matter can eventually be prevented from falling into the black hole in directions perpendicular to the rotation axis by centrifugal forces.

The matter can, however, collapse along the rotation axis of the infalling material so that a disc of matter forms about the black hole. The matter in this *accretion disc* can only fall into the black hole if it loses its angular momentum, and this is achieved by viscous forces acting in the disc. The viscosity has two effects. First, it transfers angular momentum outwards, as is illustrated by the results of an exact model calculation of the spreading of a thin viscous disc (Fig. 16.1). It can be seen that viscosity redistributes the angular momentum so that some of the matter spreads outwards taking angular momentum with it and thus allows the rest of the matter to spiral inwards (Pringle (1981)). At the same time, the viscosity acts as a frictional force, which results in the dissipation of heat. The matter in the accretion disc drifts gradually inwards until it reaches the *last stable orbit* about the black hole. At this point, the matter spirals irretrievably into the black hole (see Section 15.6.1).

Thus, the maximum energy which can be released by accretion onto black holes is given by the energy which has to be dissipated in order to reach the last stable orbit about the black hole. These binding energies were given in Sections 15.6.2 and 15.6.3. For Schwarzschild, or spherically symmetric, black holes, $\xi = 0.06$, whereas, for a maximally rotating Kerr black hole, $\xi = 0.426$. Thus, black holes, and, in particular, maximally rotating black holes, are the most powerful energy sources we know of in the Universe, and accretion is the process by which the energy can be released.

16.2.2 *The Eddington limiting luminosity*

It might seem as though we could generate arbitrarily large luminosities by allowing matter to fall at a sufficiently great rate onto a black hole. There is, however, a limit to the luminosity, and this is determined by the fact that, if the luminosity is too great, radiation pressure blows away the infalling matter. This limiting luminosity, which is known as the *Eddington luminosity*, is found by balancing the inward force of gravity against the outward pressure of the radiation.

We can find this limit by a simple calculation. We assume that the infalling matter is fully ionised and that the radiation pressure force is provided by Thomson scattering of the radiation by the electrons in the plasma. Notice that, by assuming that the pressure is due to Thomson scattering, we adopt the smallest cross-section for the processes which impede the loss of radiation from the system. Consider the forces acting on an electron–proton pair at distance r from the source. Then, the inward gravitational force is given by

$$f_{\text{grav}} = \frac{GM}{r^2}(m_p + m_e) \approx \frac{GMm_p}{r^2}$$

The radiation pressure acts upon the electron, but, because the plasma must remain neutral, this pressure is communicated to the protons by the electrostatic forces between the proton and the electron. To express this process in terms of plasma physics, the electron and proton cannot be separated by more than a Debye length $\lambda_D = v_e/\omega_p = 70(T_e/N_e)^{1/2}$ m, where v_e is the thermal speed of the electrons, ω_p is the plasma angular frequency and N_e is the number density of the electrons, before electrostatic forces in the plasma restore charge neutrality. Each photon gives up a momentum $p = h\nu/c$ to the electron in each collision. Therefore, the force acting on the electron is just the momentum communicated to it per second by the incident flux of photons N_{ph}. Thus, $f = \sigma_T N_{\text{ph}} p$, where $\sigma_T = 6.653 \times 10^{-29}$ m^{-2} is the Thomson cross-section. The flux of photons at distance r from the source is $N_{\text{ph}} = L/4\pi r^2 h\nu$, where L is the luminosity of the source, and so the outward force on the electron is $f = \sigma_T L/4\pi r^2 c$. Equating the inward and outward forces, we find

$$\frac{\sigma_T L}{4\pi r^2 c} = \frac{GMm_p}{r^2}$$

$$L_E = \frac{4\pi GMm_p c}{\sigma_T}$$

L_E is known as the *Eddington luminosity*, and is the maximum luminosity which a spherically symmetric source of mass M can emit in a steady state. Notice that the limit is independent of the radius r and depends only upon the mass M of the source. This limiting luminosity is closely related to the vibrational instability, which leads to the upper mass limit for main sequence stars (see Section 14.5). Stars with mass $M \sim 60M_\odot$ are essentially radiating at the

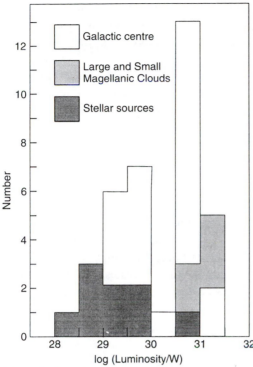

Figure 16.2. A histogram of the X-ray luminosities of bright X-ray sources in the 1–10 keV energy band in our Galaxy and the Magellanic Clouds. The Eddington limiting luminosity for a $1M_\odot$ object is 1.3×10^{31} W. (After B. Margon and J.P. Ostriker (1973). *Astrophys. J.*, **186**, 91.)

Eddington luminosity, as may be confirmed by comparing the mass–luminosity relation for main-sequence stars (Section 13.3) with the Eddington limit.

We can rewrite the expression for the Eddington luminosity in the following way by introducing the Schwarzschild radius $r_\mathrm{g} = 2GM/c^2$:

$$L_\mathrm{E} = \frac{2\pi r_\mathrm{g} m_\mathrm{p} c^3}{\sigma_\mathrm{T}} = 1.3 \times 10^{31} \left(\frac{M}{M_\odot} \right) \quad \text{W} \qquad (16.4)$$

This is a very important result. It is interesting to compare the X-ray luminosities of binary X-ray sources in our Galaxy and in the Magellanic Clouds with this limit for typical stellar masses (Fig. 16.2). It can be seen that the X-ray luminosities extend up to more or less the maximum allowed values for sources with masses $M \sim 1$–$10M_\odot$. It is possible to exceed the Eddington limit by adopting different geometries for the source, but not by a large factor. Notice also that, in non-steady-state situations, such as in supernova explosions, the Eddington limit can be exceeded by a large margin. For many of our purposes, we will find that the above spherically symmetric result provides important constraints for models of high energy astrophysical sources.

The observation of pulsed X-ray emission from many of the binary X-ray sources with pulse periods similar to those of radio pulsars, is convincing evidence that they contain rotating, magnetised neutron stars. In this case, we can work out a lower limit to the temperature of the source if it is radiating at the Eddington limit. Any source loses energy most efficiently if it radiates as a black body at temperature T. Therefore, we can equate the black-body luminosity of a neutron star of radius $r_{NS} = 10$ km to the Eddington limit for a star of mass $1M_\odot$:

$$4\pi r_{NS}^2 a T^4 = L_{Edd} = 1.3 \times 10^{31} \text{ W}; \quad T = 2 \times 10^7 \text{ K} \qquad (16.5)$$

Thus, for solar-mass neutron stars, we find $T \geq 10^7$ K. It is therefore entirely natural that neutron stars accreting at close to the Eddington limit should emit most of their energy at X-ray wavelengths. The same calculation can be performed for accretion onto white dwarfs, which have radii about 1000 times those of neutron stars. Then the typical black-body temperatures are about 3×10^5 K. Thus, accreting white dwarfs should be strong emitters of ultraviolet radiation. This expectation is in good agreement with the observed properties of *cataclysmic variables*, which are binary systems in which the compact companion is a white dwarf. Thus, these stars provide a paradigm for the processes which may take place in the more extreme cases of accreting neutron stars and black holes.

16.2.3   *Black holes in X-ray binaries and active galactic nuclei*

We can use the above results to investigate the case for the presence of accreting black holes in certain X-ray binaries and in active galactic nuclei. In the case of the X-ray pulsars, the evidence is overwhelming that the energy source is accretion – the X-ray pulse periods less than about one second are convincing evidence for rotating, magnetised neutron stars. In the case of black holes, there is not expected to be such a clear signature of a compact object, but there might well be variability on a timescale roughly equal to the light-travel time across the last stable orbit about the black hole. The light-travel time across the Schwarzschild radius of a black hole is

$$T_{min} \approx \frac{r_g}{c} = 10^{-5} \left(\frac{M}{M_\odot} \right) \quad \text{s} \qquad (16.6)$$

This is the absolute minimum timescale which we can associate with an object of mass M unless we appeal to phenomena, such as relativistic beaming, which allow the causality limit, $\tau > l/c$, to be violated. In practice, we would expect variations on timescales somewhat longer than this value. For example, the last stable orbit about a spherically symmetric black hole lies at $3r_g$ from the centre of the hole.

For solar-mass black holes, the expected temperatures of the radiating matter are similar to those given by the expression (16.5) because the last stable orbit occurs at a radius $r = 3r_g = 9(M/M_\odot)$ km, and so, if $M \sim M_\odot$, temperatures $T \sim 10^7$ K are expected at the inner edge of the accretion disc. In fact, in the black hole candidates discussed in Section 15.6.4, 'flickering' of the X-ray intensity is observed on a timescale of milliseconds (Fig. 16.3(a)). The light curve of Cygnus X-1 (Cyg X-1), which is one of the best candidates for a stellar-mass

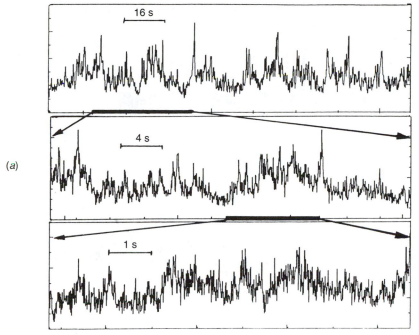

Figure 16.3 (*a*). Examples of the 'flickering' of the X-ray emission from GINGA data for the X-ray binary source Cyg X-1. The three panels show the same data set, plotted with different timescales. (K. Makishima (1988). In *Physics of neutron stars and black holes*, ed. Y. Tamaka, p. 177. Tokyo: Universal Academic Press, Inc.).

black hole, does not show regular pulsations but a characteristic flickering pattern of variation. Similar random flickering is observed in the variable X-ray emission observed from some active galactic nuclei. The light curves of the active galaxies NGC 4051, NCG -6-30-15 and NGC 5506 bear more than a passing resemblance to that of Cyg X-1 but on very much longer timescales, as expected according to expression (16.6), and it is tempting to suppose that these flickering light curves are the signatures of black holes on both the stellar and galactic scale. These remarkable observations do not prove that there are black holes in these systems, but they are certainly entirely consistent with that hypothesis.

The same type of casuality argument may be used to investigate the masses of active galactic nuclei in another way. As is apparent from Fig. 16.3, active galactic nuclei exhibit rapid variability at X-ray and optical wavelengths as well as enormous luminosities. Wandel and Mushotzky (1986) have estimated the masses of the active galactic nuclei by two independent methods. First, they used a version of expression (16.6) to find masses from the shortest variability timescales observed. Secondly, they made dynamical estimates of the masses of active galactic nuclei from the properties of their broad emission line regions. The clouds responsible for these emission lines are in motion about the nucleus and, by studying the processes of ionisation and excitation of the emission line

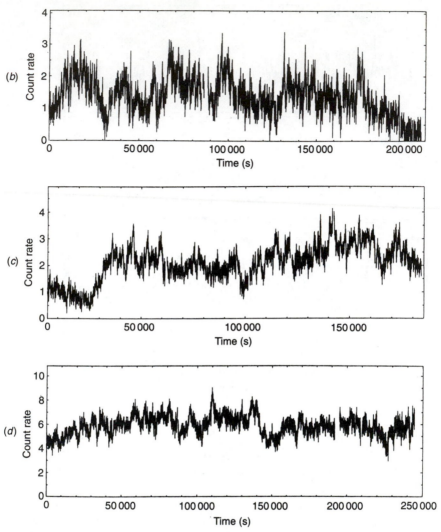

Figure 16.3 (*b*)–(*d*). The EXOSAT X-ray light curves of the active galaxies (*b*) NGC 4051, (*c*) NCG -6-30-15 and (*d*) NGC 5506 (I.M. McHardy and K.A. Pounds (1988). In *Physics of neutron stars and black holes*, ed. Y. Tamaka, p. 288. Tokyo: Universal Academic Press, Inc.)

regions, estimates of the distances of the clouds from the nucleus can be made. Since both the distances of the clouds from the nucleus and their velocities are known, mass estimates can be found by balancing the centrifugal forces acting on the clouds with their gravitational attraction for the nucleus (see Volume 3 for more details):

$$\frac{Gm_{\mathrm{cl}}M}{r^2} = \frac{m_{\mathrm{cl}}v^2}{r}, \qquad M = \frac{v^2 r}{G}$$

Wandel and Mushotzky found that, for the sample of active galactic nuclei which

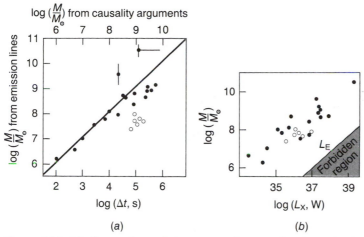

Figure 16.4. (*a*) Comparison of the mass estimates of active galactic nuclei from the variability of their X-ray emission and from dynamical estimates. The filled circles are quasars and Type I Seyfert galaxies and the open circles are Type II Seyfert galaxies. (*b*) Comparison of the inferred masses and luminosities with the Eddington limiting luminosity, $L_E = 1.3 \times 10^{31}(M/M_\odot)$. It can be seen that all of these objects lie well below the Eddington limit. (After A. Wandel and R. F. Mushotzky (1986). *Astrophys. J.*, **306**, L63–4.)

they studied, these independent estimates of the masses of the central objects were in good agreement (Fig. 16.4 (*a*)). It can be seen that the masses of the black holes inferred to be present in these active galactic nuclei range from about 10^6 to $10^9 M_\odot$. Furthermore, when they compared the luminosities of the sources with the corresponding Eddington luminosities, none of the nuclei was found to exceed that limit (Fig. 16.4 (*b*)). Indeed, even the most luminous active galactic nuclei could be accounted for by this model. Now, this argument does not prove that active galactic nuclei are powered by the accretion of matter onto supermassive black holes, but it does show that even the most extreme examples of active galactic nuclei known can be accounted for, in principle, by credible physical processes and that, although these properties are extreme, they are well within the realm of known physics. These arguments suggest that we can use these objects to elucidate the behaviour of matter in the strongest gravitational fields we have acccess to anywhere in the Universe.

There is another interesting development of the simple calculations presented above. We can work out the typical temperature of the emission from gas at roughly the last stable orbit by equating the Eddington luminosity to the thermal emission of the disc. For simplicity, we assume that the radiation originates from a spherical surface of radius $r = 3r_g = 9(M/M_\odot)$ km about the black hole. Substituting r for r_{NS} in the expression (16.5), we find

$$T \approx 2 \times 10^7 \left(\frac{M}{M_\odot}\right)^{-1/4} \quad K$$

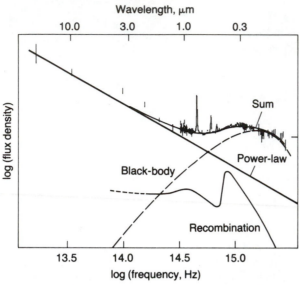

Figure 16.5. The optical–ultraviolet spectrum of the quasar 3C 273. The continuum has been decomposed into a 'power-law' component, a component associated with recombination radiation and a 'blue-bump' component, which has been represented by a black-body curve. The prominent Balmer series in the optical waveband which led to the discovery of the large redshift of 3C 273 can be seen. (From M. Malkan and W. L. W. Sargent (1982). *Astrophys. J.*, **254**, 33.)

This suggests that, for supermassive black holes with, say, $M = 10^8 M_\odot$, the thermal emission would have temperature $T \sim 2 \times 10^5$ K. It is intriguing that many active galactic nuclei have strong ultraviolet continua, 3C 273 being a good example of a galaxy with a 'blue bump' (Fig. 16.5). This is, however, a very rough argument, and it certainly cannot be the whole story since these nuclei are just as powerful emitters in the X-ray waveband.

16.3   Thin accretion discs

The simplest cases of disc accretion are thin accretion discs. Many of their essential features are described by Pringle (1981) and in the monograph with the same title as this chapter, *Accretion power in astrophysics* by Frank, King and Raine (1992). In the simplest picture, we consider steady-state discs with a constant rate of mass accretion $\dot{m}$ into the disc. The matter in the disc would take up Keplerian orbits if there were no viscous forces present. We know, however, that these forces are essential in order to transfer the angular momentum of the accreted material outwards and so allow the gas to move inwards to more tightly bound orbits.

First of all, let us work out the conditions under which the thin disc approximation is valid.

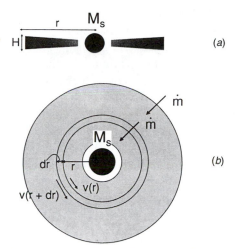

Figure 16.6. A schematic diagram illustrating the geometry of a thin accretion disc. (*a*) A side view; (*b*) a view from above the disc.

16.3.1   *Conditions for thin accretion discs*

The geometry of the thin accretion disc is shown in Fig. 16.6. It is assumed that the mass of material in the disc is much less than the mass of the central star, $M_{\mathrm{disc}} \ll M_{\mathrm{s}}$. First, we write down the condition for hydrostatic support in the direction perpendicular to the plane of the disc, that is, in the z direction:

$$\frac{\partial p}{\partial z} = -\frac{GM_{\mathrm{s}}\rho \sin \theta}{r^2}$$

But $\sin \theta \approx z/r$, and, to order of magnitude, we can write $\partial p/\partial z \approx p/H$, where H is the half-thickness of the disc. Therefore, the condition for hydrostatic support in the z direction is

$$\frac{p}{H} = \frac{GM_{\mathrm{s}}\rho H}{r^3} \tag{16.7}$$

Now, although the gas is slowly drifting into the centre, the gas moves in roughly Keplerian orbits about the star and so is in centrifugal equilibrium, that is, at any radius,

$$\frac{v_\phi^2}{r} = \frac{GM}{r^2} \qquad \frac{GM}{r} = v_\phi^2 \tag{16.8}$$

Substituting for GM/r from equation (16.8) into equation (16.7), we find

$$\frac{p}{\rho} \approx v_\phi^2 \frac{H^2}{r^2}$$

But the speed of sound in the material comprising the disc is $c_{\mathrm{s}}^2 \approx p/\rho$, and therefore

$$\frac{H}{r} \approx \frac{c_{\mathrm{s}}}{v_\phi} = \frac{1}{\mathcal{M}} \tag{16.9}$$

where $\mathcal{M}$ is the *Mach number* of the rotation velocity of material of the disc relative to the local sound speed in the disc. Thus, the condition that the thin disc approximation can be adopted is that the rotation velocity of the disc be very much greater than the sound speed; in other words, internal pressure gradients should not inflate the disc. This is exactly the same condition which is found for the confinement of the cool gas in the plane of the disc of a spiral galaxy such as our own.

16.3.2   The role of viscosity – the α parameter

Viscosity plays a central role in determining the structure of accretion discs. We need expressions for the viscous forces in a differentially rotating fluid. Let us recall the definition of viscosity in the case of fluid flow. The simplest case is unidirectional flow of the fluid in the positive x direction but with a velocity gradient in, say, the y direction. Then, the force acting on unit area in the x–z plane, that is the shear stress, is given by the expression

$$f_x(y) = \eta \frac{\partial v_x(y)}{\partial y} \tag{16.10}$$

where η is defined to be the *dynamic* or *shear viscosity*. In the case of two-dimensional flow, the expression for the shear stress acting at the point (x, y) is given by the more general form

$$f_{xy} = \eta \left[\frac{\partial v_x}{\partial y} + \frac{\partial v_y}{\partial x} \right] \tag{16.11}$$

because of the symmetry of the stress tensor, $f_{xy} = f_{yx}$ (see, for example, Landau and Lifshiftz, (1959), section 15). For the case of an accretion disc in which the gas moves in circular orbits, it is simplest to convert this expression into polar coordinates, in which $x = r \cos \phi$ and $y = r \sin \phi$. Then, we write the components of the velocity in cylindrical coordinates as

$$v_x = v_r \cos \phi - v_\phi \sin \phi \qquad v_y = v_r \sin \phi + v_\phi \cos \phi$$

and convert the differentials into polar coordinates through the relations

$$\frac{\partial}{\partial x} = \cos \phi \frac{\partial}{\partial r} - \frac{\sin \phi}{r} \frac{\partial}{\partial \phi} \qquad \frac{\partial}{\partial y} = \sin \phi \frac{\partial}{\partial r} + \frac{\cos \phi}{r} \frac{\partial}{\partial \phi}$$

Because of the cylindrical symmetry of the problem, it is convenient to evaluate the differentials at $\phi = 0$, in which case we find

$$f_{xy} = \eta \left[\frac{\partial v_x}{\partial y} + \frac{\partial v_y}{\partial x} \right] = \frac{1}{r} \left(\frac{\partial v_x}{\partial \phi} \right)_{\phi=0} + \left(\frac{\partial v_y}{\partial r} \right)_{\phi=0}$$

$$\frac{1}{r} \left(\frac{\partial v_x}{\partial \phi} \right)_{\phi=0} = -\frac{v_\phi}{r} \qquad \left(\frac{\partial v_y}{\partial r} \right)_{\phi=0} = \frac{\partial v_\phi}{\partial r}$$

Hence,

$$f = \eta \left(-\frac{v_\phi}{r} + \frac{\partial v_\phi}{\partial r} \right) = \eta r \frac{\partial}{\partial r} \left(\frac{v_\phi}{r} \right) = \eta r \frac{\partial \Omega}{\partial r} \tag{16.12}$$

Notice that this result shows correctly that there is no shear stress if the fluid rotates as a solid body, $\Omega = $ constant, justifying the choice of the expression (16.11) for the stress tensor. We can now work out the torque acting on an annular section of the disc of thickness dr at radius r from the centre. If we assume the disc has thickness H, the torque acting on the inner edge of the annulus at r is

$$G = \eta r (2\pi r^2 H) \frac{\partial \Omega}{\partial r}$$

The torque on the outer edge is

$$G(r + dr) = G(r) + \frac{\partial G}{\partial r} dr$$

and so the net torque acting on the annulus is the difference of these torques $(\partial G / \partial r) dr$. The equation of motion of the annulus is therefore

$$\frac{\partial \mathscr{L}}{\partial t} = \frac{\partial}{\partial r} \left[\eta r (2\pi r^2 H) \frac{\partial \Omega}{\partial r} \right] dr$$

where $\mathscr{L}$ is the angular momentum of the annulus. Since $\mathscr{L} = 2\pi r^2 H \rho v_\phi \, dr$, we find

$$r^2 \frac{\partial v_\phi}{\partial t} = \frac{\eta}{\rho} \frac{\partial}{\partial r} \left[r^3 \frac{\partial}{\partial r} \left(\frac{v_\phi}{r} \right) \right]$$

It is conventional to work in terms of the *kinematic viscosity* $v = \eta / \rho$, and so the equation of motion for the circumferential velocity is

$$\frac{\partial v_\phi}{\partial t} = \frac{v}{r^2} \frac{\partial}{\partial r} \left[r^3 \frac{\partial}{\partial r} \left(\frac{v_\phi}{r} \right) \right] \tag{16.13}$$

This is the basic relation we have been seeking because it indicates the role of viscosity in determining the structure of the accretion disc. There is, however, a basic problem: the viscosity is very low and the *Reynolds number* is very large. For any flow, the Reynolds number, $\mathscr{R}$, is defined to be the quantity

$$\mathscr{R} \approx \frac{L^2}{v T} = \frac{V L}{v} \tag{16.14}$$

where L, T and V are typical dimensions of length, time and velocity for the flow. In the typical situations in which viscous forces play an important role, $\mathscr{R} \sim 1$, as, for example, in the flow of a viscous fluid about a cylinder. At large Reynolds numbers, $\mathscr{R} \geq 10^3$, the flow becomes turbulent (see, for example, Landau and Lifshitz (1959), Feynman (1962), Batchelor (1967)).

Let us evaluate the Reynolds number for the material of the disc. In this case, V is the velocity of rotation of the material of the disc and L is the characteristic scale over which the velocity changes, which is just the typical radial distance from the centre of the disc. According to classical kinetic theory, the dynamic viscosity is given by $\eta = \rho v = \frac{1}{3} \rho v_s \lambda$, where ρ is the mass density of the gas, v_s is the internal sound speed, which is roughly the same as the typical speed of the particles in the gas, and λ is the mean free path of the particles. To order of magnitude, we therefore find that

$$\mathscr{R} \sim \frac{V L}{v_s \lambda} \tag{16.15}$$

We can use the results we have already derived to make some simple estimates. Let us consider the case of accretion onto a neutron star of mass $1 M_\odot$ and suppose that the luminosity is close to the Eddington limit. Therefore, the luminosity of the source is 1.3×10^{31} W, and we can equate this to the accretion luminosity $L = \frac{1}{2}\dot{m}c^2(r_g/r)$, where r is the radius of the region from which most of the luminosity is generated. For illustrative purposes, let us adopt $r = 10r_g$, corresponding to roughly three times the radius of the neutron star. We can then find the accretion rate $\dot{m} \sim 10^{15}$ kg s^{-1}. Now, for the thin disc, the accretion rate $\dot{m}$ is

$$\dot{m} = (2\pi r) \times (2H) \times \rho v_r \qquad (16.16)$$

where H is the half-width of the disc at radius r and v_r is the radial inward drift velocity. We can make a rough estimate for H from the expression (16.9) so that $H = r/\mathcal{M}$, where $\mathcal{M}$ is the Mach number of the rotation velocity of the disc relative to the sound speed in the disc. We can find $\mathcal{M}$ from the Keplerian velocity of the disc at $R = 10r_g$ and the sound speed from the fact that the temperature of the disc must be about 10^7 K from the arguments presented in Section 16.2.2. From these, we find $\mathcal{M} \sim 200$. The only remaining uncertainty is the radial drift velocity v_r. This velocity is certainly sub-sonic with respect to the speed of sound in the disc, and so let us write $v_r = \beta v_s$, where $\beta \ll 1$. Inserting these values into the expression (16.16), we find that the mass density in the disc $\rho \sim 0.1\beta^{-1}$ kg m^{-3}. Thus, if, say, $\beta = 0.01$, a typical value which comes out of more detailed models of accretion discs, we find that the mass densities are large, $\rho \sim 10$ kg m^{-3}. This mass density is in good agreement with more detailed models of the structure of accretion discs responsible for the X-ray emission from neutron stars in close binary systems.

The reason for carrying out this calculation is that we need the mass density, ρ, to work out the mean free path, λ, of the particles in the disc. We have already performed this calculation for the case of protons in Solar Wind, and exactly the same physics applies in this case. According to the expression (10.4), the mean free path of a proton is

$$\lambda \approx v_s t_c = 11.4 \times 10^6 \frac{T^{3/2} A^{1/2} v_s}{N Z^4 \ln \Lambda} \quad \text{m} \qquad (16.17)$$

Inserting the values we have derived into the expression for the Reynolds number, we find $\mathcal{R} \sim 10^{12}$.

This is a key result in the astrophysics of accretion discs. A similar answer is found for the accretion discs about white dwarfs. The result has two implications. First, ordinary viscosity associated with the deflection of charged particles in the plasma (Section 10.4) cannot play a major role in determining the structure of the disc. According to this analysis, the flow would be strongly turbulent. The result also provides a possible solution to the problem. The generation of turbulence within the disc results in a *turbulent viscosity*, which can perform all the functions of normal molecular viscosity but now the transport of momentum is associated with the motion of turbulent eddies in the plasma. There is also likely to be a magnetic field in the disc, and this provides a further means of transporting momentum on a large scale. Probably we should be talking about *magnetohydrodynamic turbulence*, but this is a very difficult area of applied mathematics and all that can be done is to adopt a simple empirical prescription to describe the processes of turbulent convection and viscosity. For example, we can write down an expression for the turbulent viscosity as $v_{turb} \sim \lambda_{turb} v_{turb}$, where λ_{turb} is the scale of the eddies and v_{turb} is their rotational or 'turn-over' velocity.

To overcome this problem, Sunyaev and Shakura (1972) introduced the following prescription for the turbulent viscosity, $v = \alpha v_s H$, where v_s is the speed of sound in the disc and H is its scale height in the z direction. The turbulent eddies must have dimension less than the scale height of the disc, and the turn-over velocities must be less than the speed of sound. Therefore, we expect α to be less than or equal to one. There is, therefore, little physics involved in this prescription of the viscosity to be used in the study of accretion discs, and there is little theoretical understanding of how α should vary through the disc as a function of density and temperature. The advantage of this formalism is that analytic solutions can be found for the structure of thin accretion discs in terms of the single parameter α – these solutions are often referred to as α *discs*. Once thin disc models have been found which account for the observed properties of X-ray sources, empirical values for the parameter α can be found.

16.3.3   *The structure of thin discs*

We can now derive further properties of thin discs by estimating the rate at which mass drifts in through the disc, that is, the radial drift velocity, v_r. Let us consider only steady-state discs in which the inflow velocity v_r is determined by the viscosity v. We have already used the equation of conservation of mass, which simply states that, in the steady state, the mass flow through any radius is a constant:

$$\dot{m} = 2\pi r v_r \int \rho \, dz = \text{constant}$$

where the integral takes account of the inflow through the full thickness of the disc in the z direction. It is convenient to work in terms of the *surface density* of the disc, which is defined to be $\Sigma = \int \rho dz$. Therefore,

$$\dot{m} = 2\pi r v_r \Sigma = \text{constant} \tag{16.18}$$

First of all, we need an expression for the torque G acting on the cylindrical surface at radius r from the centre of the disc. According to equation (16.12),

$$f = \eta r \frac{\partial \Omega}{\partial r}$$

The total shear force acting on the cylindrical surface at radius r is therefore fA, and the torque $G = fAr$. Setting $v_\phi = \Omega r$, we find

$$G = 2\pi r^3 v \frac{d\Omega}{dr} \int \rho \, dz = 2\pi r^3 v \Sigma \frac{d\Omega}{dr} \tag{16.19}$$

Now we consider the transport of angular momentum through an annular region of the disc between radii r and $r + \Delta r$. In the steady state, the change of angular momentum is entirely associated with the viscous torques acting on either side of the cylindrical annulus, that is, $G = d\mathscr{L}/dt$. As discussed above, the net torque acting on the annulus is just the difference between the torques at

r and $r + \Delta r$, that is,

$$G(r) - G(r + \Delta r) = -\frac{\partial G}{\partial r}\Delta r \qquad (16.20)$$

where G is given by the expression (16.19).

Now let us work out the rate of transport of angular momentum through the surface at r. The mass transfer per second is given by the expression (16.18), and therefore the angular momentum transport is

$$\dot{m}v_\phi r = 2\pi r^3 \Sigma v_r \Omega \qquad (16.21)$$

In exactly the same way, the angular momentum transport at radius $r + \Delta r$ is

$$(\dot{m}v_\phi r)_{r+\Delta r} = (2\pi r^3 \Sigma v_r \Omega)_{r+\Delta r}$$

We can now make a Taylor expansion and subtract from the expression (16.21). We find

$$\Delta \mathscr{L} = 2\pi \frac{\mathrm{d}}{\mathrm{d}r}\left(r^3 \Sigma v_r \Omega\right)\Delta r \qquad (16.22)$$

per unit time. Equating equations (16.20) and (16.22), we obtain

$$\frac{\mathrm{d}G}{\mathrm{d}r} = 2\pi \frac{\mathrm{d}}{\mathrm{d}r}\left(r^3 \Sigma v_r \Omega\right) \qquad (16.23)$$

We can immediately take the integral of this equation to find

$$G = 2\pi r^3 \Sigma v_r \Omega + C$$

or, using the expression (16.19) for G,

$$v\Sigma\frac{\mathrm{d}\Omega}{\mathrm{d}r} = \Sigma v_r \Omega + \frac{C}{2\pi r^3} \qquad (16.24)$$

where C is a constant. This constant is to be found from the boundary conditions, in particular, by the matching of the rotational velocity at the surface of the star to the velocity of rotation at the inner edge of the accretion disc. The matching of these velocities takes place through a boundary layer of thickness b, and this slightly complicates the analysis. We will, however, press on with the relevant approximations for thin discs in which the boundary layer is assumed to be thin compared with the radius of the star, r_*. These complications are clearly described by Frank, King and Raine (1992). Within the boundary layer, the matter is dragged round the star in solid-body rotation and so out to radius $r_* + b$, $\mathrm{d}\Omega/\mathrm{d}r = 0$. Hence, at the radius $r_* + b$,

$$C = -2\pi(r^3 \Sigma v_r \Omega)_{r_*+b} \qquad (16.25)$$

Now, at this radius, the velocity of rotation must also be the rotation velocity of material in a Keplerian orbit about the star, $\Omega^2 = GM_*/r^3$. Hence,

$$C = -2\pi[r^{3/2}\Sigma v_r(GM_*)^{1/2}]_{r_*+b}$$

But $\dot{m} = 2\pi r v_r \Sigma = $ constant at any radius, and so

$$C = -\dot{m}[GM_*(r_* + b)]^{1/2}$$

Thus, if $b \ll r_*$,

$$C = -\dot{m}(GM_* r_*)^{1/2} \qquad (16.26)$$

Notice that the constant C is just the rate of transfer of angular momentum into the boundary layer.

In the disc itself, the velocities are close to Keplerian, and so, substituting equations (16.26) and (16.18) into equation (16.24), we find the pleasant result

$$\nu\Sigma = \frac{\dot{m}}{3\pi}\left[1-\left(\frac{r_*}{r}\right)^{1/2}\right] \tag{16.27}$$

The next result we need is the rate of dissipation of energy by the viscous forces acting in the disc. The expression found in the standard text-books for the heat generated per unit volume (for example, Landau and Lifshitz (1959), section 16) is given in Cartesian coordinates as:

$$-\left(\frac{dE}{dt}\right) = \frac{1}{2}\eta\sum_{i,j}\left(\frac{\partial v_i}{\partial x_j}+\frac{\partial v_j}{\partial x_i}\right)^2 \tag{16.28}$$

where the vector $\mathbf{v} = (v_x, v_y, v_z)$ is the fluid velocity. It is convenient to convert this expression into cylindrical polar coordinates using the usual relations $x = r\cos\phi$; $y = r\sin\phi$; $z = z$ and $v_x = v_r\cos\phi - v_\phi\sin\phi$; $v_y = v_r\sin\phi + v_\phi\cos\phi$; $v_z = 0$. In the case of axial symmetry, the dissipation rate takes the simple form

$$-\left(\frac{dE}{dt}\right) = \eta r^2\left(\frac{d\Omega}{dr}\right)^2 \tag{16.29}$$

As usual, we integrate over the z coordinate, and then the dissipation rate is

$$-\left(\frac{dE}{dt}\right) = \int \eta r^2\left(\frac{d\Omega}{dr}\right)^2 dz = \nu\Sigma r^2\left(\frac{d\Omega}{dr}\right)^2 \tag{16.30}$$

We can now substitute for $\nu\Sigma$ from the expression (16.27), and hence, assuming the orbits of the matter in the disc are closely Keplerian,

$$-\left(\frac{dE}{dt}\right) = \frac{3G\dot{m}M_*}{4\pi r^3}\left[1-\left(\frac{r_*}{r}\right)^{1/2}\right] \tag{16.31}$$

This is the elegant result we have been seeking and which makes the study of thin discs so attractive – the energy dissipation rate does not depend explicitly upon the viscosity ν. The viscosity has been absorbed into the requirement of steady-state accretion at a constant rate $\dot{m}$. On the other hand, more detailed properties of the disc, such as the surface density, do depend upon ν. We assume that, in the steady state, this heat energy is carried away by radiation, and so the luminosity of the disc is found by integrating the heat dissipation rate (16.31) from r_* to infinity:

$$L = \int_{r_*}^{\infty}\left(-\frac{dE}{dt}\right)2\pi r\,dr = \frac{G\dot{m}M_*}{2r_*} \tag{16.32}$$

This is a very sensible result. The matter falling in from infinity passes through a series of bound Keplerian orbits for which the kinetic energy is equal to half of the gravitational potential energy. Thus, the matter has to dissipate half of the total potential energy which it would acquire in falling from infinity to that radius. This is the source of the luminosity of the disc. The matter which has just reached the boundary layer has only liberated half its gravitational potential energy. If this matter is then brought to rest on the surface of the star, the rest of the gravitational potential energy can be dissipated. This calculation indicates that the boundary layer can be just as important a source of luminosity as the disc itself.

It is interesting to look at this result in terms of the rate of dissipation of energy, or luminosity $L(r)$, in the annulus between radii r and $r + \Delta r$. Multiplying equation (16.31) by $2\pi r \Delta r$, the luminosity of the annulus is

$$L(r) = -\left(\frac{dE}{dt}\right) = \frac{3G\dot{m}M_*}{2r^2}\left[1 - \left(\frac{r_*}{r}\right)^{1/2}\right]\Delta r \qquad (16.33)$$

This is an interesting answer because it is apparent that this expression is more than simply the release of gravitational binding energy when matter moves from a Keplerian orbit at $r + \Delta r$ to one at r, which is simply $(G\dot{m}M_*/2r^2)\Delta r$. The difference between this expression and equation (16.33) represents the net flow of energy into the annulus Δr associated with the transport of angular momentum outwards. Thus, although the total energy released in reaching the surface of the star is simply half the gravitational potential energy, at any radius, the energy dissipation rate consists of both the energy loss due to angular momentum transport as well as the release of gravitational binding energy. From the expression (16.33), it can be seen that, at distances $r \gg r_*$, the energy dissipation rate is

$$L(r) = -\left(\frac{dE}{dt}\right) = \frac{3G\dot{m}M_*}{2r^2}\Delta r \qquad (16.34)$$

which is three times the rate of release of gravitational binding energy.

16.3.4   *Accretion discs about black holes*

Now let us extend the analysis to the case of black holes rather than considering objects with a solid surface. The problem is to determine what boundary condition should replace the expression (16.26). In the case of black holes, the matter drifts inwards through the accretion disc until it reaches the last stable orbit, from which it rapidly spirals into the black hole. As indicated in Section 15.6.1, the angular velocity now helps, rather than hinders, the collapse of matter into the black hole, since, crudely speaking, the rotational energy now contributes to the inertial mass of the infalling matter. Let us suppose that the last stable orbit has radius $r_{\rm I}$. We recall that, in the case of a non-rotating, spherically symmetric black hole, $r_{\rm I} = 3r_{\rm g} = 6GM/c^2$. Then, according to an entirely classical calculation, the condition that the matter can spiral into the hole is simply that the rotational energy of the matter should be less than or equal to half the gravitational potential energy. In other words, $\frac{1}{2}\mathscr{L}^2/I \leq \frac{1}{2}GMm/r_{\rm I}$, where $\mathscr{L}$ is the angular momentum of the element of mass m and $I = mr_{\rm I}^2$ is its moment of inertia about the hole in the last stable orbit. It is convenient to work in terms of the *specific angular momentum* $J = \mathscr{L}/m$. Therefore, the condition that the matter fall into the hole is

$$J \leq (GMr_{\rm I})^{1/2}$$

The inner boundary condition is therefore that the angular momentum with which the material arrives at the last stable orbit should be less than this critical value, and this angular momentum, as well as the matter, is then consumed by the black hole. It is conventional to write the angular momentum of the material as it arrives at r_{I} as $\mathscr{L} = \beta \dot{m}(GMr_{\mathrm{I}})^{1/2}$, where $\beta \leq 1$. This, then, becomes the boundary condition which replaces the expression (16.26). Therefore, for the case of black holes, we can write the expression for the luminosity of the disc betweeen r and $r + \Delta r$ as follows:

$$L(r) = -\left(\frac{dE}{dt}\right) = \frac{3G\dot{m}M_*}{2r^2}\left[1 - \beta\left(\frac{r_{\mathrm{I}}}{r}\right)^{1/2}\right]\Delta r \qquad (16.35)$$

Exactly as before, we find the total luminosity of the disc by integrating equation (16.35) from r_{I} to infinity. The result is

$$L = \left(\frac{3}{2} - \beta\right)\frac{G\dot{m}M_*}{r_{\mathrm{I}}} \qquad (16.36)$$

16.3.5   *The temperature distribution and emission spectra of thin discs*

Let us now make some simple estimates of the temperature distribution and emission spectrum of the disc. Suppose the disc is optically thick to radiation and that there is sufficient scattering to ensure that we can approximate the emission as black-body radiation at each point in the disc. The disc can radiate from its top and bottom surfaces, and so we can equate the heat dissipated between r and $r + \Delta r$, expression (16.34) for simplicity, to $2\sigma T^4 \times 2\pi r \Delta r$, where σ is the Stefan–Boltzmann constant:

$$\sigma T^4 = \frac{3G\dot{m}M_*}{8\pi r^3} \qquad T = \left(\frac{3G\dot{m}M_*}{8\pi r^3 \sigma}\right)^{1/4} \qquad (16.37)$$

Thus, in the outer regions of the disc, $r \gg r_*$, the temperature of the matter is expected to increase towards the centre as $T \propto r^{-3/4}$. A consequence of this result is that the sound speed in the disc increases towards the centre, and we need more detailed calculations to ensure that all the approximations we have made are indeed appropriate.

If we assume that each annulus of the disc radiates like a black body at the appropriate temperature given by the expression (16.37), we can derive the form of the integrated spectrum of the disc. The total intensity of the disc is proportional to the surface area at temperature T times the black-body intensity at that temperature. Thus,

$$I(\nu) \propto \int_{r_{\mathrm{I}}}^{r_{\max}} 2\pi r B[T(r), \nu]\,dr \qquad (16.38)$$

where $B(T, \nu)$ is the Planck function $B(\nu) \propto \nu^3[\exp(h\nu/kT) - 1]^{-1}$. But, we know that $T \propto r^{-3/4}$ and so $dr \propto (1/T)^{1/3}d(1/T)$. Therefore, we can convert the integral over dr into an integral over $(1/T)$. Carrying this out and preserving the dependence upon frequency, we find

$$I(\nu) \propto \int_{r_{\mathrm{I}}}^{r_{\max}} \left(\frac{1}{T}\right)^{4/3} \nu^3 \left[\exp\left(\frac{h\nu}{kT}\right) - 1\right]^{-1}\left(\frac{1}{T}\right)^{1/3}d\left(\frac{1}{T}\right) \qquad (16.39)$$

Now, change the variable of integration from $(1/T)$ to $x = (h\nu/kT)$ and then

$$I(\nu) \propto \frac{\nu^3}{\nu^{8/3}}\int_{r_{\mathrm{I}}}^{r_{\max}} x^{4/3}(\exp x - 1)^{-1}x^{1/3}dx \qquad (16.40)$$

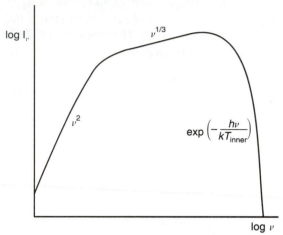

Figure 16.7. A schematic representation of the emission spectrum of an optically thick accretion disc. The exponential cut-off at high frequencies occurs at frequency $v = kT_{\text{inner}}/h$, where T_{inner} is the temperature of the innermost layers of the thin accretion disc.

But, we now notice that the integral is a definite integral, which is just a constant, and all the dependences upon v are outside the integral. Therefore,

$$I(v) \propto v^{1/3} \tag{16.41}$$

Thus, the predicted spectrum of a thin, optically thick accretion disc in the black-body approximation is that the spectrum should have the form $I(v) \propto v^{1/3}$ between the frequencies corresponding to r_{I} and r_{max}, as illustrated in Fig. 16.7. At frequencies less than that corresponding to the temperature of the disc at r_{max}, the spectrum tends towards a Rayleigh–Jeans spectrum, $I_v \propto v^2$.

16.3.6 *Detailed models of thin discs*

The analysis of Section 16.3.3 is almost as far as we can go without constructing proper models for thin discs. It was shown by Sunyaev and Shakura (1972) that, adopting the α-disc approach to the definition of the turbulent viscosity, it is possible to derive eight equations which can be solved in closed form for the structure of thin accretion discs in terms of eight unknown parameters. This analysis is straightforward, if lengthy, and the results are quoted by Frank, King and Raine (1992). It turns out that many of the properties of thin discs are only weakly dependent upon the viscosity parameter α, which is encouraging from the point of view of the confrontation with observation but disappointing from the point of view of understanding more about the nature of the viscosity in the disc.

An important aspect of these solutions is that they enable the detailed properties of realistic thin disc models to be evaluated, and, in particular, the opacity of the disc at different radii can be determined. Sunyaev and Shakura (1972) and Novikov and Thorne (1973) show that the disc can be divided into three

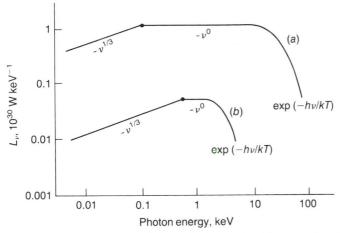

Figure 16.8. The spectra of the radiation emitted by accretion discs about a spherical, non-rotating black hole. These models are due to Sunyaev and Shakura (1972). In model (*a*), the black-hole mass is assumed to be $1M_\odot$, the accretion rate $3 \times 10^{-8}M_\odot$ year^{-1} and the accretion takes place at the Eddington limiting luminosity. In model (*b*), the mass accretion rate is less, $3 \times 10^{-10}M_\odot$ year^{-1}, and the luminosity is 10^{29} W. The radiation generated in the outer cool regions of the disc result in a power-law spectrum $I(\nu) \propto \nu^{1/3}$. In the inner regions, electron scattering is the dominant source of opacity, and the spectrum is approximately independent of frequency. The temperature of the exponential tail corresponds to the surface temperature of the inner regions of the disc modified by the effects of electron scattering. (After S.L. Shapiro and S.A. Teukolsky (1983). *Black holes, white dwarfs and neutron stars: the physics of compact objects*, p. 446. New York: Wiley Interscience.)

regions: the outer region, in which the gas pressure is much greater than the radiation pressure and the opacity is dominated by free–free (or bremsstrahlung) absorption; a middle region, in which the gas pressure is still dominant but the dominant source of opacity is electron scattering; and an inner region, in which the radiation pressure dominates and electron scattering is the most important source of opacity. These results are derived from detailed models of the structure of the accretion discs. As a result, the emitted spectrum of the disc in these regions cannot be approximated by a black-body spectrum. In the innermost regions, the disc may be optically thin to free–free absorption, even taking into account the multiple scattering of the radiation by the electrons, and then the predicted emission continuum spectrum is characteristic of optically thin bremsstrahlung, $I(\nu) \propto \nu^0$. Examples of the predicted spectra of the radiation for various assumptions about the accretion rate and the viscosity parameter α are shown in Fig. 16.8.

16.3.7   *Thick discs*

It is apparent from inspection of the expression (16.37) that a number of the approximations involved in the construction of thin disc models begin to break down towards the centre of the disc if the accretion rates become large. In

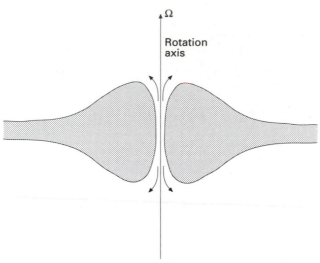

Figure 16.9. A schematic model of the structure of a thick accretion disc, showing the 'funnels' which are expected to form along the axis of the disc.

this case, the radius r approaches r_{I} and the accretion rate $\dot{m}$ can become large. In addition, if the mass of the central black hole becomes large, it can be seen that the luminosity of the disc increases, and it is no longer safe to assume that the effects of thermal pressure of the gas and the radiation pressure of the emitted radiation can be neglected in determining the structure of the inner disc. In the thin disc approximation, pressure gradients in the radial direction are neglected, although the structure in the direction perpendicular to the disc is maintained by internal pressure gradients. In the circumstance in which the rotational velocity becomes comparable to the velocity of sound, the disc inflates, so that $r \sim H$. This is a very much more complicated problem, and no fully self-consistent solution has yet been found. Fig. 16.9 is a schematic picture of a possible structure for a thick disc, and it leads to the concept of an *accretion torus* about the central compact object. Unfortunately, stability analyses which have been carried out on such structures by Papaloizou and Pringle (1984) have shown that they are globally unstable.

Nonetheless, these structures have a number of attractive features for possible accretion models for active galactic nuclei. A thick disc has 'funnels' along its rotation axis, and these may be related to the collimation of beams of particles and relativistic gases observed to be ejected at very high velocities from active galactic nuclei. We will take this subject up in much more detail in Volume 3. A discussion of some of the many problems involved in the construction of thick disc models for active galactic nuclei is given by Frank, King and Raine (1992). In the following discussion, we will consider only the simpler cases of thin accretion discs for galactic sources, in which accretion in binary systems provides a natural explanation for high energy astrophysical activity.

16.4   Accretion in binary systems

16.4.1   Feeding the accretion disc

In binary systems, the process of accretion may occur in two ways, which are illustrated in Fig. 16.10. We know that all normal stars emit stellar winds, 'quiescent' mass loss in the case of stars like the Sun and much more violent winds in the cases of luminous O and B stars, in which mass-loss rates as high as $10^{-5}M_\odot$ year^{-1} are observed (Section 14.5). Therefore, in one picture, the compact companion is embedded in an outflowing stellar wind (Fig. 16.10(a)). As in the case of the Earth's magnetosphere, there is a bow shock in the upstream direction from the compact star (Section 10.7), and accretion onto the star takes place within this cavity. In this case, the accretion may approximate more closely to the spherical case than to the disc accretion picture developed in Section 16.3.

The second model, illustrated in Fig. 16.10(b), involves what is known as *Roche lobe overflow*. The equipotential surfaces of a close binary star sytem are distorted in the rotating frame of reference when the stars fill a substantial fraction of the Roche lobe (Fig. 14.12). If the primary star fills its Roche lobe in the course of evolution, matter is transferred from the primary to the secondary through the inner Lagrangian point L_1 as the overflowing matter seeks a lower gravitational potential. As illustrated in Fig. 16.10(b), the result is a stream of plasma flowing from the inner Lagrangian point into the accretion disc. It is therefore expected that there will be a 'hot spot' at the point where the accreting matter meets the accretion disc, and there is clear evidence for such a spot in observations of certain cataclysmic variable stars.

16.4.2   The role of magnetic fields

So far, we have neglected the role of magnetic fields except to remark that they are almost certainly implicated in the turbulent viscosity within the accretion disc. The compact stars themselves possess magnetic fields, and these can strongly influence the accretion of matter onto the compact star. Direct evidence for the presence of a strong magnetic field in X-ray binaries comes from the observation of what is inferred to be an electron cyclotron absorption feature in the X-ray spectrum of Her X-1 and other X-ray binaries. The feature is observed at about 58 keV in the case of Her X-1, which would require a magnetic field in the source region of about 5×10^8 T (see Section 18.1.2). Let us assume that the magnetic field is of dipolar form, and hence that at distance r from the centre of the star the magnetic field strength is given by the expression (15.27). We can write the magnetic field strength at distance r in the form $B = (R_*/r)^3 B_s$ where B_s is the typical magnetic field strength at the surface of the compact star at radius R_*.

Let us consider the case of spherical accretion onto the compact star. The magnetic pressure at radial distance r is $p_{mag} \approx B^2/2\mu_0 \approx (B_s^2/2\mu_0)(R_*/r)^6$. The infalling matter can be considered to have fallen from rest at infinity, and so its infall velocity at radius r is $v = (2GM_*/r)^{1/2}$. The pressure which this infalling gas exerts upon the magnetic field is known as the *ram pressure*, p_{ram}, and is just the rate at which momentum is transported inwards at radius r per unit area,

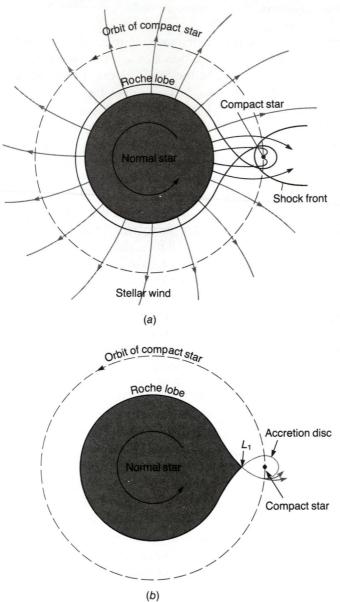

(a)

(b)

Figure 16.10. Illustrating two ways in which accretion onto stars in binary systems may take place. In (a), the primary star has a strong stellar wind and the neutron star is embedded in the strong outflow from the primary. In (b), the primary star expands to fill its Roche lobe and matter passes from the primary to the secondary star through the Lagrangian point L_1. An accretion disc is formed about the compact secondary star. (From S. L. Shapiro and S. A. Teukolsky (1983). *Black holes, white dwarfs and neutron stars: the physics of compact objects*, p. 399. New York: Wiley Interscience.)

that is, $p_{ram} = \rho v^2$. This is the same type of pressure which acts at the interface between the solar wind and the Earth's magnetosphere and which is believed to be responsible for the containment of the high energy particles in extended extragalactic radio sources. This ram pressure is balanced by the magnetic pressure of the compact star at a radius r_M, known as the *Alfvén radius*, such that

$$\rho v^2 = \frac{B_s^2}{2\mu_0} \left(\frac{R_*}{r_M} \right)^6 \tag{16.42}$$

In the case of spherically symmetric accretion, the mass accretion rate $\dot{m}$ is given by $\dot{m} = 4\pi r^2 \rho v$, and so we can reorganise equation (16.42) to find r_M in terms of the accretion rate:

$$r_M = \left(\frac{2\pi^2}{G\mu_0^2} \right)^{1/7} \left(\frac{B_s^4 R_*^{12}}{M_* \dot{m}^2} \right)^{1/7} \tag{16.43}$$

Let us insert appropriate values for a solar-mass neutron star accreting at the Eddington luminosity, that is, $L = \dot{m}\eta c^2 = 1.3 \times 10^{31}$ W, with $\eta = 0.1$, $B_s = 10^8$ T and $R_* = 10$ km. Then, $r_M = 10^3$ km, that is, about 100 times the radius of the neutron star. This is an important result – in the case of luminous X-ray sources, the immediate vicinity of the neutron star is magnetically dominated. The only way in which matter can be accreted onto the surface of the neutron star, and hence release the maximum amount of binding energy of the infalling matter, is if the matter falls down the poles of the magnetic field distribution (Fig. 16.11). It is this reasoning which leads to the concept that there is an *accretion column* associated with the infall of matter onto strongly magnetic neutron stars. Apparently, only if the magnetic field is weak can accretion take place directly onto the surface of the neutron star.

Let us repeat the calculation for white dwarfs. We assume that the radius of the star is 5000 km and that the surface magnetic field strength $B_s = 1$ T. Then, the Alfvén radius for a solar-mass white dwarf radiating at the Eddington limit is roughly the same as before, 10^3 km. Since the Alfvén radius depends upon the magnetic field strength as $B_s^{4/7}$, it follows that those white dwarfs with strong magnetic fields and lower accretion rates will be magnetically dominated, again leading to accretion along the poles of the magnetic field structure.

Let us now consider the case of *disc accretion* in the presence of a compact star with a strong magnetic field. This is a more complex problem because the interaction between the magnetic field structure and the plasma of the accretion disc has to be understood. Furthermore, the details of the interaction depend upon the angle between the magnetic axis of the compact star and the axis of the accretion disc. The radius within which the magnetic field of the star dominates the dynamics can be found by equating the torque exerted by the accretion disc on the magnetic field structure to the magnetic torque associated with the distorted magnetic field distribution at that radius. This is a non-trivial calculation, particularly when the effect of instabilities at the interface between the disc and the magnetic field structure have to be taken into account, but, in general, a result similar to that derived above for the case of spherical accretion

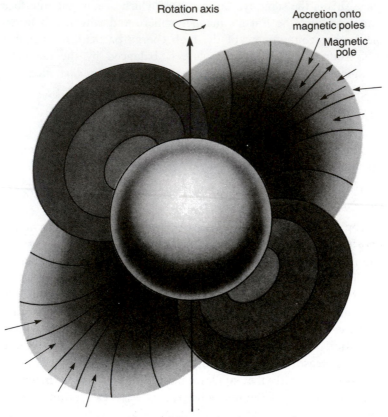

Figure 16.11. A schematic model illustrating how accretion onto a neutron star with a magnetic field nonaligned with the rotation axis can give rise to an accretion column, which, when observed at a distance, gives rise to the observation of X-ray pulses at the rotation frequency of the neutron star. The accreting matter is channelled down the magnetic field lines to the magnetic poles of the neutron star where its binding energy is deposited, resulting in strong heating of the plasma. The problems of radiative transfer under these conditions are highly complex.

is found. According to Frank, King and Raine (1992), a radius is obtained which is slightly smaller than the Alfvén radius found above, typically being about half that value. The reason why this radius is not so different from the spherically symmetric case is the very strong radial dependence of the pressure of the magnetic field, $p_{\mathrm{mag}} \propto r^{-6}$.

This picture has a number of important consequences. First, it is apparent that, once again, accretion can only take place onto the poles of the compact star, but now the accretion disc feeds matter onto the poles of the neutron star via the dipole magnetic field, as illustrated in Fig. 16.12. The net result is the formation of an accretion column with an opening angle which depends upon the angle of inclination between the axis of the magnetic field and the axis of the disc. In general, the surface area of the polar cap relative to the surface area of the star is $\sim R_*/r_{\mathrm{M}}$.

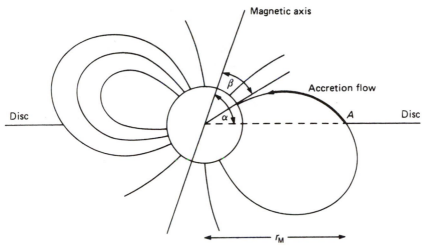

Figure 16.12. Illustrating the accretion of matter from an accretion disc onto the polar caps of a magnetised neutron star or white dwarf. (After J. Frank, A. King and D. Raine (1992). *Accretion power in astrophysics*, p. 123. Cambridge: Cambridge University Press.)

The second important point is that the accretion disc exerts a torque upon the magnetosphere of the star, which, in turn, transmits the torque to the star itself. Thus, the process of accretion leads to the speeding up of the star, and this is observed in X-ray binary stars. It is also likely to be the process which leads to the formation of the millisecond pulsars described in Section 15.4. Now, to a very good approximation, the angular velocity of the material in a thin accretion disc at radius r is just the Keplerian value, $\Omega_K^2(r) = (GM_*/r^3)$. In order for angular momentum to be transferred inwards, the Keplerian angular velocity of the disc at radius r_M should be greater than Ω, the angular velocity of the neutron star and its associated magnetosphere, which corotate within radius r_M, that is $\Omega_K(r_M) > \Omega$. Let us consider the case in which $\Omega_K(r_M) \gg \Omega$, that is, the compact star is a slow rotator. The rate of inward transfer of angular momentum at radius r_M is given by the expression (16.21), which can be rewritten $G = \dot{m} r_M^2 \Omega_K(r_M)$. This angular momentum is absorbed by the star, and so we can write

$$I\dot{\Omega} = \dot{m} r_M^2 \Omega_K(r_M) \tag{16.44}$$

where $I = \frac{2}{5} M_* R_*^2$ is the moment of inertia of the star. We can now relate the rate of speed-up of the compact star to observable properties of the binary system. If it is assumed that the luminosity of the source is due to accretion, we can relate the mass accretion rate directly to the luminosity, L, of the source through the relation $L = G\dot{m}M/r_M$. Therefore, we can find the rate of spin-up of the compact star in terms of its luminosity. Inserting this expression and the expression (16.43) into expression (16.44), we find

$$\frac{\dot{P}}{P} = -\frac{1}{2\pi I} \left(\frac{2\pi^2}{\mu_0^2 G^6} \right)^{1/14} \left(B_s^{2/7} R_*^{12/7} M_*^{-2/7} \right) \left(L^{6/7} P \right) \tag{16.45}$$

where we assume that the value of $\Omega_K(r_M)$ is given by the Keplerian value at

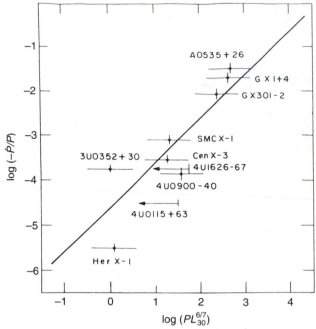

Figure 16.13. The observed spin-up rates ($\dot{P}/P$) plotted against $PL_{30}^{6/7}$ for ten pulsating X-ray binary sources. The straight line through the points is the expected spin-up rate on the basis of the simple theoretical picture developed in the text. (After J. Frank, A. King and D. Raine (1992). *Accretion power in astrophysics*, p. 126. Cambridge: Cambridge University Press; reproduced from S. Rappaport and P.C. Joss (1983). In *Accretion-driven stellar X-ray sources*, eds W.H.G. Lewin and E.P.J. van den Heuvel. Cambridge: Cambridge University Press.)

radius r_M and that P is the period of rotation of the compact star, $P = 2\pi/\Omega$. It can be seen that the speed-up rate depends principally upon the product of the rotation period of the compact star and the luminosity to the power $\frac{6}{7}$. Inserting typical values expected for a neutron star, $M_* = M_\odot$, $R_* = 10^4$ m and $B_s = 10^8$ T, we find the relation

$$\log_{10}\left(-\frac{\dot{P}}{P}\right) = -4.4 + \log_{10}\left(PL_{30}^{6/7}\right) \qquad (16.46)$$

where the luminosity L_{30} is measured in units of 10^{30} W. It can be seen from Fig. 16.13 that this simple analysis can give a good account of the observed speed-up rates of the compact stars in X-ray binaries. If further proof were needed, this argument strongly suggests that the spin-up of binary X-ray sources is associated with the accretion of matter onto a neutron star with a strong magnetic field.

We can develop another pleasant result from this simple picture of spin-up associated with the accretion of matter in binary systems. We argued above that spin-up can only occur if the Keplerian angular velocity of the matter in the disc is greater than the angular velocity of the star. Spin-up ceases when these angular velocities are equal. We can therefore find an upper limit to the angular velocity of a neutron star by equating Ω to $\Omega_K(r_M)$. Expressing this condition in terms of the period of the neutron star and using the result (16.43) for r_M, we find

$$P = 2\pi \left(\frac{r_M^3}{GM_*}\right)^{1/2} = \frac{2\pi}{(GM_*)^{1/2}} \left(\frac{2\pi^2}{G\mu_0^2}\right)^{3/14} \left(\frac{R_*^{12}}{M_*\dot{m}^2}\right)^{3/14} B_s^{6/7} \qquad (16.47)$$

Thus, the greater the mass accretion rate $\dot{m}$, the shorter the minimum period of the neutron star but the stronger the magnetic field, the less effective this process is in spinning-up neutron stars. Let us estimate the minimum period to which the neutron star could be spun-up by accretion. The maximum accretion rate is associated with the Eddington limiting luminosity, and so we can insert $L = G\dot{m}M_*/R_* = 1.3 \times 10^{31}$ W for our standard accreting neutron star with $M = M_\odot$ and $R = 10^4$ m. We then find

$$P_{\min} \approx 2B_5^{6/7} \text{ ms} \qquad (16.48)$$

where the period P is measured in milliseconds and the magnetic field strength B_s is in units of 10^5 T. This is exactly the same result given by van den Heuvel, which was quoted in Section 15.4 in relation to the minimum period which millisecond pulsars can acquire if they are spun-up by the process of accretion. The fact that many of the millisecond radio pulsars are found to be members of binary systems strongly supports the idea that they have been spun-up by accretion. Notice that the millisecond pulsars have weak magnetic fields, typically $\leq 10^5$ T, which is a distinct advantage according to the expression (16.48). To derive the line labelled 'spin-up' in Fig. 15.15, we combine the expression (16.48) with the relation between B_s and the spin-down rate of the radio pulsar, (15.28), $B_s = 3 \times 10^{15}(P\dot{P})^{1/2}$ T, and find the result $\dot{P} \leq 2 \times 10^{-15}P^{4/3}$, which is plotted on Fig. 15.15. It can be observed that even the shortest period millisecond pulsars can be accounted for by the process of spin-up by accretion torques.

16.5   Accreting binary systems

The ideas developed in the preceding two sections provide many of the tools needed to understand observations of compact objects in close binary systems. There is a wealth of data on many different types of close binary system, which are splendid objects for observation because of the many different ways in which the diagnostic tools can be used to study physical processes in the vicinity of compact stars (see, for example, the articles in the volume *X-ray binaries*, by Lewin, Paradijs and van den Heuvel (1994)).

16.5.1   *Cataclysmic variables*

The cataclysmic variables are of special interest because they display a very wide range of different types of accretion phenomena, and, in some of them, direct evidence has been found for the presence of accretion discs. These variable stars comprise a mixed bag of close binary stars, but the one common ingredient is the presence of a compact star onto which accretion takes place. The primary

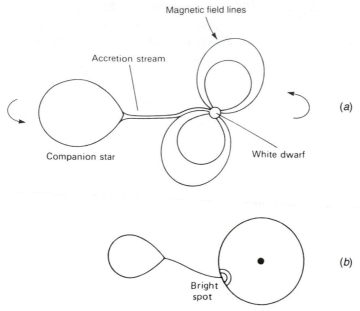

Figure 16.14. (*a*) Illustrating the accretion of matter in a polar or AM Herculis system. The magnetic field is very strong in these systems ($B \geq 10^3$ T), and so the rotation of the white dwarf is locked to the rotation of the binary system as a whole. No accretion disc forms and the matter streams more or less directly onto the surface of the white dwarf. (*b*) Illustrating the formation of a 'hot-spot' at the outer boundary of an accretion disc in a cataclysmic binary system. (From J. Frank, A. King and D. Raine (1992). *Accretion power in astrophysics*, p. 88, 127. Cambridge: Cambridge University Press.)

star is usually a late-type star near to or on the main sequence in which Roche lobe overflow takes place onto the compact star, which is normally a white dwarf. The periods of the binaries range from about 80 minutes to several hours (see Córdova (1994) for a comprehensive review of this class of variable star).

The different types of cataclysmic variable largely reflect differences in the geometries by which the accretion takes place. The best known examples of the class are the classical novae, which were the first to be discovered, and the dwarf novae. In *classical novae*, the optical brightness increases by up to about 20 magnitudes, and only one such major outburst is ever observed. This enormous release of energy is attributed to the accumulation of sufficient high temperature material on the surface of the degenerate star that *thermonuclear runaway* takes place in the envelope of the star (see Section 16.5.2). In contrast, in the *dwarf novae*, the star regularly brightens by 2 to 5 magnitudes, and it is assumed that this brightening is associated with the process of disc accretion.

Intermediate between these types are objects known as *recurrent novae*, in which the outbursts take place on the timescale of decades and which appear to comprise a variety of systems. In some cases, the companion stars are giant stars rather than main sequence stars. Close relatives of these are the *symbiotic*

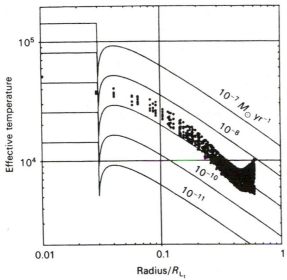

Figure 16.15. The results of eclipse mapping to determine the temperature distribution in the accretion disc in the dwarf nova Z Cha. The observations were made during an outburst so that the emission from the disc dominates the optical light. The lines show the expected temperature distributions for optically thick steady-state discs with different accretion rates. (From J. Frank, A. King and D. Raine (1992). *Accretion power in astrophysics*, p. 91. Cambridge: Cambridge University Press. After K. Horne and T. R. Marsh (1986). *Physics of accretion onto compact objects*, p. 5. Heidelberg: Springer-Verlag.)

stars, in which the accreting star is a main sequence star rather than a dwarf star. There are also systems, which are known as *novae-like stars,* which appear to be similar to novae but which are in permanent outburst.

In some of the cataclysmic variables, the white dwarfs have very strong magnetic fields, $B \geq 10^3$ T, and the rotation of the star is phase-locked to the orbit of the binary. The objects called *polars* or *AM Herculis stars* are of this type, in which the accreting matter flows directly along the magnetic field lines from the primary star onto the poles of the white dwarf (Fig. 16.14(*a*)). In other cases, the rotation period of the white dwarf is shorter than the period of the binary orbit, and these objects are referred to as *intermediate polars*. Both of these classes of magnetic cataclysmic variable exhibit high and low states of activity, presumably associated with different states of mass accretion.

Of particular interest so far as the accretion disc itself is concerned, are the dwarf novae. The reason is that, in these systems, it is possible to observe directly the optical and ultraviolet emissions from the disc. In other binary systems, the light is dominated by the emission of the primary star, as in the case of those X-ray binaries containing O and B stars, or by reprocessed X-ray light in the case of low-mass X-ray binaries. Of special interest is the fact that, in the dwarf novae, the primary star eclipses the compact star and its accretion disc. It is therefore possible to reconstruct the two-dimensional temperature distribution in the disc by making very precise high-speed photometry of the system in a number of colours. This procedure is known as *eclipse mapping*, and maximum entropy techniques can be used to reconstruct the temperature distribution in the disc. Fig. 16.15 shows the results of reconstructing the temperature distribution in the accretion disc for the dwarf nova Z Cha during an outburst (see Horne and Marsh (1986) in caption to Fig. 16.15). The reconstructed temperature distribution follows closely the relation $T \propto r^{-3/4}$, as expected from the expression (16.37). It is reassuring that the theory appears to be in good agreement with observation for this system. This result implies that during the outburst the disc is optically thick and that the steady-state accretion rate is about $10^{-9} M_\odot$ year^{-1}. In their quiescent states, however, the inferred temperature distributions of the accretion discs are not consistent with the expectations of the steady-state models, and this is likely to be due to the fact that the accretion rates are low and the discs are no longer optically thick.

In the standard disc accretion picture, the stream of matter flowing through the inner Lagrangian point L_1 impacts the disc at some point and causes a local rise in temperature of the disc in a 'hot-spot' (Fig. 16.14(b)). It has been possible to map the geometry of the hot-spot using the procedure known as *Doppler tomography*. As the binary system rotates, different projections of the disc and the hot-spot are observed, and, by taking many high resolution spectra of the system at different binary phases, it is possible to reconstruct the intensity distribution in the disc. Fig. 16.16 shows the results of carrying out this procedure for the binary system U Gem by Marsh *et al*. The emission lines of Hβ and HeII are bright in this system, and the central panels show the reconstructed brightness distribution in the disc. Most of the line emission comes from the hot-spot, although Hβ emission is also apparent from the disc. This set of observations was taken whilst the system was in its quiescent state. During an outburst, the emission from the hot-spot is swamped by the emission from the disc.

A further test of the standard thin disc model is its continuum spectrum. According to the simplest model of an optically thick steady-state disc, the continuum spectrum is expected to be of the form $I(\nu) \propto \nu^{1/3}$ (expression (16.41)). A much better approximation is to assume that the disc radiates like a stellar atmosphere from its upper and lower surfaces. Although the general trend of the observations is in agreement with this picture, in the sense that the continuum intensity increases to shorter wavelengths roughly as $\nu^{1/3}$, the models cannot reproduce the observed detailed spectra of the cataclysmic variables. It is suspected that more complete models of the processes of energy generation and radiation transfer throughout the disc are required.

In the discussion of thin accretion discs of Section 16.3.3, it was shown that only half of the gravitational potential energy of the accreted matter is dissipated in reaching the boundary layer about the star at radius $r_* + b$, the rest of the energy being dissipated in the boundary layer itself. It is therefore expected that the highest energy radiation is emitted from the boundary layer close to the stellar

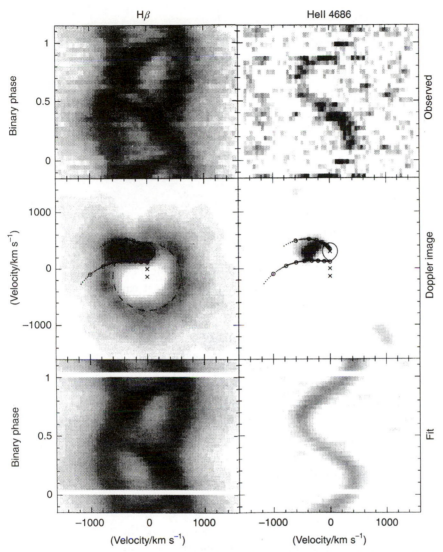

Figure 16.16. Doppler tomography of the emission lines of Hβ and HeII in the dwarf nova system U Gem during its quiescent phase. The top panels show the observed spectra as a function of orbital phase; the central panels show the reconstructed Doppler images fitted to these observations; and the lower panels show the computed velocity–phase data expected from the model-fits. The reference circles have radii corresponding to 600 km s^{-1}. (From T. Marsh, K. Horne, E.M. Schlegel, R.K. Honeycutt and R.H. Kaitchuk (1990). *Astrophys. J.*, **364**, 637.)

surface. Since the temperatures of the innermost layers are expected to be about 10^5 K, cataclysmic variables should be strong emitters in the EUV and soft X-ray wavebands. This is indeed found to be the case.

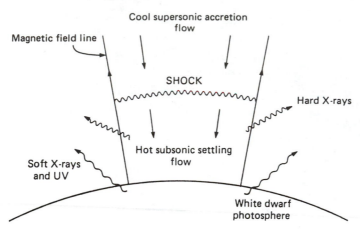

Figure 16.17. Illustrating the geometry of the accretion coloumn at the poles of a strongly magnetised white dwarf. (From J. Frank, A. King and D. Raine (1992). *Accretion power in astrophysics*, p. 137. Cambridge: Cambridge University Press.)

What was more surprising was the detection of high energy X-rays in the 1–10 keV waveband from cataclysmic variables. In most cases, the spectra can be described by thermal bremsstrahlung spectra with temperatures in the range $kT \approx 1$–5 keV. There are, however, a number of possible sources of high energy photons. One of the most important of these is associated with the accretion columns which form at the poles of an accreting white dwarf. As described above, matter is funnelled by the magnetic field onto the poles of the star where it is brought to rest, in the process dissipating its kinetic energy of collapse as thermal energy. The matter acquires a high velocity in falling onto the star, and so a shock front is formed above the stellar surface, as shown in Fig. 16.17. We can work out an upper limit to the temperature of the gas in the shocked region by equating the kinetic energy of infall of the gas to the thermal energy of the same electrons and protons once they have thermalised in the shocked region, that is,

$$\frac{1}{2}(m_{\rm p} + m_{\rm e})v^2 \approx \frac{1}{2}m_{\rm p}v^2 = \frac{1}{2}m_{\rm p}v_{\rm ff}^2 = \frac{GM_*m_{\rm p}}{R_*} = \frac{3}{2}(kT_{\rm p} + kT_{\rm e})$$

Let us assume that, on passing through the shock front, the energy of infall is shared among the electrons and protons and that thermal equilibrium is attained, that is, $T_{\rm p} = T_{\rm e}$. Thus,

$$T_{\rm e} = \frac{GM_*m_{\rm p}}{3kR_*} \tag{16.49}$$

For the case of accretion onto the poles of a solar-mass white dwarf for which $R_* = 10^4$ km, we find $T_{\rm e} = 5.4 \times 10^8$ K and $kT_{\rm e} \approx 50$ keV. This is an upper limit to the thermal energy of the plasma in the polar cap since it assumes that all the energy of infall goes into heating the plasma. As indicated in Fig. 16.17, some of the energy heats the surface layers of the white dwarf, where it is reradiated as far ultraviolet and soft X-ray emission. Thus, there is no problem, in principle, in understanding the origin of very hot gas in the vicinity of the poles of accreting white dwarfs. If the region remains optically thin to radiation, we might expect these regions to be sources of high energy bremsstrahlung emission as observed. Frank, King and Raine (1992) provide a good discussion of the many problems in constructing realistic models for the accretion columns associated with white dwarfs. To a good approximation, the problem can be reduced to a one-dimensional flow problem, with the hard X-ray emission escaping from the cylindrical surface of the shocked polar cap, as indicated in Fig. 16.17. This is unlikely to be the whole story, however. In the case of the highly inclined system OY Car, the hard X-rays seem to originate from a corona about the accretion disc since no evidence of occultation of the X-rays is observed.

A further surprise was the detection of strong winds from cataclysmic variables. The characteristic P-Cygni profiles have been observed with wind velocities up to 5000 km s^{-1}. Thus, not only is mass being accreted onto the white dwarf, but it is also being returned to the interstellar medium in the form of winds. The origin of these complications is not clear. It may well be that they are different manifestations of the processes of energy deposition in the boundary layers between the stellar surface and the accretion disc.

Finally, we must address the question of the origin of the dwarf nova phenomenon itself. There are two basic models. In one model, the enhanced accretion rate associated with the outbursts is attributed to instabilities in the disc. The question of the stability of the material of an accretion disc is a huge subject, which goes beyond the scope of this volume. It turns out to depend very sensitively upon the assumed value of the viscosity of the disc as described by the α parameter. An alternative picture is that the phenomenon is caused by short-lived mass transfer events from the primary star. Córdova (1994) provides a discussion of the merits and problems of these models.

16.5.2 *Novae and Type I supernovae*

To complete the story of accretion onto white dwarfs, we mention briefly their role in the understanding of *classical novae* and *Type I supernovae*. The principal features of the classical novae are summarised by Woosley (1986) and Starrfield (1988). As their name implies, classical novae are 'new stars' which can increase in brightness by more than 10 magnitudes in a stellar explosion. The rise to maximum light occurs over a period of a few days and can remain at that luminosity for several months. For a period of a few months, the nova radiates at roughly the Eddington luminosity for a $1M_\odot$ white dwarf, $L = 1.3 \times 10^{31}$ W and so the total energy release amounts to about 10^{38} J. There is evidence for the ejection of material from the nova with velocities between a few hundred and a few thousand kilometres per second. Of particular interest is the fact that the ejecta contain large abundances of elements synthesised by the CNO cycle at high temperatures. The event rate for classical novae is estimated to be about 30 year^{-1} in our own Galaxy and about 38 year^{-1} in M31. Thus, although they are much less powerful explosions than supernovae, they occur much more frequently.

The preferred model for classical novae involves the accretion of a critical mass of hydrogen-rich fuel onto the surface of the white dwarf. This accreted material is compressed and heated to a high temperature, $T > 10^7$ K, at the base of the accreted layer, which becomes degenerate, and, as the critical mass is approached, nuclear burning of hydrogen into helium through the CNO cycle can take place. It is estimated that the critical mass of hydrogen is about 10^{-5} to $10^{-4} M_\odot$. If the matter is sufficiently degenerate, *thermonuclear runaway* takes place because of the very strong dependence of the reaction rate of the CNO cycle upon temperature, $\epsilon \propto T^{17}$ (see Section 14.3). The CNO cycle (14.5), which results in the conversion of hydrogen into helium, involves the addition of four protons to the carbon nucleus with two intermediate β^+-decays of ^{13}N and ^{15}O. Other side-chains involve the β^+-decays of ^{14}O and ^{17}F. As described by Starrfield (1988), during the early stages of thermonuclear runaway, the timescales for the reactions involving the addition of the protons to the parent nuclei are longer than the half-lives of the above β-decay nuclei. As the temperature increases above 10^8 K, however, the timescale for the addition of protons becomes shorter than the half-life of the β-decays, with the result that these unstable nuclei become abundant and the rate at which nuclear energy generation proceeds is stabilised, since the proton capture processes have to wait until the β-decays have taken place. As a result, the thermonuclear runaway stabilises at a temperature of about 10^8 K. A further intriguing complication is that, because of the very strong temperature dependence of the energy generation rate in the CNO cycle, the layers become unstable to convection and so these unstable nuclei can be convected to the surface layers of the accreted material, where their β-decays provide a further source of heating. Once the peak temperature is reached and the envelope begins to expand, these unstable nuclei continue to provide a source of energy for the nova. These ideas have been studied in detail in computer simulations of nova explosions and can account for the main features of classical novae, including the abundances of the rarer CNO isotopes. The mass ejected in classical novae is about $10^{-4} M_\odot$, and so most of the accreted layer is expelled in the explosion.

In the case of Type I supernovae, the accretion of matter onto the surface of the white dwarf takes the star over the Chandrasekhar limit of about $1.4 M_\odot$, and so the central regions of the star begin to collapse. Woosley (1986) describes in some detail what is expected to occur in the case of carbon–oxygen white dwarfs. The central regions heat up and carbon begins to burn in a deflagration wave which passes out through the star. According to current ideas, there is no remnant left, but the whole star is disrupted, among the main products of the carbon burning being the formation of iron group elements. In an example of his models of the explosion of a white dwarf, the explosion ignites when the central temperature reaches 2.9×10^9 K and the mass of the star is $1.378 M_\odot$. A total mass of $0.86 M_\odot$ of iron is synthesised, of which roughly $0.58 M_\odot$ is initially in the form of ^{56}Ni. The energy release is 1.3×10^{44} J, which mostly goes into the kinetic energy of expansion of the supernova. The decay of the ^{56}Ni nuclei provides the characteristic exponential decline of the luminosity of the supernova, similar to the case of the Type II supernovae described in Section 15.2.1. The model in which a Type I supernova is attributed to a white dwarf exceeding the Chandrasekhar limit by mass accretion can account for the fact that the light curves and maximum luminosities of these supernovae are so similar. Because of this, they can be used as distance indicators to measure the distances of nearby galaxies.

16.5.3   X-ray binaries

It was the discovery of X-ray binaries in the early 1970s which began the remarkable upsurge of interest in accretion as an energy source for high energy astrophysical systems. The detection of pulsed X-ray emission from Her X-1 showed conclusively that the binary system contained a neutron star (Section 15.5). Much of the physics we have already discussed in relation to accretion discs about white dwarfs also applies to those about neutron stars, but the physical conditions are much more extreme. For example, the structure of the accretion columns at the poles of neutron stars is much more complicated than in the white dwarf case. The electrons become relativistic and the mean free paths for the particles become large compared with the size of the regions. It is therefore not so clear that the electrons and protons thermalise at the same temperature, nor that a standard shock wave forms in the accretion column. Almost certainly, this is another example where a collisionless shock is present, the transfer of momentum through the shock being associated with plasma processes, in particular those involving the gyration of the particles about the magnetic field in the shock. The transfer of radiation through the accretion column is a much more difficult problem as well. Nonetheless, although the physics is much more difficult, we can have confidence that the basic picture is along the correct lines. It is convenient to consider separately the *low-mass X-ray binaries*, which have many properties in common with the cataclysmic variables, and the *high-mass X-ray binaries*, in which the primary stars are massive O and B stars.

Low-mass X-ray binaries In the low-mass X-ray binaries, the primary star is normally a main sequence star with mass $M \sim M_\odot$. Members of this class are often referred to as *Galactic-bulge sources* because they lie in the general direction of the Galactic centre. A number of them have been identified with binary systems in globular clusters; the latter are members of the Galactic bulge population and are amongst the oldest systems in our Galaxy. All the stars with mass greater than about $1 M_\odot$ in these clusters have completed their evolution on the main sequence and the giant branch.

Unlike the case of the cataclysmic variables, until the last ten years it proved much more difficult to obtain definitive evidence for the binary nature of the low-mass X-ray binaries. The reason for this is now understood to be a selection effect in the sense that what is observed at X-ray wavelengths is strongly dependent upon the angle of inclination, i, of the plane of the binary to the plane of the

(a)

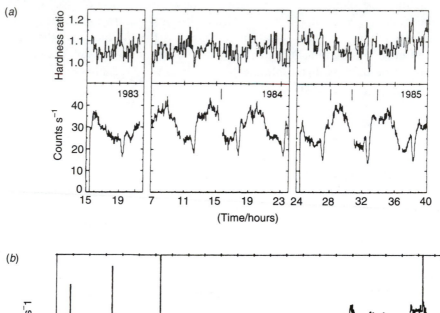

(b)

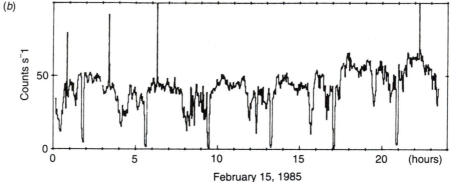

February 15, 1985

Figure 16.18. Examples of the X-ray light curves of two low-mass X-ray binary systems, illustrating a number of eclipse and absorption features. (*a*) The light curve of the 5.57-hour binary system X1822-371 determined by the EXOSAT X-ray telescope in 1983, 1984 and 1985. The upper panel shows the hardness ratio of the X-ray spectrum. (*b*) The light curve of the 3.83-hour period binary XBT0748-676. (From M.G. Watson and A.R. King (1991). In *Structure and emission properties of accretion discs*, eds C. Bertout, S. Collin, J-P. Lasota and J. Tran Thanh Van. Gif sur Yvette: Edition Frontières.)

sky. The brightest and most luminous sources are those in which the angle of inclination is small so that the orbital plane is viewed face-on. The compact X-ray source is then observed unobscured by the disc or the binary companion, but this geometry also means that it is difficult to determine the binary properties of the orbit. Lower luminosity X-ray sources display evidence of eclipses and 'dips' in their X-ray light curves, examples of some of these features being shown in Fig. 16.18. It is significant that the ratio of their X-ray to optical luminosities are much smaller than those systems inferred to be observed at small orbital inclinations. This strongly suggests that the X-ray emission is strongly attenuated when observed at large angles of inclination.

In modelling these sources, it is assumed that a standard accretion disc is the source of the X-ray luminosity, most of it originating from the inner region of the disc, but much more has to be added to account for the various observed features of their light curves. There appear to be two essential additions to the basic model. The first is related to the fact that, in a number of sources, the binary eclipses are gradual and do not completely occult the X-ray source. This is taken to be evidence for an *accretion disc corona*, which may fill a substantial fraction of the Roche lobe of the neutron star. The X-ray luminosity of this corona is only a fraction of the total luminosity of the source, and so it might be scattered radiation from the compact X-ray source in the nucleus or the emission of a hot corona within the Roche lobe about the neutron star. The physical origin of the corona is uncertain, but it could result from heating by the X-ray source or from instabilities in the accretion disc.

The second essential component is some thickening of the accretion disc to account for the statistics of absorption features seen in the low-mass X-ray binaries. This geometry is indicated schematically in Fig. 16.19(*a*). It can be seen that, unless the disc is somewhat thicker than the standard thin disc, there is only a very small likelihood of observing eclipses and absorption features in the X-ray light curves. The origin of the thickening of the accretion disc is not understood. One possibility discussed by Frank, King and Lasota (1987) is that it is associated with the fact that the stream of plasma which comes through the Lagrangian point L_1 is likely to be broader than the thickness of the thin disc at the point where the stream encounters the disc. This may lead to streams of material skimming over the surface of the disc until they reach centrifugal equilibrium at some radius. Frank, King and Lasota suggest that this process leads to the formation of cool clouds, which can provide thickening of the disc and the necessary X-ray absorption (Fig. 16.19(*b*)). There are still many uncertainties with this type of model, but it indicates the type of structure which is necessary to account for the light curves.

Another intriguing variant upon the story of accretion in low-mass X-ray binaries is provided by the *X-ray burst sources*, often referred to as *bursters*. There are two types of behaviour known, by far the more common being known as Type I bursts – there is only one example known of Type II bursts. In the Type I bursts, the X-ray luminosity typically increases by about a factor of ten with rise-times which are only about 1–10 s. The bursts last for less than a minute, and, during their decline from peak intensity, the X-ray spectra steepen to low energies, indicating that the temperature of the source region decreases as the burst evolves (Fig. 16.20). The spectra can be well-characterised by black-body spectra, which cool as the burst evolves. The Type I burst sources occur at intervals of hours to days, and so the total energy liberated in the bursts is small relative to the luminosity of the steady-state component. Typical statistics for the Type I bursts are that the interval between bursts is roughly 1000 times the duration of the burst, but during the burst the luminosity is enhanced by a factor of ten. Thus, the *average* luminosity of the steady-state component is about 100 times that of the bursts. A further clue to the nature of the bursts is provided by

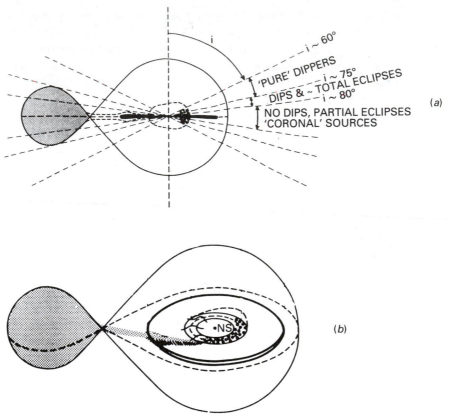

Figure 16.19. (*a*) A schematic diagram illustrating the necessity of incorporating a thick absorbing screen in the structure of the accretion disc in low-mass X-ray binaries. In this sketch, absorption features are only observed at angles of inclination greater than about 60°. It is assumed that the 'accretion disc corona' observed in X-rays fills a large fraction of the Roche lobe of the neutron star. (From J. Frank, A. King and D. Raine (1992). *Accretion discs in astrophysics*, p. 92. Cambridge: Cambridge University Press). (*b*) A schematic perspective view of the model for a low-mass X-ray binary with penetrating gas streams. (From J. Frank, A.R. King and J.P. Lasota (1987). *Astron. Astrophys.*, **178**, p. 137.)

the fact that no X-ray pulsar has ever been observed in a Type I burst source, which suggests that the magnetic fields in the neutron stars are weak or, possibly, that the magnetic field is aligned with the rotation axis. It is normally assumed that the former is the case.

These pieces of evidence can be neatly accounted for by a simple model in which the accretion disc extends in to the surface of the neutron star. The steady X-ray emission is the radiation of the accretion disc, whereas the burst emission is attributed to *thermonuclear runaway* of the accreted matter as it builds up on the surface of the neutron star. The latter process is exactly the same as that which can account for the energy release in novae (Section 16.5.2) but now applied to accretion onto neutron stars. What makes this picture appealing is that we can

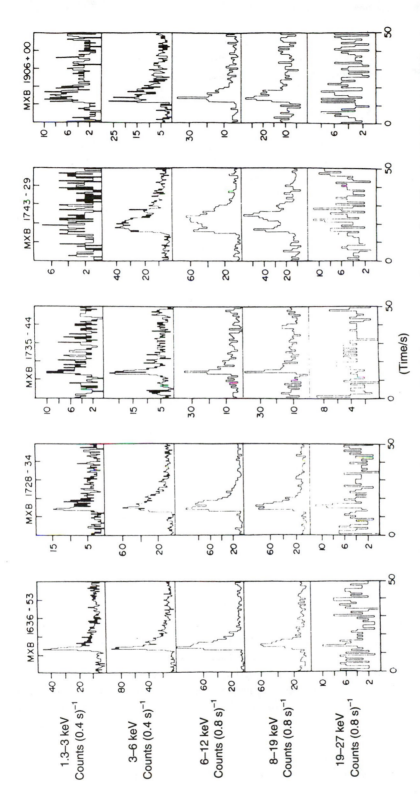

Figure 16.20. Examples of Type I X-ray bursts from five different sources. The light curves are shown in five different energy ranges, and it can be seen that the gradual decay persists longer at low energies than it does at high energies, indicating that the source cools as the source fades away. (From J. Frank, A. King and D. Raine (1992). *Accretion discs in astrophysics*, p. 162. Cambridge: Cambridge University Press.)

compare the average energy released by a parcel of matter by accretion, E_{acc}, with the energy it would release when it delivers up its available nuclear energy at the surface of the neutron star, E_{nuc}. Evidently,

$$\frac{E_{\mathrm{acc}}}{E_{\mathrm{nuc}}} \lesssim \frac{\eta mc^2}{0.007mc^2} \qquad (16.50)$$

since the maximum amount of nuclear energy which can be released is just the binding energy per nucleon of the helium nucleus, which is $0.007m_{\mathrm{p}}c^2$. For the case of neutron stars, we can take $\eta \sim 0.1$, and so the accretion disc can release about an order of magnitude more energy than the burning of the nuclear fuel on the surface of the neutron star. When account is taken of the fact that not all the nuclear energy will be emitted as black-body radiation from the surface of the neutron star, it is clear that an average luminosity ratio of about 100 in the steady-state component and the X-ray bursts can be naturally explained.

One intriguing feature of this picture of X-ray bursts is that, in the course of thermonuclear explosion, the X-ray luminosity of the neutron star approaches the Eddington limit. If the luminosity of an X-ray burster is observed to saturate at a particular value, this may be taken to correspond to the Eddington critical luminosity, and then further information may be obtained about the source region on the neutron star. This is indeed found to be the case for ten Type I burst sources for which temperatures have been measured. Assuming the sources radiate at the Eddington luminosity, the ratio of the luminosity of the burst to the surface area from which the luminosity is emitted should be a constant. The typical radius of the source is found to be about 7 km for these bursters. These are very pleasant arguments and provide further evidence, if it is needed, for the presence of neutron stars in the Type I X-ray bursters.

There is only one example known of a Type II burster, the source MXB 1730-335, which is also known as the *rapid burster*. It exhibits Type I bursts, but, in addition, there are much more rapid bursts with repetition periods of seconds to minutes. The energy of each burst is apparently proportional to the waiting time until the next burst, as if the burst had exhausted the supply of energy for the moment. The origin of this behaviour within the context of the standard accretion disc picture is not clear.

A remarkable example of the modulation of the light curve of low-mass X-ray binaries is provided by the source Her X-1. In addition to the pulsar which has period 1.24 s and a binary period of 1.7 days, the source undergoes a 35-day cycle during which it is strong for 9 days and then relatively faint for the remaining 26 days. This behaviour is accompanied by changes in the pulse profiles. The likely explanation of this behaviour is the precession of the rotation axis of the neutron star in its binary orbit so that the X-ray beam from one of the poles is pointing towards the observer when the source is bright. As precession changes the orientation of the magnetic poles with respect to the observer, the intensity observed from the bright pole decreases, but radiation from the other pole is observed. This model has important implications for the internal structure of the neutron star because the crust of the neutron star and the associated magnetic field must be decoupled (or unpinned) from the neutron superfluid since otherwise the neutron star would possess too much angular momentum and would not undergo appreciable precession. It is also necessary that the crust be sufficiently rigid to maintain a significant oblateness – otherwise there would be insufficient dipole moment for the precession torques to act continuously upon the neutron star.

Most of the bright Galactic bulge sources do not contain X-ray pulsars, but they exhibit a remarkable unstable periodic behaviour which has been termed *quasi-periodic oscillation*. Sources such as Cyg X-2, Sco X-1 and GX5-1 show rapid fluctuations in their intensities, and power spectrum analyses of the source intensities as a function of time show, not a single sharp line spectrum indicative of a stable period, but rather a broad peak spanning a range of frequencies. The mean frequency of the oscillations moves to higher frequencies as the intensity of the source increases. For example, in the source GX5-1, the central frequency increases systematically from 20 to 36 Hz as the source intensity increases by almost 50%. This behaviour is observed in most of the sources exhibiting quasi-periodic oscillations, although there are some exceptions. Almost certainly, these phenomena are associated with the flow of plasma from the accretion disc onto the neutron star. A favoured picture is one in which the oscillations are associated with 'beats' between the rotation frequency of the neutron star and the Keplerian frequency of rotation of matter at the inner edges of the accretion disc. In order that the accretion disc extend in towards the surface of the neutron star, the magnetic fields must be relatively weak, $B \leq 10^6$ T, which is consistent with the picture developed to account for the Type I X-ray bursts.

High-mass X-ray binaries Of the 100 brightest Galactic X-ray sources, about one-quarter of them are *high-mass X-ray binaries* in the sense that the primary stars are O or B giants or supergiants. Of the 26 known X-ray pulsars, 23 are associated with these classes of high-mass star. Three of the high-mass X-ray binaries are strong candidates for containing stellar-mass black holes. The O and B supergiants have large mass-loss rates, $\dot{m} \sim 10^{-6} M_\odot$ year^{-1}, and accretion takes place through the capture of a certain fraction of the outflow from the giant star by the compact star. The question of importance is whether sufficient mass is captured to power the compact X-ray source.

We can work out the *capture radius* or *accretion radius* within which matter is inevitably captured by a compact object. The compact object of mass M_X moves in its binary orbit at velocity v_X, and so, if we assume the wind from the star is radial with velocity v_W, the velocity of the wind relative to the star, v_t, is just the vector sum of these two velocities, $v_t^2 = v_X^2 + v_W^2$. We can work out the accretion rate of the compact star from the stellar wind in terms of v_t.

We showed in Section 2.2 that the impulse which a charged particle receives on passing a stationary charge is given by the inward force at the distance of closest approach b times the duration of the collision – this calculation gives

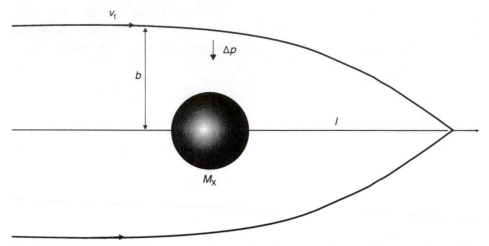

Figure 16.21. Illustrating the process of accretion by a star of mass M_X in a stellar wind of velocity v_t.

exactly the correct answer if we take the collision time to be the time it takes the moving particle to travel a distance $2b$. We can find the momentum impulse in the case of gravitational encounters in the stellar wind using exactly the same argument (Fig. 16.21). The gravitational force of attraction per unit mass at distance b is GM_X/b^2, and the duration of this force is $2b/v_t$. The momentum impulse inwards is therefore $\Delta p = 2GM_X/bv_t$. The result is that the outflowing wind is deflected towards the axis of the flow, as illustrated in Fig. 16.21. At some distance l downstream, the particles with collision parameter b collide on the axis of the flow. The perpendicular component of the velocity goes to zero, and so the necessary condition that the matter be captured by the star is that the gravitational potential energy of matter at l be greater than its kinetic energy outwards, which is approximately $\frac{1}{2}v_t^2$ per unit mass. The distance l is just $bv_t/v_\perp$, where $v_\perp = \Delta p$ since we consider unit mass. Reorganising these relations, we find that the condition $GM_X/l \geq \frac{1}{2}v_t^2$ reduces to $b \leq 2GM_X/v_t^2$. This is the result we have been seeking. This critical radius is known as the *capture radius*, R_c, and can be written

$$R_c = \frac{2GM_X}{v_X^2 + v_W^2} \tag{16.51}$$

We can now work out the X-ray luminosity of the neutron star in the steady state since $L = \eta \dot{m}c^2$, where the mass accretion rate is now $\dot{m} = (\dot{m}_P/4)(R_c/R_P)^2$, where $\dot{m}_P$ is the mass-loss rate from the primary and R_P is the distance of the neutron star from the centre of the primary star. Assuming the orbit is circular, the X-ray luminosity of the neutron star is

$$L_X \approx \frac{\eta \dot{m}_P c^2}{4} \left(\frac{2GM_X}{R_P} \right)^2 v_W^{-4} \tag{16.52}$$

where we have assumed that the wind velocity, v_W, is much greater than the orbital velocity, v_X, of the neutron star. It can be seen that the X-ray luminosity is directly proportional to the mass-loss rate of the primary star and is very sensitive to the wind velocity.

This simple model can account for the X-ray luminosities of many of the high-mass X-ray binary systems. Inverting this story, observations of the spectra of these X-ray sources can be used to study the properties of the stellar winds from O and B stars. Examples of this technique include accounting for variations of the photoelectric absorption of the X-ray source at different phases in its orbit and also variations in the X-ray spectra of those accreting X-ray sources which have elliptical rather than circular orbits (White (1985)).

16.6 γ-ray bursts

Perhaps the most extreme forms of high energy burst phenomena are those known as *γ-ray bursts*. These were discovered by accident by the Vela series of surveillance satellites, the prime task of which was to monitor nuclear explosions which might violate the nuclear test-ban treaties. γ-ray bursts are rare events, no more than about 10 to 20 being observed each year. The typical burst lasts for only a few seconds, but during that time the source becomes the strongest γ-ray source in the sky. In addition, it has been shown from simultaneous observations made with X-ray satellites that essentially all of the energy is emitted at γ-ray wavelengths. Because of the low angular resolution of γ-ray telescopes, accurate positions are not known for them and no optical identifications have been made for any of them to date. Thus, the distances of the sources are unknown.

Observations by the Gamma-Ray Observatory (GRO, see Section 7.4.3) have shown that the distribution of the γ-ray bursts is remarkably uniform over the sky. They could therefore be of local origin within our own Galaxy and, for example, be associated with the population of dead neutron stars. The problem with this interpretation is that the number of counts of γ-ray bursts is somewhat flatter than would be expected for a local uniform population of sources. The number–flux-density relation for any uniform population of sources is $N(\geq S) \propto S^{-3/2}$, whereas the counts of γ-ray bursts are found to be significantly flatter than this relation.

There is some evidence for line emission in some of the γ-ray bursts, one of which has been tentatively identified with the gravitationally redshifted electron–positron annihilation line at 512 keV and another with an electron cyclotron line. These data, if correctly interpreted, suggest that the parent object may be a neutron star. This would be consistent with the rapid variability seen in some of the bursts. Another interesting calculation is to estimate the typical luminosities of the γ-ray bursts. The total energy received per unit area in each burst, often referred to as the γ-ray *fluence*, can range between 10^{-10} and 10^{-6} J m^{-2}. If we suppose that the burst lasts 10 s and that the distance of the source is D_{kpc}, in kiloparsecs, a γ-ray burst of fluence 10^{-8} J m^{-2} would have luminosity $10^{31} D_{\text{kpc}}^2$ W, that is, roughly the Eddington luminosity of a solar-mass neutron star. Of course, there is no guarantee at all that the Eddington luminosity is of any significance in such patently non-steady situations. Ideas which have been proposed to account for the bursts, assuming that they are associated with neutron stars, include thermonuclear explosion on their surfaces, major redistributions of angular momenta within the neutron stars and collisions between comets or asteroids and neutron stars.

It cannot be excluded that the γ-ray bursts are extragalactic objects, but then their energy requirements would become very large, particularly if the flat counts of γ-ray bursts is taken to mean that they are at cosmological distances. In this case, extreme events would be needed to account for the bursts, an example being the collision of two neutron stars. This area of study is evolving very rapidly and concerns one of the most remarkable unsolved problems in high energy astrophysics.

16.7   Black holes in X-ray binaries

In the previous subsections, the evidence is overwhelming that there are neutron stars in X-ray binary systems. What about the case for the presence of black holes? The dynamical evidence for black holes in four massive X-ray binaries was discussed in Section 15.6.4, but is there any direct evidence from the X-ray observations themselves that there are black holes in these systems? Some intriguing evidence that they may be present in the X-ray sources LMC X-3, GS2000+25, LMC X-1 and GX339-4 has been obtained by the GINGA X-ray satellite (Inoue (1992)). The X-ray spectra of these sources can be decomposed into two components, a power-law component, which extends to high X-ray energies, and a soft component, which can be described by a black-body spectrum. The variability of these components was measured over a period of three years. These authors attribute the soft component to the emission of an accretion disc about the black hole. The reasoning proceeds as follows.

If it is assumed that the accretion disc is optically thick, the temperature distribution close to the black hole can be found from combining the expression (16.35) with (16.37) so that

$$T(r) = \left[\frac{3GM\dot{m}}{8\pi\sigma r^3} \left(1 - (r_i/r)^{1/2} \right) \right]^{1/4} \tag{16.53}$$

where we have assumed that there is an inner cut-off radius to the accretion disc at r_i. Assuming the disc is optically thick, its luminosity is

$$L = \int_{r_i}^{r_{max}} 4\pi r \sigma T^4(r) \, dr \approx 4\pi r_i^2 \sigma T_i^4 \tag{16.54}$$

where the radiation is assumed to be emitted from both sides of the disc and T_i is the temperature at the inner edge of the disc at r_i. The flux density observed by a distant observer is

$$S = \frac{L\cos i}{2\pi r^2} = \frac{2\cos i}{D^2} r_i^2 T_i^4 \tag{16.55}$$

where the distance of the source is D and i is the angle of inclination between the plane of the disc and the plane of the sky.

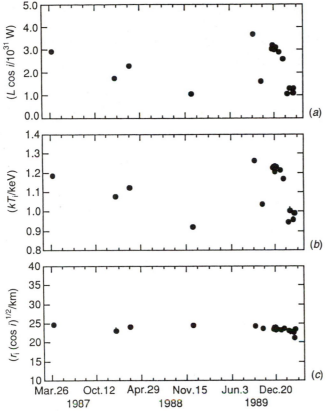

Figure 16.22. Time histories of the best-fit parameters to the soft component of the X-ray spectrum of LMC X-3 obtained by the Japanese Ginga satellite. (*a*) The bolometric luminosity of the sources; (*b*) the inferred temperature at the inner radius of the acccretion disc; (*c*) the inferred inner radius, r_i, of the accretion disc. i is the inclination angle of the plane of the orbit to the plane of the sky. (From H. Inoue (1992). *Proc. Texas/ESO-CERN Symposium on Relativistic astrophysics, cosmology and fundamental particles*, eds J.D. Barrow, L. Mestel and P.A. Thomas, pp. 86–103. New York: New York Academy of Sciences.)

Now, both the luminosity, L, and the temperature, T_i, can be measured for the soft component and so the quantity $r_i \cos^{1/2} i$ can be found. Inoue (1992) analysed variations of temperature and luminosity for these sources over a three-year period and found that the inferred value of $r_i \cos^{1/2} i$ remained remarkably constant, despite large variations in the luminosity of the soft component (Fig. 16.22). He suggested that r_i corresponds to the last stable orbit about the black hole, for example, in the case of LMC X-3, the inner radius corresponding to $r_i \cos i = 25$ km. A more complete analysis, taking account of the effects of special and general relativity has suggested that the mass of the black hole in LMC X-3 is about $5M_\odot$.

This is a remarkable result, but it is clearly dependent upon a number of assumptions, particularly that the accretion disc is optically thick. As was argued in Section 16.3.6, the inner regions of thin accretion discs are often expected to be optically thin. Nonetheless, this analysis is indicative of the type of programme which, if corrrect, provides direct evidence about the process of accretion onto black holes in relatively nearby systems.

Another possible method of investigating the inner regions of the accretion discs about black holes is to search for line emission from material close to the last stable orbit. X-ray lines such as those of FeXXV and FeXXVI are observed from high temperature plasmas, for example, from the hot intergalactic gas in clusters of galaxies. The problem with observing these lines from the inner regions of accretion discs about black holes is that they are expected to be very broad because of the large rotational velocities of the matter close to the last stable orbit and because of the effects of gravitational redshift. Searches for these broad lines require X-ray spectral observations of very high sensitivity, and these should be feasible with the next generation of large X-ray spectrometers, such as the XMM mission of the European Space Agency.

16.8   Reflections

In this chapter, we have no more than scraped the surface of one of the most exciting and important areas of contemporary high energy astrophysics. Whilst I have tried to concentrate upon those aspects of accretion phenomena which seem to be reasonably well established, there are huge areas which have been no more than mentioned. Among the most important of these are the various instabilities which must play an important role in the properties of accretion flows. A further major area which has not been addressed is the origin of the non-thermal emission seen in a number of the X-ray sources. For these and many other aspects of accretion, reference should be made to the more specialist texts listed at the end of this volume.

17

Interstellar gas and magnetic field

17.1 Introduction – a global view of the interstellar medium

Hendrik van de Hulst, the theorist who predicted the 21-cm line of neutral hydrogen, once remarked that, if you set out to detect an emission or absorption line from an atom, ion or molecule in astronomy, you are bound to discover it somewhere in the Universe. This statement is particularly true of the interstellar medium because it is now understood that it is far from equilibrium and that a very wide range of densities and temperatures are present – those found largely reflect the characteristics of the observing tools used by the astronomer. It is no surprise, therefore, that there is a great deal of physics to be studied. Astrophysically, the understanding of the nature and properties of the interstellar gas is of the first importance, since it is out of this medium that new stars are formed. It is continually replenished because of mass loss from stars, and so the medium plays a key role in the birth-to-death cycle of stars. The same astrophysics is applicable to the study of diffuse gas anywhere in the Universe, be it galaxies, the intergalactic gas or the gas clouds in the vicinity of active galactic nuclei. These diagnostic tools are essential for determining the physical conditions in which high energy astrophysical processes take place. Furthermore, interstellar gas will prove to be an essential ingredient of the fuelling mechanisms for active galactic nuclei.

The interstellar medium amounts to about 5% of the visible mass of our Galaxy. In the Galactic plane close to the Sun, the overall gas density amounts to about 10^6 particles m^{-3}, but there are very wide variations in density and temperature from place to place throughout the interstellar medium. To understand the physics of the interstellar gas, let us consider the physical processes which are important in determining its physical state.

17.1.1 Large-scale dynamics

Most of the gas in the Galaxy is confined to the Galactic plane and moves in circular orbits about the Galactic centre, the inward force of gravitational attraction being balanced by centrifugal forces. The gravitational potential in which

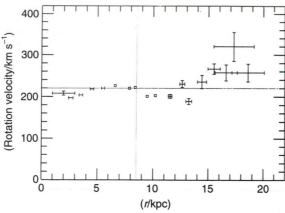

Figure 17.1. An average rotation curve for our Galaxy adopting the 1985 IAU recommended values for the Sun-centre distance of 8.5 kpc and for the mean local rotation velocity of 220 km s^{-1}. (After M. Fich and S. Tremaine (1991). *Ann. Rev. Astron. Astrophys.*, **29**, 420.)

the gas moves is defined by the mass distribution of the stars and of the Galactic dark matter. The kinematics of the interstellar gas therefore acts as a probe of the gravitational potential field of the Galaxy, and the velocities of interstellar neutral hydrogen and molecules provide among the the best information about the distribution of mass in the Galaxy.

These observations have established that the disc of the Galaxy is in a state of *differential rotation*. A recent compilation of results on the mean rotational velocity of the disc of the Galaxy as a function of distance from the Galactic centre, what is known as the *rotation curve* of the Galaxy, by M. Fich and S. Tremaine is shown in Fig. 17.1. The coordinates on the diagram have been scaled to a distance from the Galactic centre to the local standard of rest at the Sun of 8.5 kpc and a mean rotation velocity in the Solar vicinity of 220 km s^{-1}. It can be seen that the rotation velocities are remarkably constant over the radial distances from 3 to 15 kpc, with some evidence for an increase in the rotation velocity beyond 15 kpc. These results are inconsistent with solid-body rotation, for which we would expect $v_{rot} \propto r$, or Keplerian orbits, for which we expect $v_{rot} \propto r^{-1/2}$. As discussed in Volume 3, these data provide evidence for dark matter in the outer regions of the Galaxy. Similar rotation curves are found in other giant spiral galaxies.

The distribution of neutral hydrogen in the Galaxy was determined as long ago as the 1950s, and, more recently, carbon monoxide surveys have defined the distribution of the molecular gas. The picture which emerges is one in which the neutral and molecular hydrogen are closely confined to the plane of the Galaxy, the typical half-widths being about 120 and 60 pc, respectively. They have, however, very different distributions with distance from the Galactic centre. The neutral hydrogen extends from about 3 kpc to beyond 15 kpc from the centre,

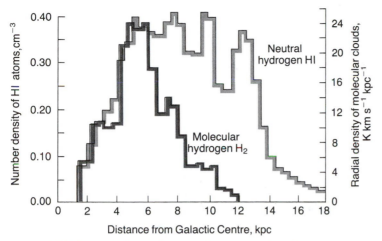

Figure 17.2. The radial distribution of atomic and molecular hydrogen as deduced from radio surveys of the Galaxy in the 21-cm line of atomic hydrogen and from millimetre surveys of the molecular emission lines of carbon monoxide, CO. (After D. Michalis and J. Binney (1981). *Galactic astronomy: structure and kinematics*, pp. 535, 554. San Francisco: W. H. Freeman and Co.)

whereas the molecular component appears to form a thick ring between radii of about 3 and 8 kpc (Fig. 17.2).

The evidence of spiral arm tracers such as O and B stars and HII regions (Fig. 17.3(*a*)) suggests that our Galaxy possesses a rather tightly wound spiral structure. Features possibly related to spiral arms have been observed in the local distribution of neutral hydrogen (Fig. 17.3(*b*)). The giant molecular clouds also tend to be found in spiral arm regions.

The importance of these observations is that they indicate that the interstellar gas is influenced by large-scale dynamical forces. Whilst the overall distribution of the gas is determined by the gravitational potential defined by the stars and the dark matter, some mechanism is needed to enhance the average gas density from about 10^6 m^{-3} to very much greater values, at least 100 to 1000 times greater, in giant molecular clouds and, in particular, to result in conditions favourable for the formation of stars in the vicinity of spiral arms. One attractive mechanism for achieving this is through a density wave set up in the distribution of stars in the Galactic disc. The *density wave theory of spiral structure* is one of the most successful attempts so far to account for the appearance of spiral arms in galaxies. The theory is based upon considerations of the stability of a disc of stars to axial perturbations. The key question is whether or not such a perturbation is stable, that is, whether or not it is not destroyed by the differential rotation of the stars about the centre of the Galaxy. It is found that the spiral density wave in the stellar distribution tends to propagate either inwards or outwards from the centre of the disc, thus destroying the perturbation. It seems that there must be some forcing mechanism which maintains the spiral pattern in the stellar

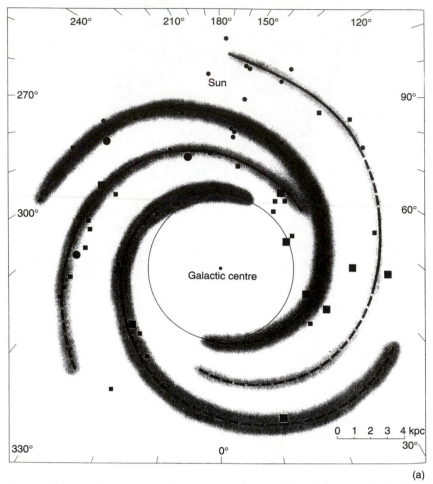

Figure 17.3 (*a*). The spiral structure of our Galaxy deduced from optical (circles) and radio (squares) observations of regions of ionised hydrogen. The dashed sections of the spiral are poorly determined. (Map by Y.M. Georgelin and Y.P. Georgelin (1988). In *The Cambridge atlas of astronomy*, eds J. Audouze and G. Israel, p. 308. Cambridge: Cambridge University Press.)

distribution. This might be due either to the interaction with companion galaxies or possibly with perturbations associated with the ellipsoidal distribution of stars in the bulge of the Galaxy. It is significant that the most recent analyses of the distribution of stars in the central bulge of the Galaxy suggest that it is ellipsoidal rather than spheroidal (Blitz *et al.* (1993)).

Assuming that the density wave in the stellar disc can be maintained, we can then ask how the gas behaves under its influence. An important point is that the sound speed in the neutral and cold gas is expected to be very low. The gas therefore tends to collect in the potential minima of the density wave, and it turns out that the velocity it acquires under its influence is such that the gas

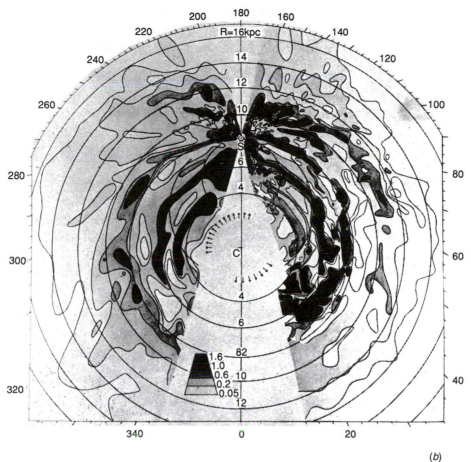

(b)

Figure 17.3 (b). The distribution of atomic hydrogen in the Galactic plane. There are prominent features which may be related to the spiral structure of our Galaxy delineated by other spiral arm tracers such as young stars and regions of ionised hydrogen. (J.H. Oort, F.J. Kerr and G. Westerhout (1958). *Mon. Not. R. Astron. Soc.*, **118**, 382.)

flow becomes supersonic. Shock waves form along the trailing edge of the stellar density wave, and a large increase in gas density behind the shock is expected because the compressed gas can cool effectively. This picture provides an attractive explanation for the formation of clouds of neutral and molecular gas in the vicinity of spiral arms, and is consistent with the observed location of young objects relative to the underlying spiral density wave defined by the old stellar populations.

Although this picture has its attractions, this is not necessarily the only process by which molecular clouds and spiral features can be formed. For example, supernova explosions lead to strong shock waves propagating through the interstellar gas, and, in the late stages of expansion, cooling of the compressed gas can lead to the formation of cool dense clouds. It has been pointed out by a number of authors that the greatest star formation rates appear to be found in the most irregular galaxies and not in those with the most beautifully developed spiral structures. One can imagine a chain reaction whereby, once the first stars are formed, the most massive of these explode over a timescale of, say, 10^6–10^7 years and trigger the next generation of star formation. This picture can be modelled as a *percolation process* occurring throughout the disc of a galaxy and has had considerable success in explaining the observation of spiral features in galaxies. It is likely that both processes are important for spiral galaxies.

17.1.2 *Heating mechanisms*

If left on its own, the interstellar gas would cool to a low temperature, but this is in conflict with the observation of many different phases at a wide range of different temperatures. In fact, there is an embarrassment of heating mechanisms for the gas, many of which are reviewed by Spitzer (1990).

The hottest gas is produced by *supernova explosions*. Supernova remnants are observed to be strong thermal X-ray sources, the gas temperature rising to 10^7 K or more (see Section 15.2.1 and Fig. 7.14). A shock wave runs ahead of the supersonically expanding shell of cooling gas and heats up the interstellar gas to high temperatures. In an elegant analysis, Cox and Smith (1974) first showed that heating by supernova explosions could lead to about 10% of the volume of the interstellar gas being heated to a high temperature. The collisions of the old shells of these supernova remnants can lead to reheating of the swept up gas as the kinetic energy of expansion is converted into heat. Thus, they predicted that the hot component would form tunnels through the interstellar gas from the overlapping of supernova shells. It is entirely plausible that at least some part of the soft X-ray emission from the plane of our Galaxy is associated with this hot gas. The most recent observations by the far ultraviolet Wide Field Camera on board the ROSAT satellite have shown that the Solar System is probably located within a large bubble of hot gas of diameter about 500 pc, which is entirely consistent with this picture. It is also probable that the hot gas, inferred to be present in the halo of our Galaxy from observations of lines of highly ionised carbon (CIV) and oxygen (OVI) by the International Ultraviolet Explorer, has attained a dynamical equilibrium in the gravitational field of the disc and halo. It is natural that the hot gas should expand to form a hot halo since its scale height is expected to be very much greater than that of the stars of the disc.

A second important heating mechanism is the *ultraviolet radiation* of young stars. The youngest of these are still embedded in the gas clouds from which they formed. The heated gas is easily recognised by the strong emission lines of hydrogen and oxygen (Fig. 13.23). In the case of heating by ultraviolet radiation, the gas temperature is determined by the balance between photoionisation of the neutral gas and recombination of the ionised component, and results in a temperature of typically 10^4 K (see, for example, Osterbrock (1989)). Older blue stars, which are no longer embedded in regions of ionised hydrogen, can ionise and heat the surrounding regions, and this form of local heating is observed in the ultraviolet spectra of certain O and B stars. The detection has also been made of a region of ionised gas about a binary X-ray source in which very high excitation species are observed, these being attributed to ionisation and heating by the source.

We have already described the role which large-scale dynamical processes play in determining the state of the gas. Wherever there are shock waves, these can lead to strong heating of the gas, for example, in the shocks associated with spiral density waves, with bipolar outflows from young stars and so on. Another dynamical source of heating is gravitational collapse, which is certainly important in the formation of stars inside molecular clouds.

There is convincing evidence that the flux of cosmic rays observed in the vicinity of the Solar System is typical of the flux of high energy particles present throughout the interstellar medium (see Sections 18.2 and 20.1). The ionisation losses of these particles are important sources of heating and ionisation of both the diffuse neutral gas and the gas in giant molecular clouds. The heating rate is poorly known because of the effects of solar modulation upon the spectra of cosmic rays of low energy observed in the vicinity of the Earth (see Section 10.3). If we simply adopt the spectrum of high energy protons observed at the top of the atmosphere, without making any allowance for the effects of solar modulation, we can find the ionisation rate of the interstellar gas by ionisation losses. This is found to amount to about $10^{-17}N$ electrons s^{-1}, the average energy of each electron being about 35 eV; N is the number density of hydrogen atoms in units of particles per cubic metre. Notice that this calculation takes account of the production of secondary electrons by the primary electrons released in the process of ionisation. Not all of this energy is available for heating the gas since much of it goes into exciting the atoms of the gas, which radiate away this energy. Notice that the heating rate could be significantly greater than the above figure once the effects of solar modulation of the local flux of cosmic rays are taken into account. On the other hand, it is unlikely to be very much greater than this figure because, as shown in Section 19.5.1, a local energy density of cosmic rays of about 1 MeV m^{-3} can be accounted for in terms of the observed energies of supernova remnants and their rate of occurrence in the Galaxy. Ionisation losses are probably the origin of the small but significant abundance of free electrons, present in molecular clouds, which are crucial for interstellar chemistry.

There are other potential sources of heating. For example, the intergalactic flux of ultraviolet ionising radiation, mass loss from all types of star, including stellar winds, infall of matter from intergalactic space and so on. It can be recognised already that there are excellent reasons why the interstellar medium should be far from equilibrium.

17.1.3   *Cooling mechanisms*

In principle, the role of cooling mechanisms should be easier to understand than the heating mechanisms because radiation is the principal means by which the thermal energy of the interstellar gas is lost, and therefore, simply by observing line and continuum emission at frequencies close to the peak of the black-body spectrum appropriate to that phase of the gas, it should be possible to observe directly the cooling processes.

For very hot ionised gas, at temperatures in excess of 10^7 K, the principal cooling mechanism is the *bremsstrahlung* or *free–free emission* of the free electrons in the plasma (Section 3.5.2 and expression (3.45)). This process is observed in supernova remnants, where, in addition, strong emission lines of 24 and 25 times ionised iron, FeXXV and FeXXVI, respectively, have been observed in the X-ray waveband at about 8 keV, confirming the high temperature of the gas.

At lower temperatures, 10^4 to 10^7 K, the emission is due to *bound–bound* and *bound–free* transitions of hydrogen, helium and heavy elements. This temperature regime is much more difficult to study observationally because most of the radiation is emitted in the unobservable ultraviolet region of the spectrum. It seems likely, however, that at least part of the soft X-ray radiation detected in the plane of the Galaxy is associated with the radiation of gas at a temperature of about 10^6 K, and the spectrum can be attributed to the bound–free emission of different elements, which, when summed, results in a smooth steep spectrum which extends to soft X-ray and far ultraviolet wavelengths.

Much of the gas observed in bright regions of ionised hydrogen is found to be at a temperature of about 10^4 K. The reason for this is that the gas is excited by radiation from hot blue stars which have strong fluxes of radiation in the ultraviolet continuum. The hard ultraviolet photons ionise the gas, and an equilibrium is set up between ionisation due to the flux of Lyman continuum radiation and recombination of the ionised gas by electron–ion collisions. This equilibrium is established at a temperature close to 10^4 K (see Osterbrock (1989)). At this temperature, the main cooling mechanism for the gas is line radiation, the resonance lines of hydrogen or the forbidden transitions of singly and doubly ionised oxygen, [OII] and [OIII], respectively (see Section 17.3.2). It is these lines which give ionised hydrogen clouds their characteristic red glow on colour photographs.

At temperatures less than 10^4 K, the ionised gas recombines, and therefore there are very few free electrons present. Between 10^3 and 10^4 K, the principal radiation loss mechanism is the line emission of neutral or singly ionised carbon, nitrogen and oxygen. The emission lines are associated with forbidden transitions of low lying energy levels. Observations from high flying aircraft, such as the Kuiper Airborne Observatory, have shown that the lines of [OI] (63 and 145 μm), [CI] (609 and 370 μm) and [CII] (157.7 μm) are particularly strong and are likely to be among the most important coolants of the interstellar gas in this temperature range.

At temperatures below about 10^3 K, *interstellar dust* can survive and plays a key role in determining the state of the gas at low temperatures. One of the more perplexing aspects of the study of the interstellar medium is that we are certain that there are large quantities of dust present, as is apparent from the patchy obscuration seen in optical images of the Galaxy (Fig. 1.8(*d*)) and regions of star formation (Fig. 13.22), but we have little definite knowledge about the exact composition and properties of the dust grains. It is known that they must contain a large fraction of the heavy elements present in the interstellar medium because the gaseous phase is very significantly underabundant in these elements. Another compelling piece of evidence is the observation of the formation of dust shells about dying stars and supernovae such as SN 1987A (Section 13.2.2), when the temperature of the ejected gas falls below about 1000 K.

Dust grains absorb electromagnetic waves efficiently at wavelengths less than or equal to their physical sizes but are transparent at longer wavelengths, and so there must be a wide range of grain sizes present in the interstellar medium to account for the fact that the absorption coefficient of the interstellar gas extends rather smoothly from ultraviolet through optical to infrared wavelengths (Fig. 17.4). Throughout the optical and infrared wavebands, the optical depth τ of the interstellar medium can be written $\tau \propto \lambda^{-1}$. This observation explains why obscuration can affect optical and ultraviolet observations very severely and yet have a relatively small effect in the infrared waveband. For example, in the direction of the Galactic centre, the attenuation in the V waveband, $\lambda = 0.55\ \mu$m, can amount to about 30 magnitudes, a factor of 10^6, in certain directions. At $2\ \mu$m, the attenutation would be only 8 magnitudes, a factor of 1600, and, at $5\ \mu$m, only 3 magnitudes, or a factor of 15.

Superimposed upon this continuum absorption curve, there are several prominent features. The most prominent is the strong, broad absorption feature observed at about 220 nm, which is present in the Galactic extinction curve. This feature corresponds rather closely with a graphite resonance band, and it is commonly assumed that this is evidence for the presence of graphite in interstellar grains. There are also dust absorption bands in the optical waveband, but these have remained unidentified despite an enormous amount of work by many authors. Dust absorption features have also been discovered in the infrared waveband, and some of these can be associated with vibrational transitions in solid materials. For example, prominent absorption and emission features at 9.7 and 18 μm are associated with silicates. Other features have been associated with water ice and ammonia molecules. The nature of the grains is therefore likely to be somewhat complex. In one favoured picture, the grains contain graphite or silicon cores surrounded by a water ice mantle. One of the most important roles of the dust grains is the formation of molecules. Atoms and molecules are adsorbed onto the grain surfaces, where they can migrate, combine with other species and then return to the interstellar medium. In other words, the grains act as a 'catalyst' for the formation of organic molecules. This is almost certainly the origin of many of the species listed in Table 17.2.

Returning to the question of the cooling of the interstellar medium, at temperatures greater than 10^3 K, the dust grains are evaporated by collisions, but, below this temperature, they survive and perform a number of different functions. First of all, dust absorbs ultraviolet and optical radiation, and therefore, within dust clouds, molecules are protected from the interstellar flux of dissociating radiation. These dust clouds are present throughout the Galactic disc, with some concentration towards spiral arms. Within the dust clouds, there are two impor-

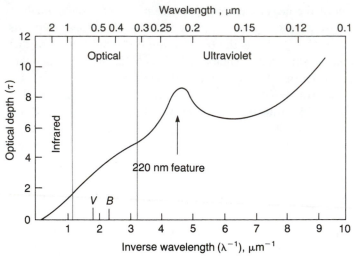

Figure 17.4. The absorption coefficient of interstellar dust grains as a function of wavelength. Superimposed upon the relatively smooth continuum absorption curve are a number of absorption bands, the strongest of these being the broad feature at about 220 nm. (After J.M. Greenberg (1982). *Laboratory and observational spectra of interstellar dust*, eds R.D. Wolstencroft and J.M. Greenberg, p. 2. Edinburgh: Royal Observatory Edinburgh Publications. Also after L. Spitzer (1978). *Physical processes in the interstellar medium*, p. 158. New York: Wiley-Interscience Publications.)

tant cooling processes. The first is molecular line emission associated either with rotational transitions of asymmetric molecules such as carbon monoxide, CO, and water vapour, H_2O, or, in some cases, with the infrared forbidden rotational and rotational–vibrational transitions of molecular hydrogen, H_2. In some regions, these lines are so strong that they must be the dominant cooling mechanism.

The second process is the *reradiation* of the radiation absorbed by the dust grains. This is almost certainly the most efficient energy loss mechanism for stars which are in the process of formation or which have just formed. Stars form in the densest regions of giant molecular clouds, and the ultraviolet radiation emitted by them is absorbed by the dust grains. The grains are heated to a temperature which is determined by the balance between the energy absorbed from the radiation field and their rate of radiation. They radiate more or less like little black bodies, the Planck distribution being modified by the emissivity function of the material of the grains. Thus, the emissivity of the grains can be written $\epsilon(v) = \kappa(v)B(v)$, where $B(v)$ is the Planck distribution and $\kappa(v)$ is the emissivity of the grains. Hildebrand (1983) has shown that, to a good approximation, $\kappa(v) \propto v$ at wavelengths $\lambda < 100$ μm and $\kappa(v) \propto v^2$ at much longer wavelengths, $\lambda > 1$ mm. A key point is that the grains radiate away the absorbed energy very rapidly at roughly the temperature to which they are heated, which is typically about 30 to 100 K for the far infrared sources found in dense molecular clouds. This means that the energy is reradiated at wavelengths of about 30 to 100 μm, at

Table 17.1. *The principal phases of the interstellar gas*

Names	Main constituent	Detected by	Volume of interstellar medium	Fraction by mass	N (m^{-3})	Temp. (K)
'Molecular clouds'	H_2, CO CS, etc.	Molecular lines; dust emission	$\sim 0.5\%$	40%	$\geq 10^9$	10–30
'Diffuse clouds'; 'HI clouds'; 'cold neutral medium'	H, C, O with some ions, C^+, Ca^+	21-cm emission & absorption	5%	40%	10^6–10^8	80
'Intercloud medium'	H, H^+, e^-; ionisation fraction 10–20%	21-cm emission & absorption; $H\alpha$ emission	40%	20%	10^5–10^6	8000
'Coronal gas'	H^+, e^-; highly ionised species, O^{5+}, C^{+3}, etc.	OVI; soft X-rays 0.1–2 keV	$\sim 50\%$	0.1%	$\sim 10^3$	$\sim 10^6$

Courtesy of Dr John Richer.

which the dust is transparent, and therefore the energy of the star can be radiated away very efficiently. This picture explains convincingly why intense far infrared emission is the signature of sites of star formation (Fig. 13.21). In addition, many galaxies, particularly those in which there is active star formation, such as the late type spiral and irregular galaxies as well as the colliding galaxies, show extreme far infrared luminosites with the characteristic emission spectra of heated dust.

The picture which emerges is one in which there are many different processes by which the interstellar gas can be heated and cooled under different circumstances. The term *violent interstellar medium* is often used, reflecting the fact that the medium is far from stationary, being constantly buffeted by supernova explosions and the winds from young stars as well as the large-scale dynamical phenomena described in Section 17.1.1.

17.1.4 *The overall state of the interstellar gas*

In spite of the complexity of the interstellar medium, it is useful to have some rough global figures to describe the various phases outlined above. These are listed in Table 17.1. It is of interest that the diffuse phases have roughly the same pressure, $p = NkT$, and so they must be more or less in pressure equilibrium. This is reassuring since they represent diffuse gas present throughout much of the interstellar medium. Notice that within the molecular clouds densities much greater than 10^9 m^{-3} are found.

Why is it that these phases are conspicuously present while others are not? The reason is almost certainly due to *thermal instabilities* in the diffuse gas. The condition for a phase of the gas to be unstable thermally was first given by Field (1965) in terms of a generalised heat-loss function $\mathcal{L}$, which is defined as the energy losses minus energy gains per unit mass of material per second. In the analysis of the stability of the diffuse interstellar gas, it is assumed that the energy losses are by radiation and that the gas is optically thin. In the classic analysis of Field, Goldsmith and Habing (1969), it was assumed that the heating is due to the ionisation losses of low energy cosmic rays (see Section 2.4). Thus, the generalised loss rate can be written

$$\mathcal{L}(N, T) = \Lambda(N, T) - \Gamma$$

where $\Lambda(N, T)$ is the cooling rate of the gas and Γ is the total heating rate. In the equilibrium state, there is balance between the heating and cooling rates so that $\mathcal{L} = 0$ and the gas is in pressure equilibrium. Field (1965) showed that the equilibrium state is unstable if $(\partial \mathcal{L}/\partial T)_p < 0$. The origin of this instability is clearly described by Shu (1992). The phases are required to remain in pressure equilibrium. Suppose in some region the density increases so that the rate of energy loss also increases. The region will contract and the decrease in thermal energy will be partly or wholly offset by the work done by the surrounding medium on the perturbed cloud. The system will be stable if the resulting pressure is more than sufficient to maintain pressure equilibrium, but, if it is not, the perturbation continues to collapse until a new equilibrium state is attained at a higher density and lower temperature.

The analysis of Field, Goldsmith and Habing (1969) showed that there are two stable phases of the interstellar medium at temperatures less than 10^4 K, one at about 8000 K and the other at a lower temperature of about 80 K, corresponding to two of the entries in Table 17.1. Between these temperatures, cooling due to the atomic and ionic lines described in Section 17.1.3 causes the gas to be unstable. This picture gave rise to the *two-phase model* of the interstellar medium. Extending the analysis to higher temperatures can account for the existence of the coronal gas as well. Thus, although there is every reason to expect the interstellar medium to be in a state of continual flux, it seems that the main components listed in Table 17.1 are in approximate pressure equilibrium.

17.2   Diagnostic tools – neutral interstellar gas

An important feature of a number of the techniques for studying the interstellar medium is that what is observed provides information about the *weighted integral* of various properties of the medium along the line of sight through the Galaxy. For example, if we measure the intensity of bremsstrahlung, we obtain information about $\int N_e^2 \, T^{-1/2} dl$ (Section 3.5.2). Clearly, the interpretation of the integral depends on how the hot gas is distributed along the line of sight as well as on its temperature distribution. Thus, interpreting integral measurements requires some care since irregularities and clumpiness in the distribution of matter have to be taken into account.

17.2.1   *Neutral hydrogen: 21-cm line emission and absorption*

Neutral hydrogen emits line radiation at a frequency $v_0 = 1420.4058$ MHz ($\lambda_0 = 21.1$ cm) through an almost totally forbidden hyperfine transition in which the spins of the electron and proton change from being parallel to antiparallel. The spontaneous transition probability for this process is $A = 2.85 \times 10^{-15}$ s^{-1} for the ground state of hydrogen, that is, about once every 10^7 years. Because there are two possible orientations of the spins of both the electron and the proton, there are four stationary states, three being degenerate in the upper state and one in the lower state. Because of the very small transition probability, collisions and other processes have time to establish an equilibrium distribution in which the populations of the upper and lower levels (labelled 2 and 1, respectively) are given by the Boltzmann distribution $N_2/N_1 = (g_2/g_1)\exp(-hv_0/kT)$, where T is the equilibrium excitation temperature and g_2 and g_1 are the statistical weights of the upper and lower levels, $g_2/g_1 = 3$. In the case of 21-cm (or HI) emission, the excitation temperature is called the *spin temperature* and is written T_s. Under all cosmic conditions, $hv_0/k = 7 \times 10^{-2}$ K $\ll T_s$, and therefore $N_2/N_1 = 3$.

When the emitting region is optically thin, only spontaneous emission need be considered, and so the emissivity of the gas is

$$\kappa = \frac{g_2}{g_2 + g_1} N_H A h v_0 = \frac{3}{4} N_H A h v_0 \qquad (17.1)$$

where N_H is the number density of neutral hydrogen atoms.

If the neutral hydrogen is distributed along the line of sight from the observer, the flux density received within solid angle Ω, say, the beam of the radio telescope, is

$$S = \int \frac{\kappa(l)}{4\pi l^2} \Omega l^2 dl \qquad\qquad I = \frac{S}{\Omega} = \frac{3}{16\pi} A h v_0 \int N_H dl \qquad (17.2)$$

$I = S/\Omega$ is the intensity of radiation in that particular direction and is a measure of the *total column density* of neutral hydrogen along the line of sight, $\int N_H dl$. In this calculation, I is measured in watts per square metre and is equal to the integral of the intensity of radiation per unit bandwidth, I_v, over the line profile, $I = \int I_v dv$.

Because of its very small transition probability, the natural line width of the 21-cm line is very narrow. If the neutral hydrogen is in motion relative to the observer, Doppler shifts of the 21-cm line radiation can be readily measured by making observations with a multi-channel receiver, which has very narrow channels on either side of the line centre, that is, a radio spectrometer. This provides a very powerful tool for investigating the dynamics of neutral hydrogen in our own and in other galaxies (see Figs 17.2 and 17.3(b)).

Non-thermal radio sources, such as supernova remnants and extragalactic radio sources, have smooth continuum spectra at radio wavelengths due to synchrotron radiation, and therefore, if neutral hydrogen clouds lie along the line of sight to the radio source, absorption features in the radio spectrum of the source are expected. We can work out the absorption coefficient for 21-cm line absorption using the same technique described in the discussion of thermal bremsstrahlung absorption at radio wavelengths in Section 3.5.3. The relation (3.60) can be used in the low frequency limit $h\nu \ll kT$, in which case the black-body intensity becomes $I_\nu = 2kT/\lambda^2$, and so

$$\chi_\nu I_\nu = \chi_\nu \frac{2kT}{\lambda^2} = \frac{\kappa_\nu}{4\pi} \tag{17.3}$$

Now, if $\Delta\nu$ is the line width of the neutral hydrogen profile, the emissivity per unit frequency interval is

$$\kappa_\nu = \frac{3}{4} N_H A h \frac{\nu_0}{\Delta\nu}$$

Therefore, the absorption coefficient, χ_ν, is

$$\chi_\nu = \frac{3}{32\pi} \frac{Ahc^2}{\nu_0^2 kT_s} \frac{\nu}{\Delta\nu} N_H \tag{17.4}$$

If the radio source is much brighter than the foreground HI emission, that is, the brightness temperature of the source $T_b \gg T_s$, its observed spectrum is

$$I_\nu = I_0(\nu)\exp(-\tau_\nu) \qquad \tau_\nu = \chi_\nu l \tag{17.5}$$

where l is the path length through the cloud. Evidently, the interpretation of the absorption spectrum requires knowledge of the spin temperature, T_s, of the intervening cloud. In practice, the absorption profile cannot normally be fitted by a simple Gaussian function but consists of a number of components with different velocities and line widths, which result from a combination of systematic and random velocities of the clouds along the line of sight to the radio source. The neutral hydrogen absorption measurements give important information about the small-scale structure and velocity dispersion of the neutral hydrogen along the line of sight on the scale of the angular size of the background source, whereas the emission profiles provide information on the scale of the beam width of the radio telescope.

17.2.2 *Molecular radio lines*

Long before the advent of radio astronomy, it was known that there exist significant abundances of molecules in interstellar space. The molecules CH, CH$^+$ and CN possess electronic transitions in the optical waveband, and absorption features associated with these were well known features of the spectra of bright stars. The great advantage of observing molecules at radio frequencies is that, unlike the optical waveband, there is no obscuration due to interstellar dust. The first interstellar molecule to be detected at radio wavelengths was the hydroxyl radical OH, which was observed in absorption against the bright radio source Cassiopaeia A in 1963. Soon afterwards, the hydroxyl lines were observed in emission, and the surprise was that the sources were very compact and variable in intensity. The corresponding brightness temperatures were very great indeed, $T_b \geq 10^9$ K, implying that the emission process must involve some form of maser action. Five years later, in 1968, ammonia, NH_3, was detected, and in the following year

Table 17.2. *Interstellar molecules*

This list of known interstellar molecules is arranged in columns showing the numbers of atoms which make up each molecule. Different isotopic species have not been included; for example, no deuterated molecules are listed. Those molecules indicated by an asterisk in brackets are ring molecules.

2	3	4	5	6	7	8	9
H_2	H_2O	NH_3	SiH_4	CH_3OH	CH_3CHO	$CHOOCH_3$	CH_3CH_2OH
OH	H_2S	H_3O^+	CH_4	NH_2CHO	CH_3NH_2		$(CH_3)_2O$
SO	SO_2	H_2CO	$CHOOH$	CH_3CN	CH_3CCH		CH_3CH_2CN
SO^+	HN_2^+	H_2CS	$HC{\equiv}CCN$	CH_3NC	CH_2CHCN		$H(C{\equiv}C)_3CN$
SiO	$HNO?$	$HNCO$	CH_2NH	CH_3SH	$H(C{\equiv}C)_2CN$		$H(C{\equiv}C)_2CH_3$
SiS	$H_2D^+?$	$HNCS$	NH_2CN	C_5H	CH_3CCN		
NO	HCN	$CCCN$	H_2CCO	HC_2CHO	C_6H		
NS	HNC	HCO_2^+	C_4H	$CH_2{=}CH_2$			
HCl	HCO	$HSiCC$	C_3H_2	H_2CCCC			
$NaCl$	HCO^+	$CCCH$	CH_2CN				
KCl	$HOC^+?$	$c{-}CCCH$	C_5				
$AlCl$	OCS	$CCCO$	SiC_4				
AlF	CCH	$CCCS$	H_2CCC				
PN	HCS^+	$HCCH$	$HCCCO?$				
SiN	$CCO?$	$HCNH^+$	$HCCNC$				
CH	CCS	$HCCN$	$HNCCC$				
CH^+	C_3						
CN	$SiC_2(*)$						
CO	$H_3^+(*)$						
CS							
C_2							
SiC							
CP							

In addition, there are molecules with 10 atoms, $CH_3(C{\equiv}C)_2CN$, 11 atoms, $H(C{\equiv}C)_4CN$, and 13 atoms, $H(C{\equiv}C)_5CN$.

Table courtesy of Dr P. Thaddeus.

water vapour, H_2O, and formaldehyde, H_2CO, were discovered. A key discovery was the great intensity of the carbon monoxide molecule, CO, which was first observed in 1970. Since that date, the number of detected molecular species has multiplied rapidly (see Table 17.2). In obscured regions of interstellar space, where the environment is protected from dissociating optical and ultraviolet radiation, complex organic molecules have been discovered with up to 13 constituent atoms. The molecules are composed of the most abundant elements: hydrogen (and deuterium), nitrogen, carbon, sulphur, silicon and oxygen and their isotopes. In fact, in some sources, the molecular line spectra are so rich that the noise in the spectra is due to the superposition of a myriad of weak molecular emission lines.

Molecules can emit line radiation associated with transitions between electronic, vibrational and rotational levels. The highest energy transitions are those associated with *electronic transitions*, and normally these lie in the optical region of the spectrum. *Vibrational transitions* are associated with the molecular binding between atoms of the molecule, which can be represented by a simple harmonic oscillator; transitions between these vibrational levels typically lie in the infrared spectral region $h\nu \approx 0.2$ eV.

The lowest energy transitions are those between rotational energy levels. The frequencies of these *rotational transitions* can easily be worked out using the rules of quantisation of angular momentum. According to quantum mechanics, the angular momentum $\mathbf{J}$ is quantised such that it can only take discrete values given by the relation $\mathbf{J}^2 = j(j+1)\hbar^2$, where the angular momentum quantum number j takes integral values, $j = 0, 1, 2, \ldots$. The energy of each of these stationary states is given by exactly the same formula which relates energy and angular momentum in classical mechanics, $E = \mathbf{J}^2/2I$, where I is the moment of inertia of the molecule about its rotation axis. When a photon is emitted or absorbed, one unit of angular momentum has to be created or absorbed, and hence j changes by one unit, so that the selection rule for these electric dipole transitions is $\Delta j = \pm 1$. The energy of the photon emitted in the rotational transition from the stationary state j to that corresponding to $j - 1$ is therefore

$$h\nu = E(j) - E(j-1) = [j(j+1) - (j-1)j]\hbar^2/2I = j\hbar^2/I \qquad (17.6)$$

For a diatomic molecule composed of atoms of masses M_1 and M_2, the moment of inertia I can be written $I = \mu r_0^2$, where μ is the reduced mass of the molecule, $\mu = M_1 M_2/(M_1 + M_2)$, and r_0 is the equilibrium spacing of the atomic nuclei. Therefore, $\nu = jh/4\pi^2\mu r_0^2$. This illustrates a very useful feature of the rotational spectrum of molecules – the rotational lines are equally spaced in frequency. This is often referred to as the *rotational ladder* of the molecule's spectrum. For example, for CO, $\mu = 6.859$ atomic mass units $= 1.11 \times 10^{-26}$ kg and $r_0 = 1.128 \times 10^{-10}$ m. Therefore, the lowest frequency rotational transition, $j = 1 \to 0$, is 115 GHz, or $\lambda = 2.6$ mm. The next transitions in the rotational ladder have frequencies 230 GHz ($j = 2 \to 1$), 345 GHz ($j = 3 \to 2$) and so on. Corresponding results are found for more complex molecules involving more than two atoms. The transition probabilities depend upon the net electric dipole moment of the molecule, and so symmetrical molecules such as hydrogen, H_2, do not emit electric dipole radiation, but asymmetrical molecules, such as CO and $HC_{11}N$, are sources of millimetre line emission.

Other molecules, such as the hydroxyl radical, OH, and formaldehyde, H_2CO, have permitted transitions in the radio waveband through molecular doubling processes. In the case of a diatomic molecule such as OH, the doubling results from the interaction between the electronic motions in the molecule and the rotation of the molecule as a whole.

Generally speaking, molecular line emission provides information about rather denser regions of the interstellar gas than 21-cm line emission. This is because molecules are fragile and can be dissociated by optical and ultraviolet photons. They are therefore predominantly found in dense molecular clouds with densities $N_H \approx 10^9$–10^{10} m^{-3}, within which the molecules are shielded from the interstellar flux of high energy photons by dust and also by *self-shielding* by the molecular hydrogen at the peripheries of the clouds. In addition, the higher frequency transitions of a particular rotational ladder have larger transition probabilities and can be used to determine much higher molecular densities within the clouds. These emission lines may be used to map the distribution of molecular species throughout our own and other galaxies.

The most common molecule is expected to be molecular hydrogen, H_2, but, as noted above, because it has no electric dipole moment, no rotational transitions are observed. Molecular hydrogen has, however, been detected by the Copernicus satellite in absorption in the ultraviolet region of the spectrum through its electronic transitions. These observations confirm that H_2 is present in large quantitites in the interstellar gas. The next most abundant molecule is expected to be carbon monoxide, CO, and, as shown above, it emits strong permitted line radiation at 2.6 mm, corresponding to the $j = 1 \rightarrow 0$ rotational transition. Strong CO radiation has been detected throughout the Galaxy and provides important information complementary to that provided by surveys of the 21-cm line of neutral hydrogen (see Fig. 17.2). The importance of the CO observations is that it can be assumed that wherever there exist CO molecules there must also exist H_2 molecules, and hence CO acts as a tracer for H_2 molecules. Indeed, the excitation mechanism for the CO molecules is collisions with hydrogen molecules, and so the CO observations provide a measure of the number density of H_2 molecules in these clouds.

Table 17.2 contains a wide variety of different types of molecule – organic molecules, inorganic molecules, free radicals and molecular ions. There is also a great range in the size of the molecules. Many consist of two atoms, but very much larger ones are observed, the record holder being the acetylenic chain molecule $HC_{11}N$ with 13 atoms. Several important patterns are discernible in Table 17.2. For example, there is the remarkable sequence of acetylenic chain molecules HCN, HC_3N, HC_5N, HC_7N, HC_9N, $HC_{11}N$ – there must be some simple mechanism for lengthening a pre-existing chain. There is also a conspicuous absence of benzene ring molecules and their derivatives, as well as biological molecules, such as glycine, the simplest amino acid, despite specific searches for these molecules. There is some evidence that some of the unidentified lines observed in the infrared waveband may be associated with fullerenes, C_{60}, or 'football' molecules.

The Universe contains an overwhelming majority of hydrogen atoms, and so the existence of many unsaturated species, that is, species containing double and triple bonds, is remarkable. If a giant molecular cloud were in thermodynamic equilibrium at a temperature of, say, 50 K, the only species expected would be saturated molecules such as CH_4, NH_3, H_2O and so on. There would be no CO nor any of the unsaturated multiply-bonded species such as $HC_{11}N$. The inference is that the interstellar medium must be very far from thermodynamic equilibrium. The principal reactions which determine the abundances of the different molecular species are gas-phase reactions and chemical reactions taking place on grain surfaces. Besides their obvious interest for the relatively new discipline of *interstellar chemistry*, the existence of these molecules provides an important tool for probing the physical conditions and velocity fields deep inside star-forming regions.

17.2.3   *Optical and ultraviolet absorption lines*

Atoms producing absorption lines within the optical waveband, 310–800 nm, must possess excited states within about 4 eV of the ground state, and there are relatively few of the more abundant species which satisfy this criterion, the most important being the transitions of NaI, CaII, CaI, KI, TiII and FeI. All these absorption lines have been observed in stellar spectra, the strongest being those of CaII and NaI, which are both doublets, the pairs of lines being known as the H and K lines of calcium (396.85 and 393.37 nm, respectively) and the D lines of sodium (DI 589.59 and D2 589.00 nm). The ultraviolet region of the spectrum, 100–300 nm, corresponds to higher energy transitions, and a very much wider range of interstellar atoms and molecules can be studied, in particular atomic and molecular hydrogen and essentially all the common heavy elements. The series of Orbital Astronomical Observatories (OAO-II and Copernicus) and the International Ultraviolet Observatory (IUE) have revolutionised studies of the interstellar medium, and absorption lines associated with all the common elements in various stages of ionisation have been detected.

The interpretation of interstellar absorption spectra requires knowledge of atomic absorption cross-sections as a function of frequency, $\sigma(v)$. For an atom at rest, the absorption cross-section may be calculated quantum mechanically in the case of simple atoms or, in most cases, may be derived from laboratory experiments. The frequency dependence of the absorption cross-section depends upon the mechanism of line broadening. The most important processes for interstellar absorption lines are *Doppler broadening*, which may result either from the random motions of the absorbing atoms in the gas or from the bulk motions of clouds, and *radiation damping*, or *natural broadening*, which results from the fact that the atom remains only a finite time, Δt, in an excited state. A rough estimate of the natural line width may be found from Heisenberg's uncertainty principle $\Delta E \approx h/\Delta t$, that is, $\Delta v \approx \Delta t^{-1}$.

We have no wish to go into the complexities of the interpretation of atomic spectra, but at least, in the optically thin case, it is evident that the optical depth of the line, τ_v, is a measure of the total column density of the atomic species, $\tau_v = \int \sigma_i N_i \mathrm{d}l$. The story becomes more complicated when τ_v is very large because natural broadening of the lines becomes important. Astronomers work in terms of the equivalent width, W, of the absorption lines, which is just the amount of energy extracted from the continuum expressed as a linewidth,

$$W = \int \left(1 - \frac{I_\nu}{I_{\nu c}}\right) d\nu \tag{17.7}$$

where $I_{\nu c}$ is the continuum spectrum expected in the absence of the absorption line. The relation between W and the column density of the species is known as the *curve of growth*.

Ultraviolet observations of this type have resulted in a number of important discoveries about the nature of the interstellar gas. Among these, the following highlights are of special significance:

1 Molecular hydrogen, H_2, has been discovered in large quantities in the interstellar gas, but there are wide variations in its abundance relative to atomic hydrogen. As discussed above, H_2 molecules can only survive if they are shielded from optical and ultraviolet photons in regions with density $N_H \geq 10^9$ m^{-3}.

2 The interstellar abundances of the heavy elements are less than their cosmic values by factors of up to $10^3 - 10^4$. A considerable fraction of these 'missing' elements is likely to be locked up in interstellar dust grains.

3 Atomic deuterium has been detected with abundance relative to neutral hydrogen of about 1.5×10^{-5}. This value is remarkably constant wherever deuterium has been detected in the interstellar gas. This is a very high abundance for such a fragile element. A convincing case can be made that this abundance of deuterium was synthesised in the first three minutes of the Hot Big Bang (see, for example, Audouze (1986), Longair (1989)).

4 Highly ionised oxygen, OVI, has been detected as a broad absorption feature in the spectra of the majority of hot stars studied by Copernicus. This is direct evidence for a hot component of the intersteilar gas having $2 \times 10^5 \leq T \leq 10^6$ K. Similar broad features have been observed in the lines of CIV in the spectra of halo stars and of B stars in the Magellanic Clouds. These are attributed to absorption in a highly ionised, hot gaseous halo about our Galaxy.

17.2.4   *X-ray absorption*

We have already described the process of photoelectric absorption in Section 4.2. We merely repeat here that, if a strong cut-off is observed in the spectrum of an X-ray source at energies $h\nu \leq 1$ keV, there is a strong probability that it is due to photoelectric absorption. If the standard cosmic abundances of the elements are assumed, the absorption coefficient shown in Fig. 4.2 is obtained, which shows the characteristic K-absorption edges of the common elements. A useful approximation to that absorption curve is

$$\tau_x = 2 \times 10^{-26} \left(\frac{h\nu}{1 \text{ keV}}\right)^{-8/3} \int N_H dl \tag{17.8}$$

where $\int N_H dl$ is referred to as the *column depth*, expressed in hydrogen atoms per square metre, and $h\nu$ in kilo-electron-volts. Notice that the absorption may take place within the source itself or in the intervening medium, for example, in our own Galaxy.

17.3   Ionised interstellar gas

17.3.1   *Thermal bremsstrahlung*

We have already discussed this mechanism in some detail (Section 3.4). We need only recall here the characteristic signature of the emission process that the spectrum is flat up to frequencies $h\nu \approx kT$, above which there is an exponential cut-off and that the intensity of radiation per unit bandwidth depends upon the combination of parameters, $N_e^2 T^{-1/2}$. Thus, the total bremsstrahlung observed along a given line of sight is

$$I_\nu = A \int N_e^2 T^{-1/2} \mathrm{d}l \tag{17.9}$$

For diffuse emission regions, this mechanism is of most importance at radio and X-ray wavelengths. At radio wavelengths, diffuse regions of ionised hydrogen at $T \approx 10^4$ K are strong sources of bremsstrahlung. If the region is compact, the region becomes optically thick and the absorption coefficient can be derived using Kirchhoff's law (Section 3.5.3). In the most compact regions of ionised hydrogen, which are found in the vicinity of regions of star formation, the radio spectrum has the form $I_\nu \propto \nu^2$ at centimetre wavelengths, the signature of bremsstrahlung absorption (see Fig. 3.4). If the form of spectrum shown in Fig. 3.4 is observed, both T and N_e can be found, provided the source is homogeneous. At the very lowest radio frequencies, $\nu \leq 10$ MHz, thermal bremsstrahlung absorption by the diffuse interstellar gas becomes important, and then the Galactic plane is observed in absorption against the background of Galactic non-thermal radio emission (Purton (1966), Bridle and Purton (1968)).

At X-ray wavelengths, bremsstrahlung has been observed from the diffuse intergalactic gas in rich clusters of galaxies (Fig. 17.5) and from the shells of supernova remnants. What makes these assertions particularly convincing is that the emission lines of very highly ionised species, such as FeXXV, are also observed in these sources, confirming the presence of a very hot gas with $T \approx 10^7 - 10^8$ K. There is also a strong possibility that part of the soft X-ray emission from the plane of the Galaxy is diffuse thermal bremsstrahlung from the hot component of the interstellar gas, which is also responsible for the ultraviolet OVI line, the temperature of which would have to lie in the range $(1 - 3) \times 10^6$ K.

17.3.2   *Permitted and forbidden transitions in gaseous nebulae*

Strong emission lines are observed from high-density regions of the interstellar gas which are excited by the ultraviolet emission of hot stars. These may be either regions in which massive young stars have formed or the vicinity of hot dying stars, such as the central stars in planetary nebulae. The mechanisms of heating and ionising the gas are photoexcitation and photoionisation, that is,

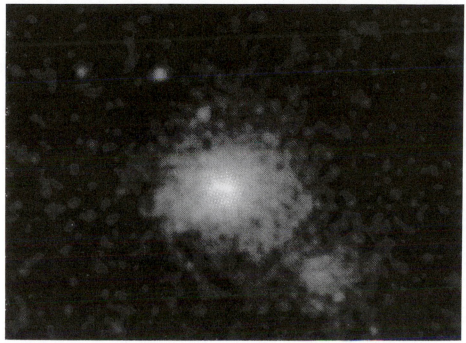

Figure 17.5. The X-ray emission from the Coma cluster of galaxies observed by the ROSAT X-ray Observatory. The observations were made in the energy band 0.5–2.4 keV and show the extended diffuse emission from the hot intracluster gas. (Courtesy of Prof. J. Trümper, Max-Planck-Institut für Extraterrestrische Physik, Garching, Germany.)

exactly the same processes described in Section 4.2 but at much lower energies, specifically at energies $h\nu \geq 13.6$ eV $= E_\mathrm{I}$, the ionisation potential of hydrogen. In the process of photoionisation, photons in the high energy tail of the Planck distribution with energy $h\nu \geq E_\mathrm{I}$ are responsible for the ionisation of the gas. The reason for this is the large cross-section of hydrogen atoms for photoionisation by photons with energies $h\nu \geq E_\mathrm{I}$. The resulting temperature of the ionised gas about the hot star is very much less than E_I/k, partly because, in a simple approximation, it can be shown that $T_\mathrm{gas} \approx T_\star$, where $T_\star$ is the effective temperature of the stellar atmosphere, and partly because of cooling by line emission. Thus, typical temperatures in the gas are about 5000–20000 K, compared with $T_\mathrm{gas} = 10^5$ K, which would be necessary for collisional ionisation of neutral hydrogen, that is, $kT_\mathrm{gas} \approx E_\mathrm{I}$. Osterbrock's book *The astrophysics of gaseous nebulae and active galactic nuclei* (1989) can be strongly recommended, both for its clear exposition of the basic atomic physics involved and of how emission lines can be used as diagnostic tools to measure physical conditions in gaseous nebulae, such as regions of ionised hydrogen, planetary nebulae, the shells of supernova remnants and the environments of active galactic nuclei.

Hydrogen recombination lines are among the strongest lines observed in the spectra of gaseous nebulae and are responsible for a large part of their cooling.

The ratio of intensities of the Balmer lines is known as the *Balmer decrement*, and it is relatively insensitive to physical conditions, unless the particle densities are very high, $N_e \geq 10^{14}$ m^{-3}, when the effects of self-absorption and collisional excitation of the Balmer series become important. The intensities of the hydrogen recombination lines do not provide direct information about the particle densities in the line-emitting regions. For example, the Hβ line of the Balmer series, in which the principal quantum number, n, changes from 4 to 2 and which has wavelength $\lambda = 486.1$ nm, is one of the strongest lines in the spectra of regions of ionised hydrogen. The line intensity is

$$L(H\beta) = N_e N_p \alpha h \nu_{H\beta} V \epsilon$$
$$= 2.28 \times 10^{-26} N_e^2 T_e^{-3/2} b_4 \epsilon V \exp(9800/T_e) \quad \text{W} \qquad (17.10)$$

where α is the recombination coefficient appropriate to the Hβ transition, V is the volume of the source, b_4 is a factor representing the departure of the population of the upper level of the Hβ transition from thermal equilibrium, T_e is the electron temperature of the gas and ϵ is the *filling factor*, which is the fraction of the volume of the source which is filled with gas; if the volume is uniformly filled with gas, $\epsilon = 1$. The intensity of the Hβ line thus measures $\int N_e^2 T^{-3/2} dl$ through the source region. Values for b_4 are given in tables by Pengelly (1964). For a temperature $T \approx (1 - 2) \times 10^4$ K, b_4 lies in the range 0.1–0.4 depending upon the precise physical conditions. There is no direct way of disentangling N_e from this study without further physical considerations.

Hydrogen recombination lines have been observed from the diffuse warm component of the interstellar gas. According to Reynolds (1990), the diffuse Hα emission appears to cover the entire sky and, at galactic latitudes $|b| > 10°$, follows the cosec$|b|$ law expected of the emission from a thin disc. The intensity of this emission provides a measure of $\int N_e^2 dl$, whereas the dispersion measures of pulsars determine $\int N_e dl$ (see Section 17.3.3) so that the clumpiness of the ionised gas can be found. Further information on the temperature and density of the diffuse ionised gas is obtained from the observation of forbidden lines of [NII], [SII] and [OIII]. The properties of the diffuse warm gas responsible for these lines and the diffuse Hα emission are similar to those labelled 'Intercloud medium' in Table 17.1.

Another application of hydrogen recombination lines is in the study of very high order transitions $n \rightarrow n - 1$, $n \geq 100$, which result in photons with energies in the radio waveband. These have been detected from many diffuse regions of ionised hydrogen and provide a further probe of physical conditions. Because the radio emission is not attenuated by interstellar dust, it provides a valuable tool for studying distant regions of ionised hydrogen, the presence of which are only known from their radio bremsstrahlung. Since the line widths are narrow, the radio recombination line velocities can be used as spiral arm tracers in the more distant parts of the Galaxy (Fig. 17.3(*a*)). Remarkably, similar recombination lines have been observed at low radio frequencies, $\nu \sim 15$–30 MHz associated

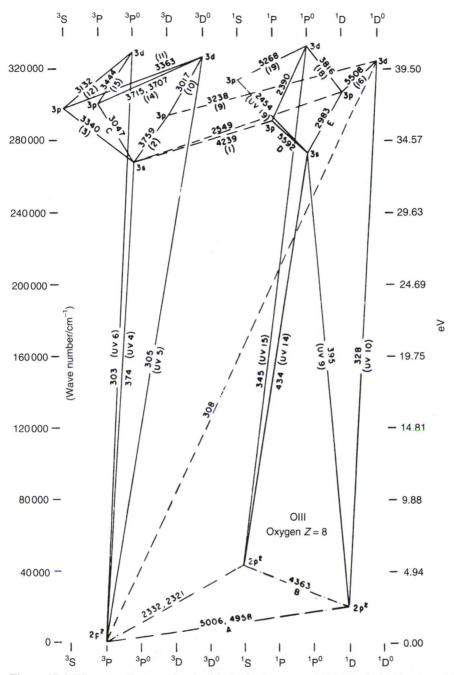

Figure 17.6. The term diagram for doubly ionised oxygen, OIII. The forbidden transitions observed in the optical waveband originate from low-lying levels associated with the ^{1}S and ^{1}D configurations of the $2p^2$ shell electrons. (From C.E. Moore and P.W. Merrill (1968). *Partial Grotrian diagrams of astrophysical interest*, p. 24. US Dept of Commerce, National Bureau of Standards.)

Table 17.3. *The critical densities for collisional de-excitation of some common ions*
All values are calculated for $T = 10\,000$ K.

Ion	Level	Critical density (N_e/m^{-3})
CII	$^2P_{3/2}$	8.5×10^7
CIII	$^3P_2$	5.4×10^{11}
NII	$^1D_2$	8.6×10^{10}
NII	$^3P_2$	3.1×10^8
NII	$^3P_1$	1.8×10^8
NIII	$^2P_{3/2}$	3.2×10^9
NIV	$^3P_2$	1.4×10^{12}
OII	$^2D_{3/2}$	1.6×10^{10}
OII	$^2D_{5/2}$	3.1×10^9
OIII	$^1D_2$	7.0×10^{11}
OIII	$^3P_2$	3.8×10^9
OIII	$^3P_1$	1.7×10^9
NeII	$^2P_{1/2}$	6.6×10^{11}
NeIII	$^1D_2$	7.9×10^{12}
NeIII	$^3P_0$	2.0×10^{10}
NeIII	$^3P_1$	1.8×10^{11}
NeV	$^1D_2$	1.6×10^{13}
NeV	$^3P_2$	3.8×10^{11}
NeV	$^3P_1$	1.8×10^{11}

From D.E. Osterbrock (1989). *The astrophysics of gaseous nebulae and active galactic nuclei*, p. 65. Mill Valley, CA: University Science Books.

with the recombination of carbon atoms but with very large principal quantum numbers, for example, $n = 631$ at 26.12 MHz and $n = 768$ at 14.7 MHz.

The other strong emission lines observed in the optical spectra of gaseous nebulae are the *forbidden lines*. Because the gas in gaseous nebulae is relatively cool ($T_e \approx 5000$–$20\,000$ K), collisions can only excite those energy levels within a few electron-volts of the ground state. For the common elements, such as C, N, O, Ne, S, the only accessible levels are various metastable levels which have excitation potentials less than about 5 eV. In these elements, the low-lying levels are associated with two, three or four electrons in incomplete p shells. An example of such a term diagram, that of doubly ionised oxygen, O^{++} or

OIII, is shown in Fig. 17.6, in which there are two $2p^2$ states within 5 eV of the ground state. The only way in which electrons in these levels can return to the ground state by a radiative transition is through the transitions, shown on the Grotrian diagram, which violate the rules for electric dipole transitions, that is, they are *forbidden* transitions. These levels above the ground state can become highly populated by electron collisions in a low density plasma because there are no selection rules for the collisional excitation of an atom or ion. This large population of ions in these metastable states is more than enough to compensate for the small spontaneous transition probability for magnetic dipole or electric quadrupole transitions between these levels and accounts for the high intensities of the forbidden emission lines.

Another class of transition which violates the selection rules for electric dipole transitions, but which is important in studies of quasar spectra, is the class of *semi-forbidden transitions*, which are less highly forbidden than the above examples. These transitions result in *intercombination lines*, in which only a single selection rule is violated. A well-known example is the semi-forbidden transition associated with doubly ionised carbon, which is denoted CIII] λ190.9 nm.

Forbidden lines provide sensitive diagnostic tools for determining densities and temperatures in emission line regions because the observability of the lines is strongly dependent upon these parameters. The strengths of the lines are determined by the competing processes by which de-excitation takes place following excitation by electron collisions. If the density is low, *radiative de-excitation* results in the emission of a photon and then the intensity of the line is proportional to the rate of collisional excitation. If, however, the density is high, de-excitation by electron collisions is more important, and this leads to suppression of the strength of the emission line. There is thus a *critical density*, above which the strength of the forbidden line emission is rapidly quenched – critical densities for a number of the common ions are listed in Table 17.3 (from Osterbrock (1989)). Critical densities can also be evaluated for the semi-forbidden lines, and, because of their greater spontaneous transition probabilities, much greater electron densities can be studied. For example, for C III], the critical density is $N_e \approx 10^{16}$ m^{-3}. In order to make estimates of parameters such as the electron density and electron temperature, it is essential to measure the *ratios* of different forbidden lines originating from the same region. The more lines with different critical densities which can be used, the better the physical conditions within the emission line region can be determined. More detailed studies involve using line ratios among the low-level forbidden lines of a particular ion which are sensitive to both density and temperature. Osterbrock (1989) provides an elegant description of the techniques by which this can be achieved. Notice that, in contrast to other techniques, this method enables particle densities to be determined directly in the regions under study. Normally these techniques are used to investigate the properties of regions with relatively high densities, $N_e \geq 10^9$ m^{-3} but Reynolds' investigations (1990) show that the same procedures can be used to study the warm diffuse interstellar medium.

17.3.3   *The dispersion measure of pulsars*

Estimates of the column density of free electrons in the Galaxy, $\int N_e dl$, may be obtained from the delay time in the arrival of radio signals as a function of frequency. In a plasma, a wave packet propagates at the group velocity v_{gr}, which is a function of frequency. At frequencies well above the gyrofrequency of the electrons in the plasma, $v \gg v_g$, the group velocity depends only upon the plasma frequency, v_p, and is given by $v_{gr} = c[1 - (v_p/v)^2]^{1/2}$, where

$$v_p = \left(\frac{e^2 N_e}{4\pi^2 \epsilon_0 m_e} \right)^{1/2} = 8.98 N_e^{1/2} \text{ Hz} \qquad (17.11)$$

where N_e is measured in electrons per cubic metre. At radio wavelengths, $v \approx 10^2 - 10^3$ MHz, $v_p/v \ll 1$ and hence

$$v_{gr} = c \left[1 - \frac{1}{2} \left(\frac{v_p}{v} \right)^2 \right] \qquad (17.12)$$

Therefore, if a pulse of radio waves is emitted at time $t = 0$, the arrival time of the signals, T_a, is a function of frequency; that is,

$$T_a = \int_0^l \frac{dl}{v_{gr}} = \int_0^l \frac{dl}{c} \left[1 + \frac{1}{2} \left(\frac{v_p}{v} \right)^2 \right] = \frac{l}{c} + \frac{e^2}{8\pi^2 \epsilon_0 m_e c} \frac{1}{v^2} \int_0^l N_e dl \qquad (17.13)$$

Thus, by measuring the arrival time of the pulse, T_a, as a function of frequency, v, the electron column density along the line of sight to the source, $\int N_e dl$, can be found. Inserting the numerical values of the constants into equation (17.13), we find

$$T_a = 4.15 \times 10^9 \frac{1}{v^2} \int N_e dl \quad \text{s}$$

where the electron density is measured in electrons per cubic metre, the distance l in parsecs and v in hertz.

For the test to be practicable, it is essential to observe sources which emit sharp pulses of radiation over a wide range of frequencies. Pulsars (Section 15.4) are ideal for this purpose, and estimates of $\int N_e dl$, which is known as the *dispersion measure*, are readily made for all of them. These data provide estimates of $\int N_e dl$ in about 500 separate directions through the interstellar gas. If it is assumed that the electron density is uniform in the plane of the Galaxy, the dispersion measure provides an estimate of the distance of the pulsar. Improved distances can be found by adopting a more detailed picture for the distribution of the ionised gas in the Galaxy (Taylor and Cordes (1993)).

17.3.4   *Faraday rotation of linearly polarised radio signals*

The partially ionised interstellar gas is permeated by the Galactic magnetic field and hence constitutes a magnetised plasma or *magnetoactive medium*. Under typical interstellar conditions, both plasma frequency, $v_p = 8.98 N_e^{1/2}$ Hz, and the gyrofrequency, $v_g = 2.8 \times 10^{10} B$ Hz, where B is measured in tesla, are much less than typical radio frequencies, $10^7 \geq v \geq 10^{11}$ Hz. Under these conditions, the position angle of the electric vector of linearly polarised radio emission is rotated on propagating through a region in which the magnetic field is uniform. This phenomenon is known as *Faraday rotation*.

This phenomonon results from the fact that the modes of propagation of radio waves in a magnetised plasma are elliptically polarised in opposite senses, that is, they are right- and left-hand elliptically polarised waves. This arises because, under the influence of a perturbing electric field, the electrons are constrained to move in spiral paths about the magnetic field direction (see Section 11.1). Therefore, when a linearly polarised signal is incident upon a magnetoactive medium, it can be resolved into equal components of oppositely handed elliptically polarised radiation. In the limit $v_g/v \ll 1$, the refractive indices, n, of the two modes are

$$n^2 = 1 - \frac{(v_p/v)^2}{1 \pm (v_g/v)\cos\theta} \tag{17.14}$$

where θ is the angle between the direction of wave propagation and the magnetic field direction. Therefore, the phase velocities of the two modes are different and one sense of polarisation runs ahead of the other. When the elliptically polarised components are added together at depth l through the region, they result in a linearly polarised wave, rotated with respect to the initial direction of polarisation. From the above dispersion relation, the difference in refractive indices under the conditions $v_p/v \ll 1, v_g/v \ll 1$, is

$$\Delta n = \frac{v_p^2 v_g}{v^3}\cos\theta \tag{17.15}$$

On propagating a distance $\mathrm{d}l$ through the region, the phase difference between the two modes is

$$\Delta\phi = \frac{2\pi v \Delta n}{c}\,\mathrm{d}l$$

On summing the two elliptically polarised waves, the direction of the linearly polarised electric vector is rotated through an angle $\Delta\theta = \Delta\phi/2$, that is,

$$\Delta\theta = \frac{\pi v_p^2 v_g}{cv^2}\cos\theta\,\mathrm{d}l$$

For $v_g \cos\theta$, we may write $2.8 \times 10^{10} B_\parallel$ Hz, where $B_\parallel$ is the component of B parallel to the line of sight, measured in tesla. Therefore, we find

$$\theta = \frac{\pi}{cv^2}\int_0^l v_p^2 v_g \cos\theta\,\mathrm{d}l \tag{17.16}$$

or, rewriting the formula in more convenient astronomical units,

$$\theta = 8.12 \times 10^3 \lambda^2 \int_0^l N_e B_\parallel\,\mathrm{d}l \tag{17.17}$$

where θ is measured in radians, λ in metres, N_e in particles per cubic metre, $B_\parallel$ in tesla and l in parsecs. Thus, measurement of the quantity θ/λ^2, which is called the *rotation measure* (in radians per square metre) gives information about the integral of $N_e B_\parallel$ along the line of sight. In addition, the sign of the rotation gives information about the weighted mean direction of the magnetic field along the line of sight. If θ/λ^2 is negative, the magnetic field is directed away from the observer; if θ/λ^2 is positive, the field is directed towards the observer.

It is fortunate that many Galactic and extragalactic radio sources emit linearly polarised radio emission, and therefore, by measuring the variation of the position angle of the electric vector with frequency, estimates of $\int N_e B_\parallel dl$ may be obtained for many different lines of sight through the Galaxy. An attractive method for making a rough estimate of the strength of the Galactic magnetic field is to combine observations of the Faraday rotation of the linearly polarised emission of pulsars with measurements of their dispersion measures. The former gives an estimate of $\int N_e B_\parallel dl$ and the latter $\int N_e dl$. We can therefore obtain a weighted estimate of the strength of the magnetic field along the line of sight:

$$\langle B_\parallel \rangle \propto \frac{\text{rotation measure}}{\text{dispersion measure}} \propto \frac{\int N_e B_\parallel dl}{\int N_e dl} \tag{17.18}$$

In addition to rotation of the plane of polarisation, the radio emission is expected to become *depolarised* as the wavelength increases. If the radio emission originates from a region of size l, in which the magnetic field B and the plasma density N_e are uniform, the radiation will remain fully polarised at high enough frequencies because internal Faraday rotation within the region is proportional to λ^2, which tends to zero as the wavelength tends to zero. At long wavelengths, however, because there is substantial rotation of the plane of polarisation through the source region, the polarisation vectors originating from different depths through the region add up at different angles as the radiation leaves the source. Evidently, when the plane of polarisation of the radiation is rotated by $\theta = \pi$ radians through the source region, the net degree of polarisation decreases. In this simple picture, the frequency at which significant depolarisation is observed provides information about the integral of $N_e B_\parallel l$ through the source region. Notice that, whereas the rotation of the plane of polarisation provides information about the $\int N_e B_\parallel dl$ from the source to the Earth, the depolarisation provides information about the source regions themselves. This process is often referred to as *Faraday depolarisation*.

Unfortunately, the above analysis only applies to the very simplest magnetic field configurations, and it becomes complicated when account is taken of the real magnetic field structure within the source region. For example, if there are irregularities or fine structure in the magnetic field and plasma distribution, we have to add the contribution of each region to the total polarisation. In addition, if the magnetic field distribution is stretched in some direction, we may obtain polarised emission, but the depolarisation would be model dependent in the sense that what is observed would depend upon how the electric field vectors are rotated on passing through different regions within the source. We will take this story up in more detail when we discuss the interpretation of the polarised radio emission observed in extragalactic radio sources (Volume 3). Evidently, the higher the angular resolution of the observations, the better the intrinsic magnetic field structure can be determined. It is generally true that, with increasing angular resolution, the percentage polarisation per resolution element increases as the effects of irregularities within the beam become smaller.

17.4   The Galactic magnetic field

17.4.1   *Faraday rotation in the interstellar medium*

We have already shown in Section 17.3.4 how measurements of the Faraday rotation of the plane of polarisation of polarised radio waves gives information about the rotation measure, which is proportional to the quantity $\int N_e B_{\parallel} dl$ along the line of sight to a radio source. A plot of the magnitude of the rotation measure as a function of galactic latitude, b, shows that the rotation measures of extragalactic radio sources increase towards low galactic latitudes (Fig. 17.7(a)). If the Galactic magnetic field were uniform and ran parallel to the plane of the Galaxy, and if the electron density were uniform, the path length through the Galactic disc would be proportional to cosec b and the component of the magnetic field along the line of sight would be proportional to $\cos b$. Therefore, it would be expected that the rotation measure would vary as $\cot |b|$. This relation provides a reasonable upper envelope to the distribution of points in Fig. 17.7(a). This indicates that most of the Faraday rotation of extragalactic radio sources originates in our own Galaxy rather than in the sources themselves. Note, however, that there is a large scatter in the values of the rotation measures at a given galactic latitude; in particular, even at low galactic latitudes, there are some sources with very small rotation measures. There must, therefore, be considerable irregularities in the distribution of the product $N_e B_{\parallel}$ along the line of sight.

If the magnitudes and signs of the rotation measures are plotted in galactic coordinates (Fig. 17.7(b)), there is general clustering of rotation measures of the same sign in different directions, which is good evidence that there is some overall order in the Galactic magnetic field. The signs of the rotation measures seem to change about galactic longitude 180°, particularly in the southern Galactic hemisphere, suggesting that the parallel component of magnetic field changes direction at this longitude. This evidence is consistent with a model in which the magnetic field runs predominantly parallel to the plane of the Galaxy in the direction of the local spiral arm. The sense of the field is such that it points away from the Earth in the direction of galactic longitude roughly 90°. That the magnetic fields in some spiral galaxies run parallel to the spiral structure is beautifully illustrated by high sensitivity radio observations of the galaxy M51 by N. Neininger (Fig. 17.8). These observations were made at a wavelength of 2.8 cm at which the Faraday rotation of the electric vectors is negligible. The figure shows the direction of the magnetic field derived from the polarisation observations, and the magnetic field structure follows very closely the spiral pattern of M51.

Another use of this technique is to combine the rotation measures of pulsars with their dispersion measures, which provide measures of $\int N_e dl$. If attention is restricted to pulsars at distances less than 2 kpc in the Galactic plane, it is found that they are consistent with a uniform magnetic field of strength 2.5×10^{-10} T

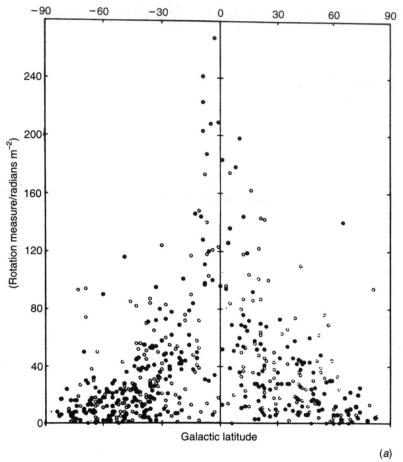

Figure 17.7 (*a*). The variation of the rotation measures of extragalactic radio sources with galactic latitude. It can be seen that the largest rotation measures are found close to the Galactic plane. (From J.B. Whiteoak (1972). In *Galactic astronomy*, eds F.J. Kerr and S.C. Simonson, p. 139. Dordrecht: D. Reidel and Co.)

running parallel to the Galactic plane in the direction of longitude $l = 90°$ (Heiles (1976)). It is apparent, however, that there are also large-scale irregularities in the field on large and small angular scales.

17.4.2   *Polarisation of starlight*

The optical polarisation of stars increases with distance through the Galactic plane and is strongly correlated with extinction by interstellar dust. Dust grains between a star and the Earth can polarise starlight if they are elongated and aligned over large regions of space. The elongated grains scatter preferentially that polarisation of the incident light waves which has its electric field vector parallel to the long axis of the grain. Therefore, the transmitted radiation is polarised parallel to the minor axis of the grains.

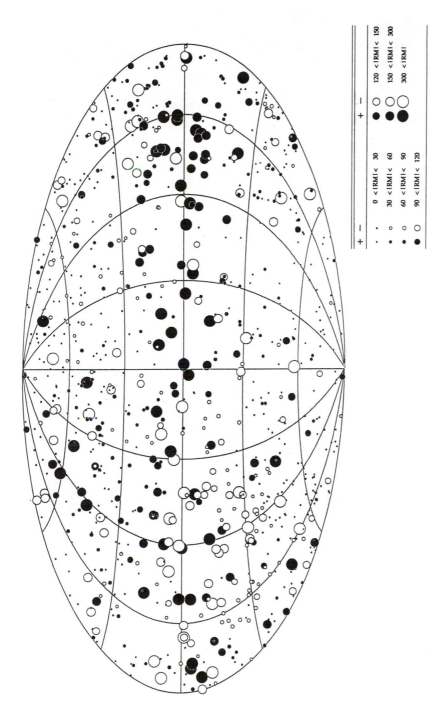

−	+			−	+	
·	·	0 <|RM| < 30		○	●	120 <|RM| < 150
◦	•	30 <|RM| < 60		○	●	150 <|RM| < 300
○	●	60 <|RM| < 90		○	●	300 <|RM|
○	●	90 <|RM| < 120				

(b)

Figure 17.7 (b). The magnitudes and signs of the rotation measures of 976 extragalactic radio sources plotted in galactic coordinates. (Data assembled by P.P. Kronberg and published by R. Wielebinski (1993). *The cosmic dynamo*, ed. F. Krause. Dordrecht: Kluwer Academic Publishers.)

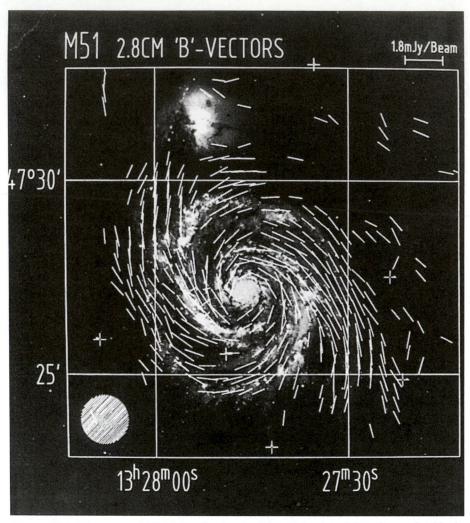

Figure 17.8. The magnetic field distribution in the spiral galaxy M51 observed with the Effelsberg 100-m telescope at a wavelength of 2.8 cm. The angular resolution of the observations is 72 arcsec and is indicated by the circle at the bottom left of the map. (From N. Neininger (1992). *Astron. Astrophys.*, **263**, 30.)

The magnetic field of the Galaxy is the vehicle for aligning the dust grains over large regions of space. Following Spitzer (1968), we can distinguish two separate phenomena which contribute to the alignment mechanism. First, if we describe elongated dust grains by prolate spheroids with principal axes $a_1 > a_2 = a_3$, in thermodynamic equilibrium their angular momentum vectors tend to become aligned with the minor axes of the grains. This may be understood as follows. If the principal moments of inertia about the axes of the grain are I_1, I_2 and I_3, the moment of inertia about the major axis, I_1, is smaller than those about the minor axes, I_2 and I_3. Let $I_2 = I_3 = \gamma I_1$, where $\gamma > 1$. In statistical equilibrium, the rotational energy about each principal axis is the same, $\frac{1}{2}I_1\omega_1^2 = \frac{1}{2}I_2\omega_2^2 = \frac{1}{2}I_3\omega_3^2$, and therefore $I_2\omega_2 = I_3\omega_3 = \gamma^{1/2}I_1\omega_1$. Therefore, the angular momentum vectors of the rotating grains in equilibrium lie preferentially perpendicular to the major axis of the grain.

In the second effect, rotation about an axis parallel to a magnetic field is energetically more favourable than rotation about an axis perpendicular to the field. In the former case, during rotation, the magnetisation of a paramagnetic material does not change, whereas, in the latter case, the direction of magnetisation changes continuously, and there are internal couples which resist such changes. Thus, rotation about an axis parallel to the magnetic field is energetically preferable. The exact efficiency of alignment depends upon the chemical composition and magnetic properties of the grains.

The combination of these two effects is that elongated grains rotate with their minor axes parallel to the magnetic field direction. Therefore the electric vector of the transmitted radiation is parallel to the magnetic field direction. From a study of the polarisation properties of about 6000 stars, D.S. Mathewson and V.L. Ford derived the map of their polarisation vectors shown in Fig. 17.9, the lengths of the lines indicating the percentage polarisation. All stars plotted lie within 3 kpc of the Sun. It is evident that the magnetic field runs predominantly parallel to the Galactic plane. These observations have suggested that the uniform magnetic field component runs in the general direction of the local spiral arm, $l \approx 50°$–$80°$, and there is reasonable agreement with radio observations of rotation measures and of the polarised Galactic radio background emission. There are also large-scale irregularities in the magnetic field distribution, some of which are associated with Galactic loops such as the North Polar Spur, the prominent feature which extends towards the North Galactic Pole from $l \approx 30°$ (see also Fig. 1.8(a)).

To estimate the magnetic field strength from optical polarisation studies, a detailed theory of the alignment mechanism is required, which, in turn, depends upon the magnetic properties of the grains. Early models of paramagnetic grain alignment required larger magnetic field strengths, $B \sim 10^{-9}$ T, than those inferred from observations of the Faraday rotation of extragalactic radio sources, but more recent modifications of the theory, including the possibility of ferromagnetic grains, suggest that smaller fields, about 3×10^{-10} T, can produce the necessary alignment.

17.4.3 *Zeeman splitting of 21-cm line radiation*

The Galactic magnetic field strength may also be estimated from the Zeeman splitting of the 21-cm neutral hydrogen line. The observational problem is formidable since the splitting amounts to only 28 GHz T^{-1} and the expected magnetic field strengths are only about 10^{-9}–10^{-10} T. Thus, the radio spectrometers must be sensitive enough to detect splittings of about 10 Hz in 1420 MHz. Fortunately, when the magnetic field runs parallel to the line of sight, Zeeman splitting results in two circularly polarised components with opposite senses of circular polarisation on opposite sides of the line centre. The splitting is always much less

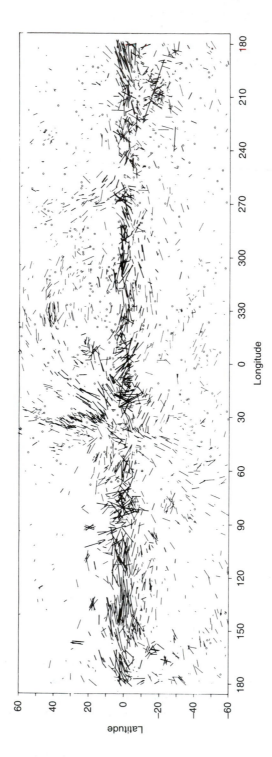

Figure 17.9. The polarisation of stars as a function of galactic coordinates. The magnitudes of the vectors indicate the strength of the polarisation, and the direction of the vectors indicate the planes of polarisation of the light. (From D.S. Matthewson and V.L. Ford (1970). *Mem. R. Astron. Soc.*, **74**, 143.)

than the width of the absorption line, and therefore the technique adopted is to observe an intense radio or millimetre absorption line and to search for an excess of oppositely circularly polarised radiation on either side of the line centre.

Magnetic field strengths have been measured by this technique using the 21-cm line of neutral hydrogen in the direction of a number of intense radio sources and are found to be greater than 10^{-9} T. It is probable that these strong magnetic fields are associated with the high density gas clouds responsible for the formation of the absorption line rather than with the general interstellar medium. Similar observations have been made of OH absorption lines, and even stronger magnetic field strengths, about 10^{-8} T, have been found. These high magnetic field strengths are likely to be associated with the dense clouds in which the OH absorption takes place.

17.4.4   *The radio emission from the Galaxy*

The final technique involves analysis of the diffuse Galactic radio emission and its polarisation. These are attributed to the synchrotron emission of ultrarelativistic electrons spiralling in the Galactic magnetic field. This topic will be studied in detail in Chapter 18. It will be shown that there are problems in deriving a unique value for the magnetic field strength from these observations, but the values are in broad agreement with other independent pieces of evidence.

17.4.5   *Summary of the information on the Galactic magnetic field*

The various techniques described above provide complementary information about different aspects of the Galactic magnetic field. The distribution of the rotation measures of pulsars and extragalactic radio sources and of the optical polarisation vectors are convincing evidence that there exists some large-scale order. In the vicinity of the Sun, the uniform component of the field runs roughly in the direction $l = 90°$ along the local spiral arm. There are, however, random components on a wide range of scales. All estimates of the magnetic field strength lie in the range $10^{-9} - 10^{-10}$ T, and a compromise value of $(2–3) \times 10^{-10}$ T is consistent with much of the evidence. In clouds, the Zeeman splitting experiments indicate that rather stronger magnetic fields are present.

17.5   Star formation

Star formation is perhaps the most important unsolved problem in astronomy, and pervades much of the astrophysics of the origin and evolution of galaxies. It is also important for high energy astrophysics because the star formation rate is related to the rate of formation of the heavy elements and to the frequency of supernovae. The explosions of supernovae in the vicinity of molecular clouds may also stimulate the star formation process.

The *initial mass function*, $\xi(M)$, describes the birth rate of stars of different masses. It is not trivial to determine this function observationally because stars are observed at widely differing stages of their evolution. The *luminosity function*

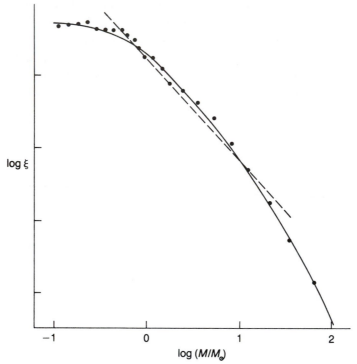

Figure 17.10. An estimate of the initial mass function of stars derived by Miller and Scalo and their best-fitting log-normal distribution. Also shown as a dashed line is the initial mass function of power-law form proposed by Salpeter (1955). (After G.E. Miller and J.M. Scalo (1979). *Astrophys. J. Suppl.*, **41**, 513–47.)

of stars, which describes the number of stars of different luminosities, can be determined and then converted into a mass function if the mass–luminosity relation is known. This function, however, underestimates the birth rate of stars more massive than $1M_\odot$ since they have lifetimes which are shorter than the age of the Galaxy, and so these statistics have to be corrected for the lifetimes of stars of different masses. A determination of the initial mass function for stars in the solar neighbourhood is shown in Fig. 17.10, from which it is apparent that it is a monotonically decreasing function of increasing mass. It is often convenient to adopt the *Salpeter initial mass function*, $\xi(M)\, \mathrm{d}M \propto M^{-2.35}\, \mathrm{d}M$, first derived by Salpeter (1955), which is a reasonable approximation for stars with masses roughly that of the Sun (dashed line in Fig. 17.10). More recent determinations have suggested that the function can be described by the log-normal distribution function proposed by G.E. Miller and J.M. Scalo:

$$\xi(\log M)\, \mathrm{d}M \propto \exp[-C_1(\log M - C_2)^2]\, \mathrm{d}M$$

where C_1 and C_2 are constants (see Fig. 17.10). This form of distribution has the intriguing feature that it describes random multiplicative processes. It is uncertain whether or not this function applies universally throughout the Galaxy and in other galaxies. With the advent of infrared array detectors, it is now becoming possible to determine the luminosity functions of stars in young clusters directly, for example, for the stars in the vicinity of the Orion Nebula (Fig. 8.17).

It is not known how the rate of star formation depends upon local physical conditions within a galaxy, for example, the temperature and density of the gas. As we show below, there is convincing evidence that stars form in cool giant molecular clouds. Ideally, one would like to determine the sequence of events which take place during the formation of stars empirically by observation. The basic problem is that, for massive stars, many of the processes involved in star formation take place over short timescales, $\leq 10^5$–10^6 years, compared with the lifetime of typical main sequence stars. This means that statistically there will always be relatively few star forming regions which can be observed at different stages in their evolution. To make matters worse, the key processes take place in the most obscured regions inside giant molecular clouds, on angular scales significantly less than the angular resolution of current telesopes. There has, nonetheless, been remarkable progress in studying these regions in the infrared and millimetre wavebands, in which the dust becomes transparent. The development of interferometers for the sub-millimetre waveband with very high angular resolution holds out the promise of being able to study many of the key processes directly.

17.5.1 *The contents of regions of star formation*

Fig. 17.11 shows a sketch of the typical contents of a region of star formation. Stars form within giant molecular clouds, the typical properties of which are given in Table 17.1. The giant molecular clouds have vastly greater sizes than the prominent regions of ionised hydrogen, which used to be thought of as the regions in which stars form. This is clearly illustrated in Fig. 13.21, in which the Orion Nebula is dwarfed by the Orion Molecular Cloud, which extends over about 16° on the sky, roughly the same size as the constellation of Orion. The giant molecular clouds in Orion are shown in higher resolution in Fig. 17.12, the Orion Nebula being the most prominent region of ionised hydrogen close to the densest region of the molecular cloud. These high angular resolution observations show that there is much fine structure within the molecular clouds, each density enhancement being a potential site of star formation. There are large quantities of dust associated with the clouds, which protect the interstellar molecules from being photodissociated by the interstellar flux of ionising radiation. This means that we tend to see optically only those regions of ionised hydrogen which lie close to the front surface of the clouds. The Orion Nebula, for example, is probably a 'blister' on the front surface of the Orion giant molecular clouds. The temperature of the molecular gas in the clouds is typically about 30–100 K.

The youngest objects which have been observed in the clouds are the hot far infrared sources. Optically, the Orion Nebula is by far the most prominent feature in the region of Orion, but, at far infrared wavelengths, most of the luminosity is associated with a region to the north-west of it, an object known as the Becklin–Neugabauer object (the diffuse region to the top right of the infrared image of

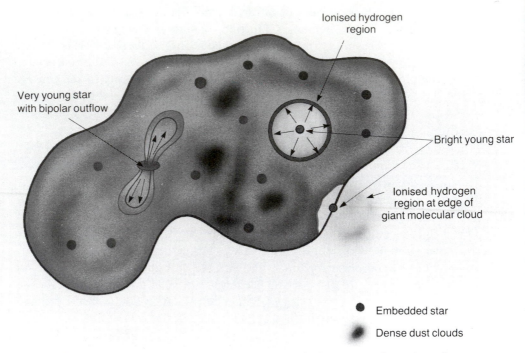

Figure 17.11. A schematic diagram illustrating the typical contents of a region of star formation. These include embedded stars, regions of ionised hydrogen, known as HII regions, protostars and bipolar outflow sources associated with protostars or very young stars. It is believed that the Orion Nebula is an example of an HII region which has been formed close to the nearside surface of the Orion giant molecular cloud. The sketch shows an example of this at the bottom right of the giant molecular cloud.

Orion in Fig. 8.17). This compact far infrared source has luminosity about 10^5 times that of the Sun. The spectrum is sharply peaked in the far infrared region of the spectrum, typical of the emission spectrum of reradiated dust (Fig. 13.21). The natural picture to account for these observations is that the source contains a very young star, or cluster of young stars, which is still embedded in the dust cloud out of which it formed. Unlike the case of the Orion Nebula, there is no region of ionised hydrogen surrounding the young stars, which suggests that the stars must be very newly formed, or even in the process of formation.

One of the intriguing questions is whether far infrared sources such as those in Orion are truly stars in the sense that hydrogen burning has begun in their cores or whether they are protostars. The distinction between the two cases is that, in the case of a *protostar*, the energy is derived from the dissipation of the gravitational binding energy of the star, that is, its energy of collapse, whereas, in the case of a young star, the energy source is the nuclear burning of hydrogen into helium. The search for protostars has been recognised as the holy grail of infrared and sub-millimetre astronomy from the time it was realised that the

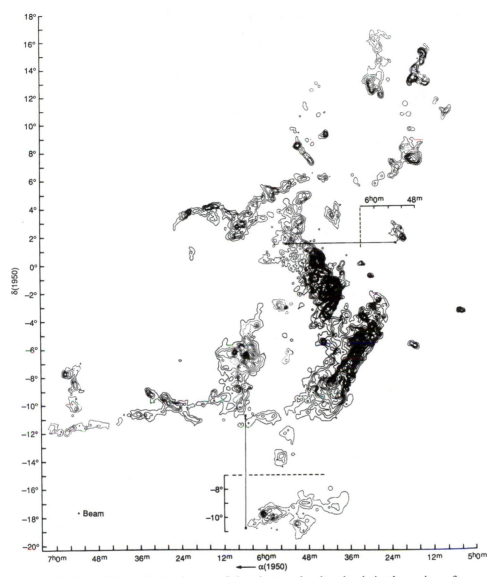

Figure 17.12. A high resolution image of the giant molecular clouds in the region of Orion. This contour map should be contrasted with that shown in Fig. 13.21. The higher resolution of this figure reveals that the giant molecular clouds contain many dense knots which may be the sites of the next generation of new stars. (R.J. Maddalena, M. Morris, J. Moscowitz and P. Thaddeus (1986). *Astrophys. J.*, **303**, 375.)

intense far infrared sources must be associated with what are among the youngest stellar systems known. No protostar has yet been identified with certainty.

The far infrared sources display a number of remarkable properties, perhaps the most surprising being the fact that virtually all of them are associated with *bipolar outflows*. Observations at millimetre and infrared wavelengths have shown that

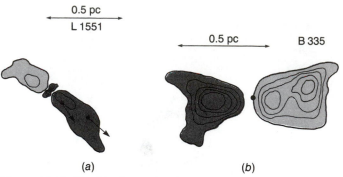

(a)   (b)

Figure 17.13. The bipolar molecular outflow sources L1551 and B 335, in which the molecular line emission originates from jets directed in opposite directions away from a central source, which is assumed to be a very young star or protostar. In both cases, one of the lobes is moving towards and the other away from the observer. The strong infrared sources associated with the sources are indicated by black filled circles. (*a*) In L1551, a molecular disc has also been observed in the CS line about the infrared source. The dots with arrows are two Herbig–Haro objects moving outwards in the direction of the outflow as measured from their proper motions. These objects are probably condensations in the outflow and are observed optically. (*b*) In B 335, the bipolar outflow appears to increase in width with distance from the centre, as is expected in a freely expanding jet. (From C.J. Lada (1985). *Ann. Rev. Astron. Astrophys.*, **23**, 307, 311.)

there are molecular outflows from these infrared sources, which are remarkably well collimated in opposite directions (Fig. 17.13). Measurements of the velocity of outflow from the Doppler shifts of the molecular lines show that the velocities are highly supersonic, velocities as large as 50 to 100 km s^{-1} being observed. Polarisation observations of the infrared molecular hydrogen emission in Orion show, in addition to a molecular hydrogen reflection nebula, polarisation vectors parallel to the molecular outflow. This is interpreted as evidence for a magnetic field in the outflow. A schematic representation of what is believed to take place in such a bipolar outflow is shown in Fig. 17.14. These outflows bear more than a passing resemblance to the jets associated with double radio sources in which the beams consist of relativistic material. Also, like the extragalactic radio sources, the jets are sometimes very narrow and, on occasion, one-sided.

Another remarkable feature of these sources has been the detection of molecular rings about the axis of the bipolar outflow (Fig. 17.13(*a*)). There has been speculation that this disc of molecular gas may be the material out of which a planetary system about the young star may be formed. Further evidence for this type of phenomenon has come from observations made with the IRAS satellite, which discovered what are believed to be dust discs about a number of the brightest stars in the sky. In the case of Vega, the third brightest star in the sky, a far infrared extended source has been observed with temperature about 80 K, which has been interpreted as a cool disc out of which a planetary system might form. Infrared speckle observations of young stars in a nearby low-mass star forming region have revealed similar phenomena. Further evidence for discs

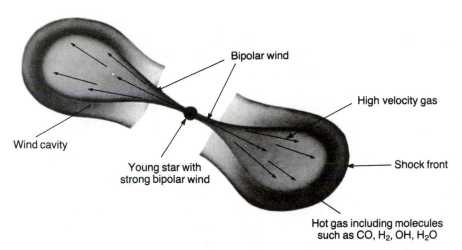

Figure 17.14. A schematic diagram illustrating the properties of a typical molecular outflow source. The outflow is supersonic and compresses the surrounding molecular gas. The heating of the molecular gas by the outflow and the cooling by molecular line emission results in a temperature of about 2000 K, at which it can be observed through its infared molecular line emission.

about young low-mass stars is provided by ^{13}CO observations of *T-Tauri stars*. These stars are observed just above the main sequence, and a convincing case can be made that they are newly formed and in the process of settling down to become normal main-sequence stars. Cool molecular discs have been observed about these stars, and in a few cases the disc has been resolved (Beckwith *et al.* (1990)). These discs extend to more than 1000 AU from the star.

17.5.2   *The problems of star formation*

There are three major problems which have to be solved in understanding how regions with densities of about 10^9 m^{-3} can collapse to form stars with about 10^{30} times greater densities. First of all, there is an *energy* problem. To form a stable star, the protostar must get rid of its gravitational binding energy. Secondly, any cloud possesses some angular momentum, and, because of conservation of angular momentum, the rotational energy increases during collapse. Unless there is some way of getting rid of angular momentum, the growth of rotational energy will halt the collapse in the equatorial plane (see Section 16.2.1). This is the *angular momentum* problem. Thirdly, if there is a *magnetic field* present in the collapsing cloud, its field strength is amplified during collapse, and this could become sufficiently strong to halt collapse in the equatorial plane. It will be recognised that these problems are remarkably similar to those encountered in the study of accretion onto compact objects (Chapter 16), and indeed there are many similarities between the two cases. These problems are discussed in detail in the review article by Shu, Adams and Lizano (1987).

Gravity ensures that, on a large enough scale, a gas cloud of any density and temperature is unstable. This instability is known as the *Jeans instability*. If a uniform medium is perturbed, the self-gravitation of the perturbation causes the region to collapse, but this is resisted by internal pressure gradients. The criterion for collapse is therefore that the gravitational force should exceed the internal pressure forces. A simple calculation shows how this comes about. The force of gravity acting on 1 m^3 of matter at the edge of the uniform cloud of mass M, radius R and density ρ is $\sim GM\rho/R^2$, while the force associated with the pressure gradient which prevents collapse is $dp/dr \sim p/R$. When the former exceeds the latter, collapse occurs. Now, since the speed of sound, c_s, is approximately $(p/\rho)^{1/2}$, and $M \sim \rho R^3$, the condition $GM\rho/R^2 > p/R$ reduces to $R \geq R_J = c_s/(G\rho)^{1/2}$. This characteristic length scale is known as the *Jeans length*, R_J, and is the largest scale a cloud can have before collapse under self-gravity becomes inevitable. Notice that the criterion depends only upon the speed of sound in the gas, c_s, and the density of the medium, ρ. We can therefore define a *Jeans mass* as the mass contained within the region which has scale R_J:

$$M_J \sim \rho R_J^3 \approx 10^5 \frac{T^{3/2}}{\mu_H^2 N^{1/2}} \quad M_\odot$$

where μ_H is the average mass of the particles contributing to the pressure relative to the mass of the hydrogen atom. The timescale for collapse of the unstable region is roughly $\tau \approx R_J/c_s \sim (G\rho)^{-1/2}$. I have given elsewhere a more formal derivation of these results, starting from the equations of gas dynamics coupled with Poisson's equation for the gravitational potential (Longair (1989)). In the case of instability in a disc, a similar result can be derived using the same arguments, Spitzer (1978) giving the criterion $R_J \geq c_s^2/G\rho(0)$, where $\rho(0)$ is the density in the midplane of the disc.

The values of the Jeans mass for the phases of the interstellar medium listed in Table 17.1 range from about 10000 to 20$M_\odot$ for the diffuse HI clouds at $T = 80$ K and the coldest molecular clouds at $T = 10$ K, respectively. Thus, the observed components of the interstellar medium are expected to condense into objects with roughly the masses needed to explain the formation of stars and star clusters. The Jeans mass decreases as the density of the cloud increases, and therefore the instability leads naturally to continued fragmentation as the cloud collapses. The fragmentation ceases when the cloud becomes optically thick to radiation, which is expected to occur for masses $M \sim 0.01M_\odot$. In fact, all the stars we observe have masses greater than $0.1M_\odot$, and this is attributed to the fact that stars with mass less than about $0.08M_\odot$ are not hot enough in their centres for nuclear burning to take place. Unfortunately, there is no direct evidence that this process of fragmentation actually occurs.

The Jeans criterion indicates that, when the region has increased in density to 10^{-16} kg m^{-3} at a temperature of 10 K, the Jeans mass corresponds to $1M_\odot$. According to small perturbation analyses, such a cloud simply collapses *en masse*, and this may well be what happens if the mass of the collapsing cloud is much greater than the Jeans mass. The rate of growth of the instability is, however, strongly dependent upon the density, and therefore, if a region of enhanced density within the collapsing cloud forms, it tends to collapse more rapidly. In the case of masses which just exceed the Jeans mass, computations of the collapse have shown that, even if the initial mass distribution is uniform, a density maximum rapidly forms in the central regions. The origin of this behaviour can be understood in terms of the propagation of rarefaction waves into the cloud which enhances the central density. The net result is that the central regions collapse more rapidly, forming a central core.

The material of the collapsing gas cloud consists of molecular gas and dust. During collapse, the material of the cloud is heated, and the energy can be lost very efficiently by radiation, so long as the cloud remains optically thin. When the central regions become optically thick to radiation, the radiation is trapped and the core begins to heat up. The core forms a pressure supported structure whilst the matter in the envelope continues to be accreted onto the core.

The evolution of the central core of the protostar is more or less independent of the behaviour of the accreting envelope. Fig. 17.15 shows what might be expected for the structure of the protostar. The core is indicated by the region labelled *hydrostatic core*, and the regions outside this are associated with the accretion flow. In the outer envelope, the matter and dust are optically thin and radiate away their thermal energy very efficiently. The collapse in this region is therefore close to isothermal. Eventually, the matter and dust densities increase to such values that the dust becomes optically thick and the region with optical depth unity is referred to as the *dust photosphere*. Inside this radius, there is a dust envelope within which the temperature increases with decreasing radius until it becomes so hot that the dust evaporates, at $T \approx 2300$ K for graphite grains, and then the radiative transfer is determined by the properties of the gas rather than the dust. The gas then falls in towards the hydrostatic core, and, since the latter acts as a 'solid body', an accretion shock is set up, which has the effect of dissipating the kinetic energy of the infalling gas and radiating away the binding energy of the gas. This radiation has to escape out through the accreting matter.

This picture makes it clear why protostars are expected to be intense infrared sources. The dissipation of the binding energy of the matter accreted onto the protostar is carried away by radiation from the accretion shock, and this energy is trapped and degraded in the dust envelope. The energy is radiated away at the temperature of the dust photosphere. Models such as those of Adams and Shu (1985) show a broad maximum corresponding to the superposition of the emission from grains at different temperatures in the regions of the photosphere with typical temperatures of about 100 K. These are similar to the spectra seen in a number of sources inferred to be protostellar objects (see Fig. 17.17).

Although this general picture can account for the observed properties of regions of star formation, the detailed evolution of stars from the collapse phase until they become hydrogen burning stars on the main sequence is complex

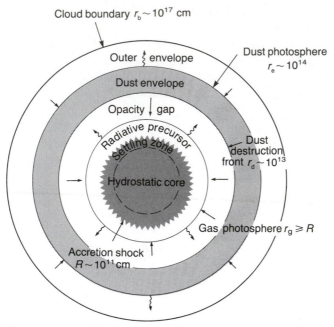

Figure 17.15. A diagram illustrating the structure of an accreting protostar according to the analysis of Shu and his colleagues. The various regions are described in the text. (From S.W. Stahler, F.H. Shu and R.E. Taam. (1980). *Astrophys. J.*, **241**, 641.)

and is strongly dependent upon the mass of the star. The origin of these differences lies in the characteristic timescales for the star. In addition to the collapse timescale, $\tau \approx R_\mathrm{J}/c_\mathrm{s} \sim (G\rho)^{-1/2}$, there is the timescale over which the star can radiate away its gravitational binding energy. The latter is known as the *Kelvin–Helmholtz timescale*, and is just $\tau_\mathrm{KH} \approx GM^2/RL$, where L is the luminosity of the star. For low-mass stars, this timescale is longer than the collapse time, and so the star undergoes a prolonged accretion phase. For a $1M_\odot$ star, the Kelvin–Helmholtz timescale is about 2×10^7 years compared with a collapse time of only about 10^6 years. On the other hand, for stars with mass greater than about $5M_\odot$, the Kelvin–Helmholtz timescale is shorter than the collapse time, and so the evolution and formation of the star is much more rapid, $\tau \sim 10^5 - 10^6$ years. Thus, for massive stars, their binding energies are rapidly radiated away and further accretion can take place directly onto the star.

It is intriguing that the timescales of formation for stars in a molecular cloud are similar to the main-sequence lifetimes of massive stars. Therefore, the massive stars can complete their evolution and explode as Type II supernovae while stars are still forming in the giant molecular cloud. This association is indeed found in galaxies in which Type II supernovae have been observed – they occur close to OB associations and giant HII regions in spiral arms. Huang and Thaddeus (1986) have shown that, in the outer regions of our own Galaxy, there is a strong correlation between supernova remnants and giant molecular clouds. They estimate that roughly half the supernova remnants in the outer Galaxy are associated with giant molecular clouds, which would be consistent with statistics which suggest that roughly half the supernovae which explode in a galaxy such as our own are of Type I and half of Type II. The intriguing possibility suggests itself that the supernova remnant itself may lead to the compression of the gas in the molecular cloud and hence to a new burst of star formation. This is referred to as *sequential star formation*. There is some evidence for this process taking place in star formation regions, but it is not by any means universal.

The problem of removing the angular momentum and magnetic fields in the collapsing cloud may be related to the formation of gaseous discs about protostars and young stars. The removal of angular momentum may take place in a manner similar to that described in Section 16.3 for accretion discs, namely, the transport of angular momentum outwards through the disc by viscous torques. The problem of expelling the magnetic fields is summarised by Shu, Adams and Lizano (1987) in their review.

A 'standard' picture of star formation has been described by Shu, Adams and Lizano (1987) which attempts to synthesise all these ideas into one general picture and which is illustrated schematically in Fig. 17.16. The process begins with the collapse of cool density enhancements within giant molecular clouds, and, in the early stages, the energy source in protostars and pre-main-sequence stars is the accretion of matter onto the core of the protostar rather than nuclear energy generation. Because the infalling matter is bound to have some angular momentum, a rotating disc forms perpendicular to the rotation axis. The removal of the gravitational binding energy of the accreted matter is effected by the reradiation of heated dust at far infrared wavelengths, at which the protostellar cloud is transparent as described above. At some stage, a stellar wind breaks out along the rotation axis of the system, creating a bipolar outflow. Finally, when the accretion phase is completed, all that is left is the newly formed star with a circumstellar disc. One of the more striking discoveries of the IRAS mission has been that objects with the spectral characteristics corresponding to each of these stages have been observed (Fig. 17.17). Objects at the earliest stages in their evolution are deeply embedded in the dust clouds out of which they formed, and these are purely far infrared sources. At later stages, the emission from the star and the disc or accretion envelope can be identified.

There are many unknowns in this picture. For example, a good case can be made that in some regions only low-mass stars are formed, as in the star forming region in the Taurus molecular cloud, while in others, such as Orion, both high- and low-mass stars are formed. Mezger and Smith (1977) argued that the sites of high-mass and low-mass stars are spatially distinct, with the OB stars forming in large cloud complexes, whereas objects like T-Tauri stars can form in large and small clouds distributed throughout the disc. This picture is supported by the analysis of Solomon, Sanders and Roivolo (1985), who find that the coldest molecular clouds do not contain stars with spectral types earlier than late B

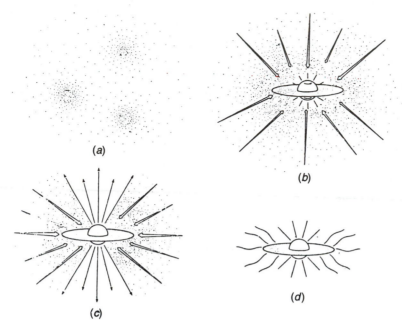

Figure 17.16. A schematic representation of a plausible scenario for the formation of stars. (*a*) Density inhomogeneities collapse under their own self-gravity. (*b*) The main accretion phase, in which an accreting core has formed and infall of matter onto that core takes place. The binding energy of the accreted matter is removed by radiation which is absorbed by dust and reradiated in the far infrared waveband. (*c*) Jets of material burst out of the accreting star along its rotation axis, producing the characteristic 'bipolar outflows' observed in most young stars. (*d*) The accretion of material ceases and the system is left with a young, hydrogen burning star and a rotating dust disc. (From F. H. Shu, F. C. Adams and S. Lizano (1987). *Ann. Rev. Astron. Astrophys.*, **25**, 23.)

and are spread uniformly throughout the Galactic disc, whereas the warm clouds contain massive stars and HII regions. Thus, the initial mass function of stars may not have a single origin, but rather consist of the superposition of the birth rate of high- and low-mass stars formed at different locations in the Galaxy.

Another problem concerns the origin of the bipolar outflows. The rotation axis of the accretion disc provides a natural axis for the ejection of matter. According to the picture of Shu, Adams and Lizano (1987), the collimation may be associated with a stellar wind as it escapes from the accreting envelope along the path of least resistance, which is along the rotation axis. Various models have been proposed for the formation of the wind close to the surface of the star. In particular, the hot wind may be associated with the dissipation of energy in the boundary layer between the accretion disc and the stellar surface or with the hot innermost layers of the accretion disc. The hot gas may be channelled by the magnetic field in the 'magnetosphere' of the accreting star along the polar

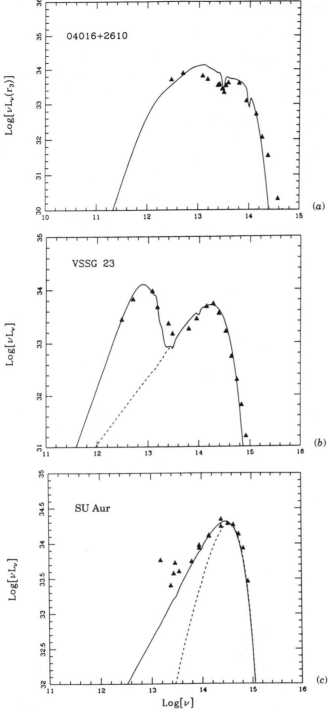

Figure 17.17. Comparison of the theoretical and observed spectra of sources in the Taurus and Ophiuchus molecular clouds. The ordinate is $\nu I(\nu)$ so that it represents the energy emitted at each frequency. All the sources have mass of order $1M_\odot$. (a) The source 04016+2610 is inferred to be a protostar during its main infall phase, that is, the star and disc are embedded in an infalling dust envelope. (b) In VSSG 23, it is inferred that an intense wind has broken out along the rotation axis revealing the newly born star surrounded by a nebular disc. (c) The source SU Aur is a T Tauri star with a small infrared excess. The disc has disappeared leaving an isolated pre-main-sequence star. (F.C. Adams, C.J. Lada and F.H. Shu (1987). *Astrophys. J.*, **312**, 788.)

directions. We will have to return to this class of problem when we tackle the origin of jets and beams in extragalactic radio sources.

There is much that remains uncertain in these pictures, but it is encouraging that new observational capabilities are being developed at infrared and millimetre wavelengths for the study of these compact infrared sources which should help elucidate how stars actually succeed in forming. The reason for dwelling upon the problem of star formation is that, besides its intrinsic interest and importance, star formation is an important component of active galaxies, and some models attribute many of the observed features of these nuclei to star formation activity.

18

Synchrotron radiation and the radio
emission of the Galaxy

18.1 Synchrotron radiation

The synchrotron radiation of relativistic and ultrarelativistic electrons is the process which dominates high energy astrophysics. It is the radiation emitted by very high energy electrons gyrating in a magnetic field. It was originally observed in early betatron experiments, in which electrons were first accelerated to ultrarelativistic energies. This same mechanism is responsible for the radio emission from the Galaxy, from supernova remnants and extragalactic radio sources. It is also responsible for the non-thermal optical emission observed in the Crab Nebula and possibly for the optical and X-ray continuum emission of quasars. The reasons for these assertions will become apparent in the course of this chapter.

The word *non-thermal* is used frequently in high energy astrophysics to describe the emission of high energy particles. I find this an unfortunate terminology, since all emission mechanisms are 'thermal' in some sense. The word is conventionally taken to mean 'continuum radiation from particles, the energy spectrum of which is not Maxwellian'. In practice, continuum emission is often referred to as 'non-thermal' if it cannot be accounted for by the spectrum of thermal bremsstrahlung or black-body radiation.

It is a very major undertaking to work out properly all the properties of synchrotron radiation, and that is beyond the scope of this book. For details, I refer the enthusiast to the books by Bekefi (1966), Pacholczyk (1970) and Rybicki and Lightman (1979) and the three review articles by Ginzburg and his colleagues (see the *References* section for this chapter). We will find that many of the most important results can be drived by simple arguments (see, for example, Scheuer (1966)). First of all, let us work out the total energy loss rate.

229

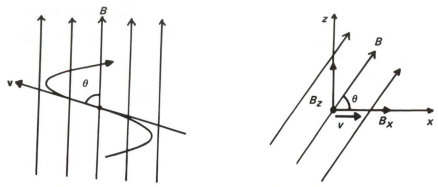

Figure 18.1. The coordinates used in working out the total radiation rate due to synchrotron radiation.

18.1.1   *The total energy loss rate*

Most of the essential tools needed in this analysis have already been derived in Sections 3.3 and 11.1. The rules for the radiation emitted by an accelerated charge are so important that we repeat them here:

1 The total radiation loss rate of an accelerated charge, q, in its instantaneous rest frame is

$$-\left(\frac{\mathrm{d}E}{\mathrm{d}t}\right)_{\mathrm{rad}} = \frac{q^2|\ddot{\mathbf{r}}|^2}{6\pi\epsilon_0 c^3} \tag{3.9}$$

where $\ddot{\mathbf{r}}$ is the acceleration of the charge.

2 The *polar diagram* of the radiation is of *dipolar* form, that is, if θ is the angle between the acceleration vector and the direction in which the radiation is emitted, the electric field strength varies as $\sin\theta$ and the power radiated per unit solid angle in the direction θ as $\sin^2\theta$ (expression (3.8)).

3 The radiation is *polarised*, with the electric field vector lying in the direction of the acceleration vector projected onto a sphere at a large distance.

These aspects of *electric dipole radiation* were derived in Sections 3.3.2 and 3.3.3.

In a uniform magnetic field, a high energy particle moves in a spiral path at a constant pitch angle, θ. Its velocity along the field lines is constant whilst it gyrates about the magnetic field direction at the relativistic gyrofrequency $v_{\mathrm{g}} = eB/2\pi\gamma m_{\mathrm{e}} = 28\gamma^{-1}$ GHz T^{-1}, where γ is the Lorentz factor of the particle, $\gamma = (1-v^2/c^2)^{-1/2}$. During this motion, the charged particle is accelerated towards the guiding centre of its orbit (see Sections 11.1 and 11.2), and then the radiation rate of the particle can be worked out using the results of Section 3.3. The object of the calculations in this section is to combine these results to work out the total radiation rate for a high energy electron gyrating in a magnetic field of field strength **B**.

There are several ways of using these tools. First of all, we can go directly to the expression (11.2) for the acceleration of the electron in its orbit and insert this into the expression for the radiation rate of a relativistic electron (3.23). In the present case, the acceleration is always perpendicular to the velocity vector of the particle and hence $a_\parallel = 0$. Therefore, the total radiation loss rate of the electron is

$$-\left(\frac{\mathrm{d}E}{\mathrm{d}t}\right) = \frac{\gamma^4 e^2}{6\pi\epsilon_0 c^3}|a_\perp|^2$$

$$= \frac{\gamma^4 e^2}{6\pi\epsilon_0 c^3}\frac{e^2 v^2 B^2 \sin^2\theta}{\gamma^2 m_e^2}$$

$$= \frac{e^4 B^2}{6\pi\epsilon_0 c m_e^2}\frac{v^2}{c^2}\gamma^2 \sin^2\theta \qquad (18.1)$$

Another way of arriving at the same result, and revising some of the rules we have already derived, is to start from the fact that, in the instantaneous rest frame of the electron, the acceleration of the particle is small, and therefore in that frame we can use the non-relativistic expression for the radiation rate. Let us choose the coordinate system shown in Fig. 18.1. The instantaneous direction of motion of the electron in the laboratory frame, the frame in which **B** is fixed, is taken as the positive x axis. Then, to find the force acting on the particle, we transform the field quantities into the instantaneous rest frame of the electron using the standard relativistic transformations for the magnetic field strength. In the frame of reference of the moving electron, S', the force on the electron is

$$\mathbf{F}' = m_e\dot{\mathbf{v}}' = e(\mathbf{E}' + \mathbf{v}' \times \mathbf{B}') = e\mathbf{E}' \qquad (18.2)$$

since the particle is instantaneously at rest in S', $\mathbf{v}' = 0$. Therefore, in transforming the magnetic field strength **B** into S', we need only consider the transformed components of the electric field, $\mathbf{E}'$:

$$\begin{cases} E_x' = E_x \\ E_y' = \gamma(E_y - vB_z) \qquad \text{and hence} \\ E_z' = \gamma(E_z + vB_y) \end{cases} \qquad \left.\begin{array}{l} E_x' = 0 \\ E_y' = -v\gamma B_z \\ \quad = -v\gamma B \sin\theta \\ E_z' = 0 \end{array}\right\} \qquad (18.3)$$

Therefore,

$$\dot{\mathbf{v}}' = -\frac{e\gamma v B \sin\theta}{m_e}$$

Consequently, in the rest frame of the electron, the loss rate by radiation is

$$-\left(\frac{\mathrm{d}E}{\mathrm{d}t}\right)' = \frac{e^2|\dot{\mathbf{v}}'|^2}{6\pi\epsilon_0 c^3} \qquad (18.4)$$

$$= \frac{e^4\gamma^2 B^2 v^2 \sin^2\theta}{6\pi\epsilon_0 c^3 m_e^2} \qquad (18.5)$$

Since $(\mathrm{d}E/\mathrm{d}t)$ is a Lorentz invariant (Section 3.3), we recover the formula (18.1). Let us rewrite this in the following way:

$$-\left(\frac{\mathrm{d}E}{\mathrm{d}t}\right) = 2\left(\frac{e^4}{6\pi\epsilon_0^2 c^4 m_e^2}\right)\left(\frac{v}{c}\right)^2 c\,\frac{B^2}{2\mu_0}\gamma^2 \sin^2\theta$$

The quantity in the first set of round brackets on the right-hand side of this expression is just the Thomson cross-section, σ_T (see Section 4.3.1). We have also used the relation $c^2 = (\mu_0\epsilon_0)^{-1}$. Therefore,

$$-\left(\frac{dE}{dt}\right) = 2\sigma_T c U_{mag} \left(\frac{v}{c}\right)^2 \gamma^2 \sin^2\theta \qquad (18.6)$$

where $U_{mag} = B^2/2\mu_0$ is the energy density of the magnetic field. In the ultrarelativistic limit, $v \to c$, we may approximate this result by

$$-\left(\frac{dE}{dt}\right) = 2\sigma_T c U_{mag} \gamma^2 \sin^2\theta$$

These results apply for electrons of a specific pitch angle, θ. Particles of a particular energy, E, or Lorentz factor, γ, are often expected to have an isotropic distribution of pitch angles, and therefore we can work out their average energy loss rate by averaging over an isotropic distribution of pitch angles, $p(\theta)d\theta = \frac{1}{2}\sin\theta d\theta$:

$$-\left(\frac{dE}{dt}\right) = 2\sigma_T c U_{mag} \gamma^2 \left(\frac{v}{c}\right)^2 \frac{1}{2} \int_0^\pi \sin^3\theta d\theta$$

$$= \frac{4}{3}\sigma_T c U_{mag} \left(\frac{v}{c}\right)^2 \gamma^2 \qquad (18.7)$$

Notice that there is a deeper sense in which the expression (18.7) is the average loss rate for an individual particle of energy E. During its lifetime, it is likely that the high energy particle is randomly scattered in pitch angle, and then the expression (18.7) is the correct expression for its average energy loss rate.

18.1.2   *Non-relativistic gyroradiation and cyclotron radiation*

Before tackling the spectral distribution of synchrotron radiation, let us make a short diversion into the non-relativistic and mildly relativistic cases, which are of considerable interest astrophysically. Consider first of all the simplest case of non-relativistic gyroradiation, in which $v \ll c$ and hence $\gamma = 1$. Then, the expression for the loss rate of the electron becomes

$$-\left(\frac{dE}{dt}\right) = 2\sigma_T c U_{mag} \left(\frac{v}{c}\right)^2 \sin^2\theta = \frac{2\sigma_T}{c} U_{mag} v_\perp^2 \qquad (18.8)$$

and the radiation is emitted at the gyrofrequency of the electron, $\nu_g = eB/2\pi m_e$.

One interesting aspect of this emission mechanism is the fact that the polarisation properties are quite distinctive. In the non-relativistic case, there are no beaming effects, and thus what is observed by the distant observer can be derived from the simple rules given at the beginning of Section 18.1.1. When the magnetic field is perpendicular to the line of sight, *linearly polarised radiation* is observed because the acceleration vector is observed to perform simple harmonic motion in a plane perpendicular to the magnetic field by the distant observer. The electric field strength varies sinusoidally at the gyrofrequency as the dipole distribution of radiation sweeps past the observer. On the other hand, when the magnetic field is parallel to the line of sight, the acceleration vector is continually changing direction as the electron moves in a circular orbit about the magnetic field lines, and therefore the radiation is observed to be 100% *circularly polarised*. When observed at an arbitrary angle θ to the magnetic field direction, the radiation is observed to be *elliptically polarised*, the ratio of axes of the polarisation ellipse being $\cos\theta$.

One of the most remarkable potential applications of gyroradiation is in the binary X-ray source Hercules X-1. Hard X-ray spectral observations have discovered what is referred to as a *cyclotron absorption feature* at 34 keV (Fig. 18.2). If this feature is attributed to absorption at the gyrofrequency by hot gas in the vicinity of the poles of the magnetised neutron star, an estimate of the magnetic field strength can be found. Inserting 34 keV into the formula for the gyrofrequency, we find a magnetic field strength of about 3×10^8 T, a very strong field indeed but within the range of values found from studies of the rate of slow-down of radio pulsars. Thus, it is entirely plausible that this is an example of absorption at the gyrofrequency in the intense magnetic fields present in magnetised neutron stars.

The other example concerns mildly relativistic cyclotron radiation, in which account has to be taken of the beaming of the radiation. Even for slowly moving electrons, $v \ll c$, not all the radiation is emitted at the gyrofrequency because there are small aberration effects which slightly distort the observed angular distribution of the intensity from a simple $\cos^2\theta$ law. From the symmetry of these aberrations, it can be seen that the observed polar diagram of the radiation may be decomposed by Fourier analysis into a sum of equivalent dipoles radiating at harmonics of the relativistic gyrofrequency, ν_r, where $\nu_r = \nu_g/\gamma$. These harmonics have frequencies

$$\nu_l = \frac{l\nu_r}{[1 - (v_\parallel/c)\cos\theta]} \tag{18.9}$$

where l takes integral values, $l = 1, 2, 3, \ldots$ and the fundamental gyrofrequency has $l = 1$. The factor $[1 - (v_\parallel/c)\cos\theta]$ in the denominator takes account of the Doppler shift of the radiation of the electron due to its translation velocity along the field lines, $v_\parallel$, projected onto the line of sight to the observer. In the limit $lv/c \ll 1$, it can be shown that the total power emitted in a given harmonic for the case $v_\parallel = 0$ is

$$-\left(\frac{dE}{dt}\right)_l = \frac{2\pi e^2 \nu_g^2}{\epsilon_0 c} \frac{(l+1)l^{2l+1}}{(2l+1)!} \left(\frac{v}{c}\right)^2 \tag{18.10}$$

and hence, to order of magnitude,

$$\frac{(dE/dt)_{l+1}}{(dE/dt)_l} \approx \left(\frac{v}{c}\right)^2$$

Thus, the energy radiated in high harmonics is small when the particle is non-relativistic. Notice that the loss rate (18.10) reduces to equation (18.8) for $l = 1$.

When the particle becomes significantly relativistic, however, for example $v/c > 0.1$, the energy radiated in the higher harmonics becomes important. The Doppler corrections to the observed frequency of the emitted radiation become significant, and a wide spread of emitted frequencies is associated with the different pitch angles of an electron of energy $E = \gamma mc^2$. The result is to broaden the width of the emission line of a given harmonic, and, for high harmonics, the lines are so

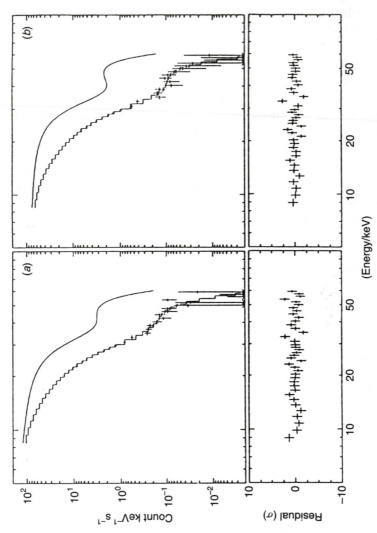

Figure 18.2. The hard X-ray spectrum of the binary X-ray source Hercules X-1, as observed by the GINGA satellite, showing the 34 keV cyclotron absorption feature. The X-ray source has a pulse period of 1.24s, and in (a) the spectrum observed at the pulse maxima is shown. In (b), the spectrum is derived from the difference between the spectra observed at pulse maximum and pulse minimum, and shows more clearly the absorption feature. The lower panels show the residuals obtained for the fit of the model to the data. (From T. Mihara, K. Makashima, T. Ohashi, T. Sakao, M. Tashiro, F. Nagase, Y. Tanaka, S. Kitamoto, S. Miyamoto, J.E. Deeter and P.E. Boynton (1990). *Nature*, **346**, 250.)

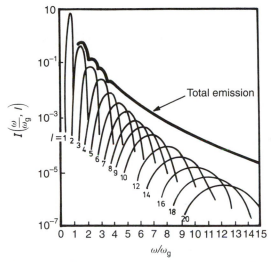

Figure 18.3. The spectrum of emission of the first 20 harmonics of mildly relativistic cyclotron radiation. The electron has $v = 0.4c$. (After G. Bekafi (1966). *Radiation processes in plasmas*, p. 203. New York: John Wiley and Sons, Inc.)

broadened that the emission spectrum becomes continuous rather than a series of harmonics at well defined frequencies. The results of calculations for a relativistic plasma having $kT_e/m_e c^2 = 0.1$, corresponding to $\gamma = 1.1$ and $v/c \approx 0.4$, are shown in Fig. 18.3. The spectra of the first 20 harmonics are shown as well as the total emission spectrum found by summing the spectra of the individual harmonics. One way of looking at synchrotron radiation is to consider it as the relativistic limit of the process illustrated in Fig. 18.3, in which all the harmonics are washed out and a smooth continuum spectrum is observed.

Just as in the case of gyroradiation, the harmonics of cyclotron radiation are circularly polarised. This means that, if the elliptical polarisation of the radiation from a celestial object can be measured in detail, it is possible to learn a great deal, not only about the strength of the magnetic field, but also about its orientation with respect to the line of sight. Circularly polarised optical emission has been discovered in the eclipsing magnetic binary stars known as *AM Herculis binaries* or *polars*, circular polarisation percentages as large as 40% being observed. In these systems, a red dwarf star orbits a white dwarf with a very strong magnetic field. Accretion of matter from the surface of the red dwarf onto the magnetic poles of the white dwarf (Fig. 16.14(a)) results in the heating of the matter to temperatures in excess of 10^7 K. Thus, in addition to radiating X-rays, these objects are strong sources of cyclotron radiation. Fields of order 2000 T have been found in these objects, and hence the fundamental gyrofrequency is expected to correspond to a wavelength of about 5μm. Very often, the individual harmonics of the cyclotron radiation are washed out, but recently it has been possible to distinguish them in the X-ray source EXO 033319-2554.2 (Fig. 18.4).

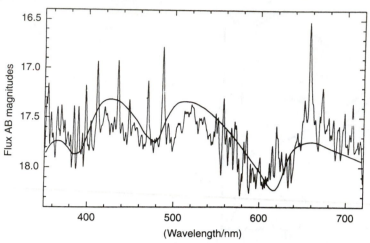

Figure 18.4. A broad-band spectrum of the AM Herculis object EXO 033319-2554.2, which is a soft X-ray source. The presence of a strong magnetic field is inferred from the observation of strongly circularly polarised emission. The solid line shows a best fit of the cyclotron emission spectrum to the broad cyclotron harmonics at 420, 520 and 655 nm. The inferred strength of the magnetic field is 5600 T. (From L. Ferrario, D. T. Wickramsinghe, J. Bailey, I. R. Tuohy and J. H. Hough (1989). *Astrophys. J.*, **337**, 832.)

The frequency spacing between the harmonics has enabled an accurate estimate of the magnetic field strength to be made. This turns out to be 5600 T, the largest known magnetic field strength in an AM Herculis system.

18.1.3   The spectral distribution of radiation from a single electron – physical arguments

The next step is to work out the spectral distribution of synchrotron radiation, but an exact analysis requires very much more effort. Let us analyse first of all some basic aspects of radiation mechanisms involving relativistic particles, which will prove to be invaluable in understanding where many of the exact results come from.

One of the basic features of the radiation of relativistic particles in general is the fact that the radiation is *beamed* in the direction of motion of the particle. Part of this effect is associated with the relativistic aberration formulae between the frame of reference of the particle and the observer's frame of reference. There are, however, subtleties about what is actually observed because, in addition to aberration, we have to consider carefully the time development of what is seen by the distant observer.

Let us consider first the simple case of a particle gyrating about the magnetic field at a pitch angle of 90°. The electron is accelerated towards its guiding centre, that is, radially inwards, and in its instantaneous rest frame it emits the usual dipole pattern with respect to the acceleration vector. This is illustrated in Fig. 18.5(*a*). We can therefore work out the radiation pattern in the laboratory frame of reference by applying the aberration formulae with the results illustrated

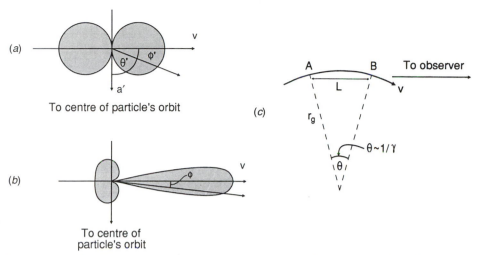

(a)

To centre of particle's orbit

(b)

To centre of
particle's orbit

(c)

A   B   To observer

$\theta \sim 1/\gamma$

Figure 18.5. Illustrating the relativistic beaming and Doppler effects associated with synchrotron radiation. (a) The polar diagram of the dipole radiation of the electron in its instantaneous rest frame, showing the strength of the electric field as a function of angle θ'. (b) The polar diagram of the radiation transformed into the laboratory frame of reference. (c) The geometry of the path of the electron when the radiation is observed by the distant observer.

schematically in Fig. 18.5(b). As discussed in Section 3.3.2, the angular distribution of radiation with respect to the velocity vector in the frame S' is $I_\nu \propto \sin^2\theta' = \cos^2\phi'$. We may think of this as being the probability distribution with which photons are emitted by the electron in its rest frame. The appropriate aberration formulae between the two frames are:

$$\sin\phi = \frac{1}{\gamma}\frac{\sin\phi'}{1 + (v/c)\cos\phi'} \qquad \cos\phi = \frac{\cos\phi' + v/c}{1 + (v/c)\cos\phi'} \qquad (18.11)$$

To illustrate the beaming of the radiation, let us consider the angles $\phi' = \pm\pi/4$, these being the angles at which the intensity of radiation falls to half its maximum value in the instantaneous rest frame of the particle. The corresponding angles, ϕ, in the laboratory frame of reference are

$$\sin\phi \approx \phi \approx 1/\gamma \qquad (18.12)$$

Thus, the radiation emitted within $-\pi/4 < \phi' < \pi/4$ is beamed in the direction of motion of the electron within an angle $-1/\gamma < \phi < 1/\gamma$. As observed in the frame S, the dipole beam pattern is very strongly distorted and the intensity of the radiation is strongly Doppler-shifted. Only when this elongated beam pattern sweeps past the observer is a significant amount of radiation observed. A large 'spike' of radiation is observed every time the electron's velocity vector lies within an angle of about $1/\gamma$ to the line of sight to the observer. This analysis shows why the observed frequency of the radiation must be very much higher than the gyrofrequency of the electron in its orbit. The spectrum of the radiation received by the distant observer is the Fourier transform of this pulse once the effects of time retardation and Doppler shifting are taken into account. This phenomenon is illustrated in Fig. 18.5(*b*), which shows how the observed radiation pattern narrows as the velocity of the particle increases.

Thus, the observer sees significant radiation from only about $1/\gamma$ of the particle's orbit, but the observed duration of the pulse is less than $1/\gamma$ times the period of the orbit because radiation emitted at the trailing edge of the pulse almost catches up with the radiation emitted at the leading edge. Let us illustrate this second effect by a simple calculation which is carried out entirely in the laboratory frame of reference and which concerns the time of arrival of the signals at the distant observer. The portion of the particle's orbit from which radiation is received by the distant observer is shown in Fig. 18.5(*c*). Consider the observer located at a distance R from the point A. The radiation from A reaches the observer at time R/c. Now consider the radiation emitted from B at time L/v later, which then has to travel a distance $(R-L)$ at the velocity of light to reach the observer. The trailing edge of the pulse therefore arrives at the observer at a time $L/v+(R-L)/c$. The duration of the pulse as measured by the observer is therefore

$$\Delta t = \left[\frac{L}{v} + \frac{(R-L)}{c}\right] - \frac{R}{c} = \frac{L}{v}\left[1 - \frac{v}{c}\right]$$

Notice that the observed duration of the pulse is much less than the value L/v which might naively have been expected. Only if light propagated at an infinite velocity would the duration of the pulse be L/v. The intriguing point about this analysis is that this factor $1 - (v/c)$ is exactly the same factor which appears in the Liénard–Weichert potentials, and takes account of the fact that the source of radiation is not stationary but is moving towards the observer (see equation (3.18)). In fact, the relativistic particle almost catches up with the radiation emitted at A since $v \approx c$. Now we can rewrite the above expression, noting that

$$\frac{L}{v} = \frac{r_g\theta}{v} \approx \frac{1}{\gamma\omega_r} = \frac{1}{\omega_g}$$

where ω_g is the non-relativistic angular gyrofrequency and $\omega_r = \omega_g/\gamma$. We also note that we can rewrite $1 - (v/c)$ as

$$1 - \frac{v}{c} = \frac{[1 - (v/c)]\,[1 + (v/c)]}{1 + (v/c)} = \frac{1 - (v^2/c^2)}{1 + (v/c)} \approx \frac{1}{2\gamma^2} \tag{18.13}$$

since $v \approx c$. Therefore, the observed duration of the pulse is roughly

$$\Delta t \approx \frac{1}{2\gamma^2\omega_g}$$

This means that the duration of the pulse as observed by a distant observer in the laboratory frame of reference is roughly $1/\gamma^2$ times shorter than the non-relativistic gyroperiod, $T_g = 2\pi/\omega_g$. The maximum Fourier component of the spectral decomposition of the observed pulse of radiation is expected to correspond to a frequency $\nu \sim \Delta t^{-1}$, that is,

$$\nu \sim \Delta t^{-1} \sim \gamma^2\nu_g \tag{18.14}$$

where ν_g is the non-relativistic gyrofrequency.

In the above analysis, it has been assumed that the particle moves in a circle about the magnetic field lines, that is, the pitch angle θ is 90°. The same calculation can be performed for any pitch angle, and then the result becomes

$$v \sim \gamma^2 v_g \sin \theta$$

The reason for performing this simple exercise in detail is that the beaming of the radiation of ultrarelativistic particles is a very general property and does not depend upon the nature of the force causing the acceleration.

Returning to an earlier part of the calculation, the observed frequency of the radiation can also be written

$$v \approx \gamma^2 v_g = \gamma^3 v_r = \frac{\gamma^3 v}{2\pi r_g}$$

where v_r is the relativistic gyrofrequency and r_g is the radius of curvature of the particle's orbit. Notice that, in general, we may interpret r_g as the instantaneous radius of curvature of the particle's orbit and v/r_g as the angular frequency associated with it. This is a useful result because it enables us to work out the frequency at which most of the radiation is emitted, provided we know the radius of curvature of the particle's orbit. The frequency of the observed radiation is roughly γ^3 times the angular frequency v/r, where r is the instantaneous radius of curvature of the particle in its orbit. This result is important in the study of *curvature radiation*, which has important applications in the emission of radiation from the magnetic poles of pulsars (Section 15.4).

For many order of magnitude calculations, it is sufficient to know that the energy loss rate of the relativistic electron is given by the expression (18.7) and that most of the radiation is emitted at a frequency which is roughly $v = \gamma^2 v_g$, where v_g is the non-relativistic gyrofrequency. However, we often have to do somewhat better than this, and that is the subject of the next subsection.

18.1.4 *The spectrum of synchrotron radiation – improved version*

I am not aware of any particularly simple way of deriving the spectral distribution of synchrotron radiation, and I do not find the present analysis particularly appealing. The outline given below follows closely the presentation of Rybicki and Lightman (1979). The analysis proceeds by the following steps:

1 write down the expression for the energy emitted per unit bandwidth for an arbitrarily moving electron;
2 select a suitable set of coordinates in which to work out the field components radiated by the electron spiralling in a magnetic field;
3 then battle away at the algebra to obtain the spectral distribution of the field components.

Let us outline these stages in the calculation, highlighting a number of important points as we go along.

The spectrum of radiation of an arbitrarily moving electron The first thing we need is the generalisation of the formulae for the radiation of an accelerated charge now moving at a relativistic velocity. For comparison, we will quote the formulae we have already derived for a slowly moving charge. We start with the Liénard–Weichert potentials, which are written (see expressions (3.18)):

$$\mathbf{A}(\mathbf{r}, t) = \frac{\mu_0}{4\pi r} \left[\frac{q\mathbf{v}}{1 - \dfrac{\mathbf{v} \cdot \mathbf{n}}{c}} \right]_{\text{ret}} \quad ; \quad \phi(\mathbf{r}, t) = \frac{1}{4\pi \epsilon_0 r} \left[\frac{q}{1 - \dfrac{\mathbf{v} \cdot \mathbf{n}}{c}} \right]_{\text{ret}} \tag{3.18}$$

The key differences as compared with the expression for a slowly moving charge (expression (3.17)) are the presence of the Doppler-shift factor $[1 - (\mathbf{v} \cdot \mathbf{n}/c)]$ in the denominator and the explicit recognition that retarded quantities have to be used to work out the fields at the observer. We will write $\kappa = [1 - (\mathbf{v} \cdot \mathbf{n}/c)]$.

The next expression we need is the relation between the acceleration of the electron and the spectral energy distribution of the radiation. We have already given the expression for the radiation emitted by the particle when there is no net motion, (3.27), and we repeat it here, writing out explicity the Fourier transform of the acceleration.

$$I(\omega) = \frac{e^2}{6\pi^2 \epsilon_0 c^3} \left| \int_{-\infty}^{\infty} \dot{\mathbf{v}}(t) \exp(i\omega t) \mathrm{d}t \right|^2 \tag{18.15}$$

The corresponding result for the case of a moving particle can be written

$$\frac{\mathrm{d}I(\omega)}{\mathrm{d}\Omega} = \frac{e^2}{16\pi^3 \epsilon_0 c} \left| \int_{-\infty}^{\infty} \left\{ \mathbf{n} \times \left[\left(\mathbf{n} - \frac{\mathbf{v}}{c} \right) \times \frac{\dot{\mathbf{v}}}{c} \right] \kappa^{-3} \right\}_{\text{ret}} \exp(i\omega t) \mathrm{d}t \right|^2 \tag{18.16}$$

where the angular dependence of the emitted radiation has been preserved (Rybicki and Lightman (1979), after Jackson (1975)). The vector $\mathbf{n}$ is the unit vector from the particle to the point of observation, $\mathbf{n} = \mathbf{R}/|R|$. Integrating equation (18.16) over solid angle $\mathrm{d}\Omega = 2\pi \sin \theta \, \mathrm{d}\theta$ in the non-relativistic limit gives the expression (18.15). The key differences between equations (18.15) and (18.16) are the inclusion of the Doppler-shift factor κ^3 in the denominator and the fact that the expression in square brackets has to be evaluated at *retarded time* t', where $t' = t - R(t')/c$ in the usual way.

The next step is to manipulate the expression (18.16) into a more manageable form. First of all, we change the integration from an integral over $\mathrm{d}t$ to one over $\mathrm{d}t'$. This is done by noting that $t' = t - R(t')/c$. Differentiating both sides and noting that the unit vector $\mathbf{n}$ points towards the observer, we find

$$\mathrm{d}t' = \mathrm{d}t - \frac{1}{c} \frac{\mathrm{d}R(t')}{\mathrm{d}t'} \mathrm{d}t'$$

$$\mathrm{d}t = \mathrm{d}t' \left(1 - \frac{\mathbf{n} \cdot \mathbf{v}}{c} \right) = \kappa \mathrm{d}t'$$

Therefore, the integral over t becomes one over retarded time t'. A further simplification is introduced if we write the distance to the particle as $R(t') = |\mathbf{r}| - \mathbf{n} \cdot \mathbf{r}_0(t')$, where $\mathbf{r}_0(t')$ is the position vector which describes the position of the particle relative to an origin at $\mathbf{r}$. Note that, in all our calculations, $\mathbf{r}_0(t') \ll \mathbf{r}$.

Therefore, expression (18.16) reduces to

$$\frac{\mathrm{d}I(\omega)}{\mathrm{d}\Omega} = \frac{e^2}{16\pi^3\epsilon_0 c}\left|\int_{-\infty}^{\infty} \mathbf{n}\times\left[\left(\mathbf{n}-\frac{\mathbf{v}(t')}{c}\right)\times\frac{\dot{\mathbf{v}}(t')}{c}\right]\right.$$

$$\left.\times\kappa^{-2}\exp\left[\mathrm{i}\omega\left(t'-\frac{\mathbf{n}\cdot\mathbf{r}_0(t')}{c}\right)\right]\mathrm{d}t'\right|^2 \tag{18.17}$$

The next step is to simplify the vector triple product inside the integral using the identity

$$\mathbf{n}\times\left[\left(\mathbf{n}-\frac{\mathbf{v}}{c}\right)\times\frac{\dot{\mathbf{v}}}{c}\right]\kappa^{-2} = \frac{\mathrm{d}}{\mathrm{d}t'}\left\{\kappa^{-1}\left[\mathbf{n}\times\left(\mathbf{n}\times\frac{\mathbf{v}}{c}\right)\right]\right\} \tag{18.18}$$

This is found by differentiating $\kappa^{-1}[\mathbf{n}\times(\mathbf{n}\times\mathbf{v}/c)]$ with respect to t' and then using the vector triple product rule $\mathbf{a}\times(\mathbf{b}\times\mathbf{c}) = (\mathbf{a}\cdot\mathbf{c})\mathbf{b}-(\mathbf{a}\cdot\mathbf{b})\mathbf{c}$. Substituting expression (18.17) into expression (18.16) and integrating by parts, we find

$$\frac{\mathrm{d}I(\omega)}{\mathrm{d}\Omega} = \frac{e^2\omega^2}{16\pi^3\epsilon_0 c}\left|\int_{-\infty}^{\infty}\mathbf{n}\times\left(\mathbf{n}\times\frac{\mathbf{v}}{c}\right)\exp\left[\mathrm{i}\omega\left(t'-\frac{\mathbf{n}\cdot\mathbf{r}_0(t')}{c}\right)\right]\mathrm{d}t'\right|^2 \tag{18.19}$$

Notice that, in the process of using the identity (18.18), we have apparently eliminated the acceleration of the charge – it is now simply the dynamics of the particle which appears in expression (18.19).

The system of coordinates We now look at the dynamics of the relativistic electron in its orbit and choose the most convenient set of coordinates for evaluating the integrals in the expression (18.19). The particle spirals about the magnetic field lines at an angular frequency $\omega_{\mathrm{r}} = eB/\gamma m_{\mathrm{e}}$ and at a pitch angle α with respect to the magnetic field direction. At any time, the orbit has a certain radius of curvature, a, and we take the instantaneous plane of its orbit to be the x–y plane. We simplify the calculations considerably if we take the x axis to have its origin at the point where the velocity vector, $\mathbf{v}$, of the particle lies in the x–z plane, which includes the observer (see Fig. 18.6), and the y axis to be the direction of the instantaneous radius vector of the electron. Thus, the unit vector, $\mathbf{n}$, pointing from the origin of the system of coordinates to the observer lies in the x–z plane. The importance of this choice of coordinates is that, since $\mathbf{v}$ is tangential to the orbit of the particle at $x = y = 0$, the vector $\mathbf{n}$ is parallel to the magnetic field direction as seen in projection by the observer. This enables us to define another orthogonal set of coordinates with the same origin as x, y, z, with the $\epsilon_{\|}$ lying in the plane containing $\mathbf{n}$ and the magnetic field direction and $\epsilon_{\perp}$ lying along the y axis so that $\epsilon_{\|} = \mathbf{n}\times\epsilon_{\perp}$. The importance of this step is that the unit vectors $\epsilon_{\|}$ and $\epsilon_{\perp}$ form the natural system of coordinates for describing the observed polarisation of the radiation, the $\|$ and $\perp$ symbols referring to components parallel and perpendicular to the magnetic field direction as seen in projection by the observer.

The algebra We can deal separately with the vector triple product and the exponent in the integral (18.19). To evaluate the vector triple product, we write down the coordinates of the electron in the $(\mathbf{n}, \epsilon_{\|}, \epsilon_{\perp})$ coordinate system, taking

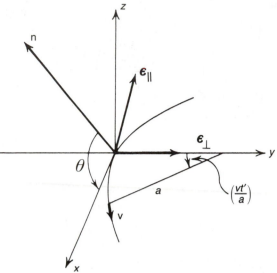

Figure 18.6. The geometry for evaluating the intensity and polarisation properties of synchrotron radiation. At $t = 0$, the particle velocity is along the x axis and a is the radius of curvature of the trajectory. (After G.B. Rybicki and A.P. Lightman (1979). *Radiative processes in astrophysics*, p. 175. New York: John Wiley and Sons.)

$x = y = z = 0$ as the point at which $t' = 0$. Therefore, after time t', the particle has moved a distance vt' round the orbit corresponding to the angle $\phi = vt'/a$, where a is the radius of curvature of the particle's orbit. From the geometry of Fig. 18.6, we see that

$$\mathbf{v} = |\mathbf{v}| \left[\mathbf{i}_x \cos \left(\frac{vt'}{a} \right) + \boldsymbol{\epsilon}_\perp \sin \left(\frac{vt'}{a} \right) \right] \tag{18.20}$$

We can now decompose this velocity into components in the $(\mathbf{n}, \boldsymbol{\epsilon}_\parallel, \boldsymbol{\epsilon}_\perp)$ coordinate system:

$$\mathbf{v} = |\mathbf{v}| \left[\boldsymbol{\epsilon}_\perp \sin \left(\frac{vt'}{a} \right) + \mathbf{n} \cos \theta \cos \left(\frac{vt'}{a} \right) - \boldsymbol{\epsilon}_\parallel \sin \theta \cos \left(\frac{vt'}{a} \right) \right] \tag{18.21}$$

where θ is the angle between the unit vector $\mathbf{n}$, which points towards the observer, and the x–y plane. Finally, we take the vector product $\mathbf{n} \times (\mathbf{n} \times \mathbf{v})$, recalling that $\boldsymbol{\epsilon}_\parallel = \mathbf{n} \times \boldsymbol{\epsilon}_\perp$ and $\boldsymbol{\epsilon}_\perp = -\mathbf{n} \times \boldsymbol{\epsilon}_\parallel$:

$$\mathbf{n} \times (\mathbf{n} \times \mathbf{v}) = |\mathbf{v}| \left[-\sin \left(\frac{vt'}{a} \right) \boldsymbol{\epsilon}_\perp + \sin \theta \cos \left(\frac{vt'}{a} \right) \boldsymbol{\epsilon}_\parallel \right] \tag{18.22}$$

The second part of the calculation is to evaluate the part of the exponent in square brackets, $[t' - \mathbf{n} \cdot \mathbf{r}_0(t')/c]$. Again we refer to Fig. 18.6 to evaluate $\mathbf{r}_0(t_0)$, which is just the position vector of the particle in its orbit. By simple geometrical arguments, we can see that

$$\mathbf{r}_0(t') = 2a \sin\left(\frac{vt'}{2a}\right) \left[\boldsymbol{\epsilon}_\perp \sin\left(\frac{vt'}{2a}\right) + \mathbf{n}\cos\theta \cos\left(\frac{vt'}{2a}\right)\right.$$
$$\left. -\boldsymbol{\epsilon}_\parallel \sin\theta \cos\left(\frac{vt'}{2a}\right)\right] \tag{18.23}$$

Then, by substitution we find

$$\left[t' - \frac{\mathbf{n}\cdot\mathbf{r}_0(t')}{c}\right] = t' - \frac{a}{c}\cos\theta \sin\left(\frac{vt'}{a}\right) \tag{18.24}$$

Now, we have to look at the origin of the main contributions to the integral
(18.19). It is evident that the greatest contributions come from the smallest values
of $[t' - \mathbf{n}\cdot\mathbf{r}_0(t')/c]$. If this were large, there would be many 'oscillations' in the
integral and these would average out to a very small value. We know from our
physical description of the synchrotron radiation process in Section 18.1.3 that
most of the radiation is strongly beamed in the direction of motion of the electron.
This means that the principal contributions to the spectral distribution of the
radiation come from small values of θ and correspondingly small values of vt'/a
(see the geometry of Fig. 18.6). Therefore, we can now expand equation (18.24)
for small vaules of θ and vt'/a. Expanding as far as third order in the small
quantities θ and vt'/a, we find

$$t' - \frac{\mathbf{n}\cdot\mathbf{r}_0(t')}{c} = t'\left(1 - \frac{v}{c}\right) + \frac{v}{c}\frac{\theta^2}{2}t' + \frac{v^3}{6ca^2}t'^3$$

Since $v \approx c$, we can again write $1 - (v/c) = 1/2\gamma^2$ (expression (18.13)) and hence

$$t' - \frac{\mathbf{n}\cdot\mathbf{r}_0(t')}{c} = \frac{1}{2\gamma^2}\left[t'\left(1 + \gamma^2\frac{v}{c}\theta^2\right) + \frac{v^3\gamma^2 t'^3}{3ca^2}\right]$$
$$= \frac{1}{2\gamma^2}\left[t'(1 + \gamma^2\theta^2) + \frac{c^2\gamma^2 t'^3}{3a^2}\right]$$

where, in the last relation, we have set $v = c$.

We now perform the same type of reduction for the quantity $\mathbf{n} \times (\mathbf{n} \times \mathbf{v}/c)$, and
find

$$\mathbf{n} \times \left(\mathbf{n} \times \frac{\mathbf{v}}{c}\right) = \frac{|\mathbf{v}|}{c}\left[-\sin\left(\frac{vt'}{a}\right)\boldsymbol{\epsilon}_\perp + \sin\theta\cos\left(\frac{vt'}{a}\right)\boldsymbol{\epsilon}_\parallel\right]$$
$$\approx \left(-\frac{vt'}{a}\boldsymbol{\epsilon}_\perp + \theta\boldsymbol{\epsilon}_\parallel\right)$$

We are now in a position to write down the integrals for the radiation components
in the $\boldsymbol{\epsilon}_\perp$ and $\boldsymbol{\epsilon}_\parallel$ directions:

$$\frac{\mathrm{d}I_\perp(\omega)}{\mathrm{d}\Omega} = \frac{e^2\omega^2}{16\pi^3\epsilon_0 c}\left|\int_{-\infty}^{\infty} \frac{vt'}{a}\exp\left\{\frac{\mathrm{i}\omega}{2\gamma^2}\left[t'(1 + \gamma^2\theta^2) + \frac{c^2\gamma^2}{3a^2}t'^3\right]\right\}\mathrm{d}t'\right|^2$$
$$\frac{\mathrm{d}I_\parallel(\omega)}{\mathrm{d}\Omega} = \frac{e^2\omega^2\theta^2}{16\pi^3\epsilon_0 c}\left|\int_{-\infty}^{\infty}\exp\left\{\frac{\mathrm{i}\omega}{2\gamma^2}\left[t'(1 + \gamma^2\theta^2) + \frac{c^2\gamma^2}{3a^2}t'^3\right]\right\}\mathrm{d}t'\right|^2 \tag{18.25}$$

We are now very close to the answers we need. We only have to make a number
of straightforward changes of variable to reduce the integrals to a standard form.

Again, because most of the power emitted by the electron is contained within small values of θ, corresponding to small values of t', there is little error in taking the limits of the integrals from $-\infty$ to $+\infty$. We therefore write

$$\theta_\gamma^2 = 1 + \gamma^2\theta^2 \quad ; \quad y = \gamma ct'/a\theta_\gamma \quad ; \quad \eta = \omega a\theta_\gamma^3/3c\gamma^3$$

and then

$$\frac{\mathrm{d}I_\perp(\omega)}{\mathrm{d}\Omega} = \frac{e^2\omega^2}{16\pi^3\epsilon_0 c}\left(\frac{a\theta_\gamma^2}{c\gamma^2}\right)^2\left|\int_{-\infty}^{\infty} y\exp\left[\mathrm{i}\eta\frac{3}{2}\left(y+\frac{y^3}{3}\right)\right]\mathrm{d}y\right|^2$$

$$\frac{\mathrm{d}I_\parallel(\omega)}{\mathrm{d}\Omega} = \frac{e^2\omega^2\theta^2}{16\pi^3\epsilon_0 c}\left(\frac{a\theta_\gamma}{c\gamma}\right)^2\left|\int_{-\infty}^{\infty} \exp\left[\mathrm{i}\eta\frac{3}{2}\left(y+\frac{y^3}{3}\right)\right]\mathrm{d}y\right|^2$$

(18.26)

The integrals can be expressed in terms of modified Bessel functions using the following relations, which can be derived from those presented by Abramovitz and Stegun (1965, see relations 10.4.22–10.4.32):

$$\int_0^\infty \cos\left[\frac{3\eta}{2}\left(x+\frac{1}{3}x^3\right)\right]\mathrm{d}x = \frac{1}{\sqrt{3}}K_{1/3}(\eta)$$

$$\int_0^\infty x\sin\left[\frac{3\eta}{2}\left(x+\frac{1}{3}x^3\right)\right]\mathrm{d}x = \frac{1}{\sqrt{3}}K_{2/3}(\eta)$$

(18.27)

where $K_{2/3}$ and $K_{1/3}$ are modified Bessel functions of order $\frac{2}{3}$ and $\frac{1}{3}$, respectively. We use the symmetry of the integrands to derive the following expressions for the integrals:

$$\frac{\mathrm{d}I_\perp(\omega)}{\mathrm{d}\Omega} = \frac{e^2\omega^2}{12\pi^3\epsilon_0 c}\left(\frac{a\theta_\gamma^2}{c\gamma^2}\right)^2 K_{2/3}^2(\eta)$$

$$\frac{\mathrm{d}I_\parallel(\omega)}{\mathrm{d}\Omega} = \frac{e^2\omega^2\theta^2}{12\pi^3\epsilon_0 c}\left(\frac{a\theta_\gamma}{c\gamma}\right)^2 K_{1/3}^2(\eta)$$

(18.28)

The final step is to integrate over the angle θ. Since most of the radiation is emitted within a very small angle, θ, with respect to the pitch angle of the particle, it is quite permissible to assume that, over one period of gyration of the particle about the magnetic field direction, the angle over which the integral is to be taken is $2\pi\sin\alpha\,\mathrm{d}\theta$ because the element of solid angle varies very little over $\mathrm{d}\theta$ whilst the radiation pattern is a strong function of θ (Fig. 18.7). We make little error in taking the limits of the integrals to $\pm\infty$ because all the power is concentrated in the angle $\mathrm{d}\theta$ about the pitch angle, α. Therefore, the integrals can be written:

$$I_\perp(\omega) = \frac{e^2\omega^2 a^2\sin\alpha}{6\pi^2\epsilon_0 c^3\gamma^4}\int_{-\infty}^{\infty}\theta_\gamma^4 K_{2/3}^2(\eta)\mathrm{d}\theta$$

$$I_\parallel(\omega) = \frac{e^2\omega^2 a^2\sin\alpha}{6\pi^2\epsilon_0 c^3\gamma^2}\int_{-\infty}^{\infty}\theta_\gamma^2\theta^2 K_{1/3}^2(\eta)\mathrm{d}\theta$$

(18.29)

These integrals have been evaluated by Westfold (1959) and by Le Roux (1961). If Westfold's paper is used, the following relations may be found from comparison

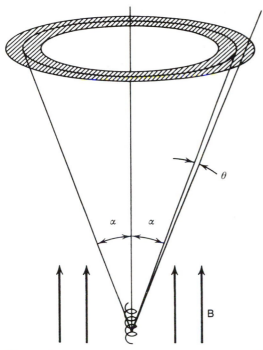

Figure 18.7. Synchrotron emission from a particle with pitch angle α. The radiation is confined to the shaded solid angle. (After G.B. Rybicki and A.P. Lightman (1979). *Radiative processes in astrophysics*, p. 178. New York: John Wiley and Sons.)

of his equations (23) and (25):

$$\int_{-\infty}^{\infty} \theta_\gamma^4 K_{2/3}^2 \left(\frac{x}{2}\theta_\gamma^3\right) d\theta = \frac{\pi}{\sqrt{3}\gamma x}\left[\int_x^{\infty} K_{5/3}(z)dz + K_{2/3}(x)\right]$$
$$\int_{-\infty}^{\infty} \gamma^2\theta^2\theta_\gamma^2 K_{1/3}^2 \left(\frac{x}{2}\theta_\gamma^3\right) d\theta = \frac{\pi}{\sqrt{3}\gamma x}\left[\int_x^{\infty} K_{5/3}(z)dz - K_{2/3}(x)\right]$$

(18.30)

It will be recalled that $\theta_\gamma = (1 + \gamma^2\theta^2)$ and $x = 2\omega a/3c\gamma^3$. It is traditional to write

$$F(x) = x \int_x^{\infty} K_{5/3}(z)dz \qquad ; \qquad G(x) = x K_{2/3}(x)$$ (18.31)

Then, using the expression $a = 3c\gamma^3 x/2\omega$ to eliminate a from the expressions (18.29), we find

$$I_\perp(\omega) = \frac{\sqrt{3}e^2\gamma \sin\alpha}{8\pi\epsilon_0 c}[F(x) + G(x)]$$
$$I_\parallel(\omega) = \frac{\sqrt{3}e^2\gamma \sin\alpha}{8\pi\epsilon_0 c}[F(x) - G(x)]$$

(18.32)

It is now convenient to define a *critical angular frequency* $\omega_c = 3c\gamma^3/2a$ so that $x = \omega/\omega_c$. We recall that a is the radius of curvature of the particle in its spiral orbit. Now, at any instant, the plane of the particle's orbit is inclined at a pitch angle α to the magnetic field. Therefore, with respect to the guiding centre of the particle's trajectory, the radius of curvature is $a = v/(\omega_r \sin \alpha)$, and hence

$$\omega_c = 2\pi\nu_c = \frac{3}{2}\left(\frac{c}{v}\right)\gamma^3\omega_r \sin \alpha \tag{18.33}$$

or, taking the limit $v \to c$ and rewriting the expression in terms of the *non-relativistic gyrofrequency* $\nu_g = eB/2\pi m_e = 28$ GHz T^{-1} (see Section 11.1), we find

$$\nu_c = \frac{3}{2}\gamma^2\nu_g \sin \alpha \tag{18.34}$$

This result is remarkably similar to that which we derived in Section 18.1.3 for the frequency at which most of the radiation is emitted, $\nu \approx \gamma^2\nu_g$.

Finally, we note that, in integrating over $2\pi \sin \theta \, d\theta$ in equations (18.29), the expressions (18.32) represent the energy emitted in the two polarisations during one period of the electron in its orbit, that is, in a time $T_r = \nu_r^{-1} = 2\pi\gamma m_e/eB$. Therefore, the emissivities of the electron in the two polarisations are

$$j_\perp(\omega) = \frac{I_\perp(\omega)}{T_r} = \frac{\sqrt{3}e^3 B \sin \alpha}{16\pi^2\epsilon_0 c m_e}[F(x) + G(x)]$$

$$j_\parallel(\omega) = \frac{I_\parallel(\omega)}{T_r} = \frac{\sqrt{3}e^3 B \sin \alpha}{16\pi^2\epsilon_0 c m_e}[F(x) - G(x)] \tag{18.35}$$

The total emissivity of a single electron by synchrotron radiation is the sum of $j_\perp(\omega)$ and $j_\parallel(\omega)$:

$$j(\omega) = j_\perp(\omega) + j_\parallel(\omega) = \frac{\sqrt{3}e^3 B \sin \alpha}{8\pi^2\epsilon_0 c m_e}F(x) \tag{18.36}$$

This is the spectral energy distribution we have been seeking. The spectrum is shown in Fig. 18.8 and the function $F(x)$ is given in tabular form in Table 18.1. This form of spectrum confirms the physical arguments given in Section 18.1.3. It has a broad maximum centred roughly at the frequency $\nu \approx \nu_c$, with $\Delta\nu/\nu \sim 1$. The maximum of the emission spectrum in fact has value $\nu_{max} = 0.29\nu_c$. The spectrum is continuous, and use is made of this feature in large synchrotron radiation machines such as the Synchrotron Radiation Source (SRS) at the Daresbury Laboratory, UK, to generate a precisely defined, high intensity, continuum spectrum at infrared, optical, ultraviolet and X-ray wavelengths.

Let us develop some further interesting features of the emission spectrum. First of all, let us take the integral of the emission spectrum to ensure that we have obtained the correct expression for the total energy loss rate:

$$-\frac{dE}{dt} = \int_0^\infty j(\omega)d\omega = \frac{\sqrt{3}e^3 B\omega_c \sin \alpha}{8\pi^2\epsilon_0 c m_e}\int_0^\infty F\left(\frac{\omega}{\omega_c}\right)d\left(\frac{\omega}{\omega_c}\right)$$

$$= \left(\frac{9\sqrt{3}}{4\pi}\right)\left(\frac{e^2}{6\pi\epsilon_0^2 c^4 m_e^2}\right)c\frac{B^2}{2\mu_0}\gamma^2 \sin^2 \alpha \int_0^\infty F(x)dx$$

$$= \sigma_T c U_{mag}\gamma^2 \sin^2 \alpha \left(\frac{9\sqrt{3}}{4\pi}\right)\int_0^\infty F(x)dx \tag{18.37}$$

We now use the integrals presented by Rybicki and Lightman (1979) to find

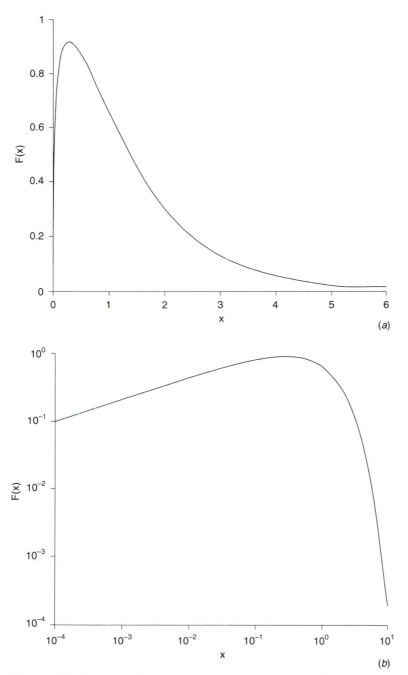

Figure 18.8. The intensity spectrum of the synchrotron radiation of a single electron shown (*a*) with linear axes and (*b*) with logarithmic axes. The function is plotted in terms of $x = \omega/\omega_{\mathrm{c}} = v/v_{\mathrm{c}}$, where ω_{c} is the critical angular frequency, $\omega_{\mathrm{c}} = 2\pi v_{\mathrm{c}} = \frac{3}{2}\left(\frac{c}{v}\right)\gamma^{3}\omega_{\mathrm{r}}\sin\alpha.$

Table 18.1. *The function $F(x)$, which describes the synchrotron radiation spectrum of a single ultrarelativistic electron (see expressions (18.36) and (18.31))*

x	$F(x)$	x	$F(x)$	x	$F(x)$
1.0×10^{-4}	0.0996	2.0×10^{-1}	0.904	1	0.655
1.0×10^{-3}	0.213	2.8×10^{-1}	0.918	2	0.301
1.0×10^{-2}	0.445	3.0×10^{-1}	0.918	3	0.130
3.0×10^{-2}	0.613	5.0×10^{-1}	0.872	5	2.14×10^{-2}
1.0×10^{-1}	0.818	8.0×10^{-1}	0.742	10	1.92×10^{-4}

the constant:

$$\int_0^\infty x^\mu F(x)\mathrm{d}x = \frac{2^{\mu+1}}{(\mu+2)}\Gamma\left(\frac{\mu}{2}+\frac{7}{3}\right)\Gamma\left(\frac{\mu}{2}+\frac{2}{3}\right) \tag{18.38}$$

$$\int_0^\infty x^\mu G(x)\mathrm{d}x = 2^\mu\Gamma\left(\frac{\mu}{2}+\frac{4}{3}\right)\Gamma\left(\frac{\mu}{2}+\frac{2}{3}\right) \tag{18.39}$$

From expression (18.38), setting $\mu = 0$, we find

$$\frac{9\sqrt{3}}{4\pi}\int_0^\infty F(x)\mathrm{d}x = \frac{9\sqrt{3}}{4\pi}\Gamma\left(\frac{7}{3}\right)\Gamma\left(\frac{2}{3}\right) = 2 \tag{18.40}$$

by use of the recurrence relations for Γ-functions given by Abramovitz and Stegun (1965, p. 255). Thus,

$$-\left(\frac{\mathrm{d}E}{\mathrm{d}t}\right) = 2\sigma_T c U_{\mathrm{mag}}\gamma^2 \sin^2\alpha \tag{18.41}$$

which is exactly the same result derived for the total energy loss rate (expression (18.6)) in Section 18.1.2.

The next interesting point is to look at the asymptotic expressions for the emissivity of the electron in the limits of high and low frequencies. The asymptotic expressions for the function $F(x)$ are quoted by Rybicki and Lightman (1979), and are as follows:

$$F(x) = \frac{4\pi}{\sqrt{3}\Gamma(1/3)}\left(\frac{x}{2}\right)^{1/3} \qquad x \ll 1 \tag{18.42}$$

$$F(x) = \left(\frac{\pi}{2}\right)^{1/2} x^{1/2}\exp(-x) \qquad x \gg 1 \tag{18.43}$$

The high frequency emissivity of the electron is therefore given by an expression of the form

$$j(v) \propto v^{1/2}\exp(-v/v_c) \tag{18.44}$$

which is dominated by the exponential cut-off at frequencies $v \gg v_c$. This simply means that there is very little power at frequencies $v > v_c$, which can be understood on the basis of the physical arguments developed in Section 18.1.3; namely, there is very little structure in the observed polar diagram of the radiation emitted by the electron at angles $\theta \ll \gamma^{-1}$.

At low frequencies, $v \ll v_c$, the spectrum is given by

$$j(\omega) = \frac{\sqrt{3}e^3 B \sin \alpha}{8\pi^2 \epsilon_0 cm_e} \frac{4\pi}{\sqrt{3}\Gamma\left(\frac{1}{3}\right)} \left(\frac{\omega}{2\omega_c}\right)^{1/3}$$

$$= \frac{e^2}{3^{1/3}\Gamma\left(\frac{1}{3}\right) 2\pi\epsilon_0 c} \left(\frac{eB \sin \alpha}{\gamma m_e}\right)^{2/3} \omega^{1/3}$$

that is, the emissivity is proportional to $v^{1/3}$. A very pleasant argument, given by Scheuer (1966), explains the origin of this dependence. Let us return to the expression for the vector potential $\mathbf{A}$, expression (3.18), which determines the intensity of the radiation field, and take the limit of small angles to the line of sight to the observer:

$$\mathbf{A} = \frac{\mu_0}{4\pi r} \frac{e\mathbf{v}}{1 - \left(\frac{v}{c}\cos\theta\right)} = \frac{\mu_0}{4\pi r} \frac{e\mathbf{v}}{1 - \frac{v}{c}\left(1 - \frac{\theta^2}{2}\right)}$$

$$= \frac{\mu_0}{4\pi r} \frac{e\mathbf{v}}{\left(1 - \frac{v}{c}\right) + \frac{v\theta^2}{2c}} \tag{18.45}$$

$$= \frac{\mu_0}{2\pi r} \frac{e\mathbf{v}}{\frac{1}{\gamma^2} + \theta^2}$$

using again the relation (18.13) and setting $v = c$. The radiation is strongly beamed in the forward direction, and, as we have shown, much of it is associated with angles $\theta < \gamma^{-1}$. This radiation is emitted at angular frequencies $\omega \sim \omega_c$, and is associated with the fact that the electron has velocity very close to the velocity of light. In this case,

$$\mathbf{A} \approx \frac{\mu_0 e\gamma^2 \mathbf{v}}{2\pi r}$$

However, at angles $\theta > \gamma^{-1}$, which correspond to Fourier components of the radiation field with frequencies less than v_c, it can be seen from expression (18.45) that the magnitude of the vector potential is determined by the angle θ rather than by how close the velocity of the electron is to the velocity of light:

$$\mathbf{A} \approx \frac{\mu_0 e\mathbf{v}}{2\pi r\theta^2}$$

To put it another way, the low frequency part of the emission spectrum should not depend upon the precise value of the Lorentz factor, γ. Another way of expressing this same result is that the intensity of emission should be independent of the rest mass of the particles responsible for the radiation. Let us therefore rewrite the expression for total energy loss rate of synchrotron radiation, eliminating the rest mass of the electron and expressing the result simply in terms of the relativistic gyrofrequency of the particle and the critical frequency, v_c. We know that

$$-\frac{dE}{dt} = \int_0^{\omega_c} j(\omega)d\omega = 2\sigma_T cU_{mag}\gamma^2 \sin^2 \alpha$$

Now, we write $\omega_c = \frac{3}{2}\gamma^3 \omega_r \sin \alpha$ and rewrite the expression for the total emission rate of the electron in terms of ω_r and ω_c. A simple calculation shows that

$$\frac{\mathrm{d}E}{\mathrm{d}t} = 2\left(\frac{e^4}{6\pi\epsilon_0^2 c^4 m_e^2}\right) c \frac{B^2}{2\mu_0} \gamma^2 \sin^2\alpha$$

$$= \frac{e^4 c^3 B^2}{6\pi\epsilon_0} \frac{\gamma^4}{E^2} \sin^2\alpha$$

(18.46)

where $E = \gamma m_e c^2$ is the total energy of the particle. Now,

$$\omega_r = eB/\gamma m_e = eBc^2/E$$

and hence

$$-\frac{\mathrm{d}E}{\mathrm{d}t} = \frac{e^2 \omega_r^2 \sin^2\alpha}{6\pi\epsilon_0 c}\gamma^4$$

Substituting for γ^4, we find

$$-\frac{\mathrm{d}E}{\mathrm{d}t} = \int_0^{\omega_c} j(\omega)\mathrm{d}\omega = \left(\frac{2}{3}\right)^{4/3} \frac{e^2(\omega_r \sin\alpha)^{2/3}}{6\pi\epsilon_0 c}\omega_c^{4/3}$$

which depends only upon the angular gyrofrequencies ω_r and ω_c. The angular gyrofrequency simply describes the trajectory of the particle and depends only upon the total energy of the particle rather than its mass since $\omega_r = eBc^2/E$. Therefore, we can differentiate the expression (18.46) and find

$$j(\omega) = \left(\frac{2}{3}\right)^{4/3} \frac{2e^2(\omega_r \sin\alpha)^{2/3}}{9\pi\epsilon_0 c}\omega^{1/3}$$

This is of exactly the same form as found above from the exact analysis, apart from a slightly different constant.

18.1.5   The synchrotron radiation of a power-law distribution of electron energies

Now that we have derived the expressions (18.35), (18.37), (18.38) and (18.39), we can readily work out many other useful formulae for the observed properties of synchrotron radiation from astrophysical sources. The first thing to do is to work out the radiation spectrum of a distribution of particle energies rather than from particles of a single energy. We have seen that the energy spectra of cosmic rays and cosmic ray electrons can be approximated by power-law distributions. Let us therefore work out the emission spectrum of an electron energy distribution of power-law form $N(E)\mathrm{d}E = \kappa E^{-p}\,\mathrm{d}E$, where we will always take $N(E)\,\mathrm{d}E$ to be the number of electrons per unit volume in the energy interval E to $E + \mathrm{d}E$. Notice that sometimes the energy spectra are written in terms of the Lorentz factors, γ, of the particles rather than their energies, $N(\gamma)\,\mathrm{d}\gamma = \kappa'\gamma^{-p}\,\mathrm{d}\gamma$ – to convert between these conventions, we see that $\kappa = \kappa'(m_e c^2)^{(p-1)}$. Let us work out the proper answer first and then look at the results from a more physical point of view.

First of all, we assume that the power-law distribution of electron energies applies to electrons with a fixed pitch angle α. We have to integrate the contributions of electrons of different energies to the intensity at angular frequency ω.

This means integrating over all values of x. We recall that

$$x = \frac{\omega}{\omega_c} = \frac{\omega}{\frac{3}{2}\gamma^2\omega_g \sin\alpha} = \frac{2\omega m_e^2 c^4}{3E^2\omega_g \sin\alpha} = \frac{A}{E^2} \tag{18.47}$$

As expected, at a particular frequency, we sum over the low energy tail of $F(x)$ for high energy electrons and the over the exponential cut-offs for low energy electrons. The emissivity per unit volume is therefore

$$J(\omega) = \int_0^\infty j(x)\kappa E^{-p}\,dE$$

From expression (18.47), we find

$$E = (A/x)^{1/2} \quad ; \quad dE = -1/2A^{1/2}x^{-3/2}dx$$

and therefore

$$J(\omega) = \frac{\kappa}{2A^{(p-1)/2}} \int_0^\infty j(x)x^{(p-3)/2}dx$$

$$= \frac{\sqrt{3}e^3 B\kappa \sin\alpha}{16\pi^2\epsilon_0 cm_e A^{(p-1)/2}} \int_0^\infty F(x)x^{(p-3)/2}dx$$

We can now use the integral (18.38) with $\mu = (p-3)/2$. We find

$$J(\omega) = \frac{\sqrt{3}e^3 B\kappa \sin\alpha}{8\pi^2\epsilon_0 cm_e(p+1)} \left(\frac{\omega m_e^3 c^4}{3eB \sin\alpha}\right)^{-(p-1)/2}$$

$$\times \; \Gamma\left(\frac{p}{4} + \frac{19}{12}\right)\Gamma\left(\frac{p}{4} - \frac{1}{12}\right) \tag{18.48}$$

To complete the analysis, we can now integrate over the pitch angle α. Notice that the emissivity of the electron at a particular frequency, ω, depends strongly upon α, as shown by the relations (18.47) and (18.48). We can, of course, use any pitch angle distribution we like, but the most important case is that of an isotropic distribution of pitch angles, for which the probability distribution of α is $\frac{1}{2}\sin\alpha \, d\alpha$. We use the result

$$\frac{1}{2}\int_0^\pi \sin^{(p+3)/2}\alpha \, d\alpha = \frac{\sqrt{\pi}\Gamma\left(\frac{p+5}{4}\right)}{2\Gamma\left(\frac{p+7}{4}\right)}$$

so that the emission per unit volume can be written

$$J(\omega) = \frac{\sqrt{3}e^3 B\kappa}{16\pi^2\epsilon_0 cm_e(p+1)} \left(\frac{\omega m_e^3 c^4}{3eB}\right)^{-(p-1)/2}$$

$$\times \; \frac{\sqrt{\pi}\Gamma\left(\frac{p}{4} + \frac{19}{12}\right)\Gamma\left(\frac{p}{4} - \frac{1}{12}\right)\Gamma\left(\frac{p}{4} + \frac{5}{4}\right)}{\Gamma\left(\frac{p}{4} + \frac{7}{4}\right)} \tag{18.49}$$

We observe that the key dependences for the emissivity in the above formulae are

$$J(\omega) \propto \kappa B^{(p+1)/2}\omega^{-(p-1)/2} \tag{18.50}$$

We note the important result that, if the electron energy spectrum has power-law index p, the spectral index of the synchrotron emission of these electrons is $\alpha = (p-1)/2$.

It is useful to provide a much simpler derivation of these dependences, which will help elucidate some physical points. We have shown that the emitted spectrum of electrons of energy E is quite sharply peaked near the critical frequency v_c (Fig. 18.8) and is certainly very much narrower than the breadth of the electron energy spectrum. Therefore, to a good approximation, it may be assumed that all the radiation of an electron of energy E is radiated at the critical frequency v_c, which we may approximate by

$$v \approx v_c \approx \gamma^2 v_g = \left(\frac{E}{m_e c^2} \right)^2 v_g \qquad v_g = \frac{eB}{2\pi m_e} \qquad (18.51)$$

Therefore, the energy radiated in the frequency range v to $v + dv$ can be attributed to electrons with energies in the range E to $E + dE$. We may therefore write

$$J(v)dv = \left(-\frac{dE}{dt} \right) N(E)dE \qquad (18.52)$$

Now,

$$\left\{ \begin{array}{c} E = \gamma m_e c^2 = \left(\dfrac{v}{v_g} \right)^{1/2} m_e c^2 \\[3mm] dE = \dfrac{m_e c^2}{2 v_g^{1/2}} v^{-1/2} dv \\[3mm] -\left(\dfrac{dE}{dt} \right) = \dfrac{4}{3} \sigma_T c \left(\dfrac{E}{m_e c^2} \right)^2 \dfrac{B^2}{2\mu_0} \end{array} \right\} \qquad (18.53)$$

Substituting these quantities into equation (18.52), we find that the emissivity may be expressed in terms of κ, B, v and fundamental constants:

$$J(v) = \text{(constants)} \; \kappa B^{(p+1)/2} v^{-(p-1)/2} \qquad (18.54)$$

The constant has the same form as in the expression (18.49). This is exactly the functional dependence which comes out of the full analysis. The physical basis of the process is perhaps more evident from the simple physical argument. The spectral shape is determined by the shape of the electron energy spectrum rather than by the shape of the emission spectrum of a single particle. The quadratic nature of the relation between emitted frequency and the energy of the electron accounts for the difference in slopes of the emission spectrum and the electron energy spectrum.

18.1.6   *The polarisation of synchrotron radiation*

As we discussed in Section 18.1.2, if we look along the magnetic field lines, the radiation of a non-relativistic electron is circularly polarised, and, in general, when looked at at any angle, the radiation is elliptically polarised. In the case of relativistic electrons, however, a significant amount of radiation is only observed if the trajectory of the electron lies within an angle $1/\gamma$ of the line of sight. To understand the polarisation properties of synchrotron radiation, it is helpful to introduce the concept of *the velocity cone*, which is the cone described by the velocity vector **v** of the electron as it spirals about the magnetic field. The axis of the cone is the magnetic field direction and the velocity vector precesses about this direction at the relativistic gyrofrequency.

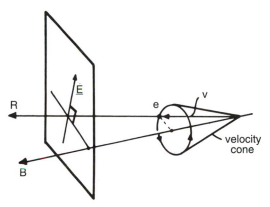

Figure 18.9. Illustrating the geometry of the velocity cone of an ultrarelativistic electron and the polarisation of the emitted radiation.

Consider first the case of those electrons whose velocity cones lie precisely along the line of sight to the observer (see Fig. 18.9). At the instant the electron points directly to the observer, its acceleration vector, $\mathbf{a}$, is in the direction $\mathbf{v} \times \mathbf{B}$, and the observed radiation is linearly polarised parallel to the direction $\mathbf{v} \times \mathbf{B}$ and lying in the plane perpendicular to the wave vector $\mathbf{k}$, as indicated by the vector $\mathbf{E}$ in Fig. 18.9. In this case, the $\mathbf{E}$ vector is perpendicular to the projection of $\mathbf{B}$ onto the plane of the sky. In fact, as we have shown in Section 18.1.3, when we integrate over the dynamical history of the particle, there is also a component parallel to the magnetic field direction associated with the radiation observed when the electron is not precisely pointing to the observer within the cone of angle $1/\gamma$. The radiation from a single electron is elliptically polarised because the component parallel to the field has a different time dependence within each pulse compared with that of the perpendicular component. This is reflected in the fact that the frequency spectra of the two polarisations are different (Fig. 18.10).

When there is a distribution of pitch angles, however, all the electrons with velocity cones within the angle $1/\gamma$ of the line of sight contribute to the intensity measured by the observer. These contributions are elliptically polarised in opposite senses on either side of the velocity cone. The total net polarisation is found by integrating over all particles which contribute to the intensity, and, because the beamwidth $1/\gamma$ on either side of the line of sight is very small when the electron is highly relativistic, the components of elliptical polarisation parallel to the projection of $\mathbf{B}$ cancel out and the resultant polarisation is linear. This means that we obtain the correct expression for the linearly polarised component of the radiation if we take averages of the $j_{\parallel}$ and $j_{\perp}$ components and neglect their time variation through the pulse.

We work out the exact results for the linear polarisation of synchrotron radiation using the formulae derived above. First of all, let us consider the emission of a single electron and work out the amount of energy in each polarisation. We

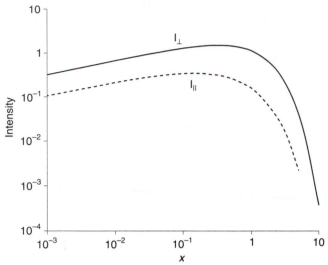

Figure 18.10. The intensity spectra of the two polarisations $I_\perp$ and $I_\parallel$ of synchrotron radiation for a single high energy electron.

use the formulae (18.35), (18.38) and (18.39) to find

$$\frac{I_\perp}{I_\parallel} = \frac{\int_0^\infty [F(x) + G(x)]\mathrm{d}x}{\int_0^\infty [F(x) - G(x)]\mathrm{d}x}$$

Using expressions (18.38) and (18.39) with $\mu = 0$, we find

$$\frac{I_\perp}{I_\parallel} = \frac{\Gamma\left(\frac{7}{3}\right)\Gamma\left(\frac{2}{3}\right) + \Gamma\left(\frac{4}{3}\right)\Gamma\left(\frac{2}{3}\right)}{\Gamma\left(\frac{7}{3}\right)\Gamma\left(\frac{2}{3}\right) - \Gamma\left(\frac{4}{3}\right)\Gamma\left(\frac{2}{3}\right)}$$

Since

$$\Gamma(n+1) = n\Gamma(n), \tag{18.55}$$

$$\frac{I_\perp}{I_\parallel} = \frac{\frac{4}{3}+1}{\frac{4}{3}-1} = 7 \tag{18.56}$$

Thus, the energy liberated in the two polarisations from a single electron is exactly in the ratio 7:1, a result which is derived at an early stage in his analysis by Le Roux (1961).

We have already derived the formulae necessary to work out the fractional polarisation as a function of frequency for a single electron. The fractional polarisation is defined to be

$$\Pi = \frac{I_\perp(\omega) - I_\parallel(\omega)}{I_\perp(\omega) + I_\parallel(\omega)} \tag{18.57}$$

Therefore, inserting the expressions for the emissivities in the two polarisations given by the expressions (18.35), we find

$$\Pi(\omega) = \frac{G(x)}{F(x)} \tag{18.58}$$

This function is displayed in Fig. 18.11.

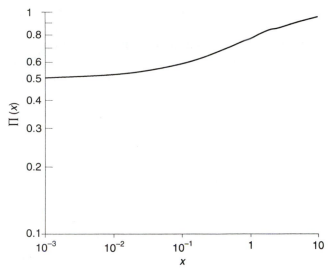

Figure 18.11. The polarisation Π of the synchrotron radiation of a single electron as a function of frequency.

The most useful result is the percentage polarisation at frequency ω from a power-law distribution of electron energies. If the particles have energy spectra $N(E) = \kappa E^{-p}\,dE$, we have to average over the range of energies which can contribute to the intensity observed at frequency ω. Performing exactly the same type of calculation as was carried out in Section 18.1.4, we find that the fractional polarisation is

$$\Pi = \frac{\int_0^\infty G(x)x^{(p-3)/2}dx}{\int_0^\infty F(x)x^{(p-3)/2}dx}$$

Using again the expressions (18.38), (18.39) and (18.55), we find

$$\Pi = \frac{p+1}{4}\frac{\Gamma\left(\frac{p}{4}+\frac{7}{12}\right)}{\Gamma\left(\frac{p}{4}+\frac{19}{12}\right)} = \frac{p+1}{4\left(\frac{p}{4}+\frac{7}{12}\right)}$$

$$= \frac{p+1}{p+\frac{7}{3}} \tag{18.59}$$

It can be seen that, for a typical value of the exponent of the energy spectrum of the electrons, $p = 2.5$, the fractional polarisation of synchrotron radiation is expected to be about 72%, that is, the synchrotron radiation of the electrons in a uniform magnetic field is expected to be highly polarised.

If the particles do not have extreme values of γ, some circular polarisation may be observable because of the inexact cancellation of the elliptically polarised components on either side of the velocity cone. There are two reasons for this. First of all, the number of particles on either side of the velocity cone are different simply because of the $\sin\alpha$ factor in the expression for the solid angle contained within $d\alpha$, $d\Omega = \frac{1}{2}\sin\alpha\,d\alpha$. Secondly, within the cone $\theta \sim 1/\gamma$, the particles which radiate with smaller values of α must have larger energies to radiate at frequency

ω because the frequency at which most of the radiation is emitted is $\omega = \gamma^2 \omega_g \sin \alpha$. Because $N(E) = \kappa E^{-p}$, different numbers of electrons radiate at frequency ω on either side of the velocity cone. These two effects mean that the cancellation of the elliptical polarisation is not exact, particularly if the values of γ are small. These somewhat lengthy calculations have been carried out by Ginzburg, Sazonov and Syrovatskii (1968) and by Legg and Westfold (1968). To order of magnitude, the fractional circular polarisation amounts to about γ^{-1} of the linear polarisation, and the effect is therefore quite small. Circular polarisation has been detected from some compact sources of radio emission at about the 1% level, and these are important observations because they provide independent information about the energies of the electrons emitting the synchrotron radiation.

18.1.7 Synchrotron self-absorption

According to the principle of detailed balance, to every emission process there is a corresponding absorption process – in the case of synchrotron radiation, this is known as *synchrotron self-absorption*. Let us give first of all some physical arguments.

Suppose a source of synchrotron radiation has a power-law spectrum, $S_\nu \propto \nu^{-\alpha}$, where the spectral index $\alpha = (p - 1)/2$. If the source has the same physical size at all frequencies, its *brightness temperature*, $T_b = (\lambda^2/2k)(S_\nu/\Omega)$, is proportional to $\nu^{-(2+\alpha)}$, where S_ν is its flux density and Ω is the solid angle it subtends at the observer (see Section 8.5.2). Therefore, at low enough frequencies, the brightness temperature of the radiation may approach the kinetic temperature of the radiating electrons. When this occurs, we expect self-absorption effects to become important since thermodynamically the radiating electrons cannot result in a brightness temperature greater than their kinetic temperature.

We have derived the expressions for the synchrotron radiation spectrum of a power-law energy distribution of relativistic electrons, $N(E)\mathrm{d}E = \kappa E^{-p}\,\mathrm{d}E$ in Section 18.1.4. Now, this energy spectrum is *not* a thermal equilibrium spectrum, which for relativistic particles would be a *relativistic Maxwellian distribution*. The concept of temperature can still be used, however, for particles of a particular energy E for the following reasons. First, the spectrum of the radiation emitted by particles of energy E is peaked about the critical frequency $\nu \approx \nu_c \approx \gamma^2 \nu_g$, where $\gamma = E/m_e c^2 \gg 1$ and $\nu_g = eB/2\pi m_e$ is the non-relativistic gyrofrequency. Thus, the emission and absorption processes at frequency ν are associated with electrons of roughly the same energy. Secondly, the characteristic timescale for the relativistic electron gas to relax to an equilibrium spectrum is very long indeed under typical cosmic conditions because the particle number densities are very small and all interaction times with matter are very long. Therefore, we can associate a temperature T_e with electrons of a given energy through the relativistic formula which relates particle energy to temperature:

$$\gamma m_e c^2 = 3k T_e \tag{18.60}$$

One way of understanding the difference between this result and the standard

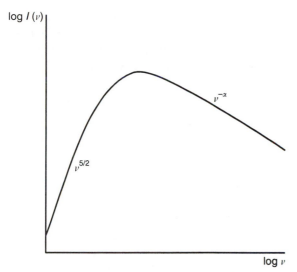

Figure 18.12. The spectrum of a source of synchrotron radiation which exhibits the phenomenon of synchrotron self-absorption.

result of kinetic theory, $E = \frac{3}{2}kT_e$, is to recall that the ratio of specific heats γ_{SH} is $\frac{4}{3}$ for a relativistic gas and $\frac{5}{3}$ for a non-relativistic gas. The internal thermal energy density of a gas is $u = NkT/(\gamma_{SH} - 1)$, where N is the number density of particles. Setting $\gamma_{SH} = \frac{5}{3}$, we obtain the classical result, and, setting $\gamma_{SH} = \frac{4}{3}$, we obtain the expression (18.60) for the energy per particle.

The important point is that the *effective temperature* of the particles now becomes a function of their energy. Since $\gamma \approx (v/v_g)^{1/2}$,

$$T_e \approx (m_e c^2/3k)(v/v_g)^{1/2}$$

For a self-absorbed source, the brightness temperature of the radiation must be equal to the kinetic temperature of the emitting particles, $T_b = T_e$, and therefore, in the Rayleigh–Jeans limit,

$$S_v = \frac{2kT}{\lambda^2}\Omega = \frac{2m_e}{3v_g^{1/2}}\Omega v^{5/2} \propto \frac{\theta^2 v^{5/2}}{B^{1/2}} \tag{18.61}$$

where Ω is the solid angle subtended by the source, $\Omega \approx \theta^2$.

This calculation demonstrates the physical origin of the steep low frequency spectrum expected in sources in which synchrotron self-absorption is important, $S_v \propto v^{5/2}$. It does not follow the standard Rayleigh–Jeans law because the effective kinetic temperature of the electrons varies with frequency. Notice that, in a self-absorbed source, the spectral form $S_v \propto v^{5/2}$ is independent of the spectrum of the emitting particles so long as the magnetic field is uniform. The typical spectrum of a self-absorbed radio source is shown in Fig. 18.12. Spectra of roughly this form are found at radio, centimetre and millimetre wavelengths in the nuclei of active galaxies and quasars, and it is conventionally assumed that synchrotron self-absorption is the process responsible for these low frequency cut-offs.

Let us now work out the absorption coefficient for synchrotron self-absorption. The simplest way of proceeding is to regard the emission of a photon of energy $h\nu$ by the synchrotron radiation process as originating in a two-level system, in which the electron makes a transition from a state with energy E and momentum $\mathbf{p}$ (level 2) to one with energy $E' = E - dE$ and momentum $\mathbf{p}' = \mathbf{p} - d\mathbf{p}$ (level 1). We have worked out classically the emission coefficient for this process, and hence we can write down a spontaneous transition probability which describes the rate of emission of photons in the frequency interval ν to $\nu + d\nu$:

$$A_{21} = \frac{j(\nu, E)}{h\nu} \qquad \text{photons Hz}^{-1}\text{s}^{-1} \tag{18.62}$$

where $j(\nu)$ is now the emissivity per unit frequency interval rather than per unit angular frequency, that is, $j(\nu, E) = 2\pi j(\omega, E)$. Notice that this expression contains no information about the directional properties of the radiation. In fact, we will derive the absorption coefficient under the assumption that the magnetic field is chaotic and the particle distribution within the source region is isotropic. In this case, it is safe to assume that the emission from the source region is isotropic. There are complexities in a more complete calculation which are discussed by Ginzburg and Syrovatskii (1969).

We can immediately write down the Einstein coefficients for absorption and induced emission using the expressions (3.54):

$$A_{21} = \frac{2h\nu^3}{c^2}B_{12} = \frac{2h\nu^3}{c^2}B_{21} \tag{18.63}$$

Now, the Einstein coefficients are defined in terms of the number density, $n(\mathbf{p})$, of particles per unit volume of phase space, $d^3\mathbf{p}$, rather than per unit energy interval. The absorption coefficient is then given by the usual expression involving the Einstein coefficients, but now for pairs of states separated in momentum by $d\mathbf{p} = (h\nu/c)\, \mathbf{i}_k$. For a particular pair of states, we can write, following the expression (3.57),

$$\chi_\nu = \frac{h\nu}{4\pi}[n(\mathbf{p} - \hbar\mathbf{k})B_{12}\, d^3\mathbf{p} - n(\mathbf{p})B_{21}\, d^3\mathbf{p}] \tag{18.64}$$

Making a Taylor expansion for small values of $h\nu/c$, we find

$$n(\mathbf{p} - \hbar\mathbf{k}) = n(\mathbf{p}) - \frac{h\nu}{c}\frac{dn}{dp}$$

$$\chi_\nu = -\frac{h^2\nu^2}{4\pi c}B_{12}\frac{dn}{dp}d^3\mathbf{p}$$

This result has to be integrated over all possible pairs of electron momenta which could be involved in the absorption process, which means that we have to integrate over all momentum states. We assume an isotropic particle distribution in momentum space, and so we can write

$$\chi_\nu = -\int_0^\infty \frac{h^2\nu^2}{4\pi c}B_{12}\frac{dn}{dp}4\pi p^2 dp \tag{18.65}$$

$$= -\frac{hc}{2\nu}\int_0^\infty A_{21}\frac{dn}{dp}p^2 dp$$

$$= -\frac{c}{2v^2} \int_0^\infty j(v, E) \frac{\mathrm{d}n}{\mathrm{d}p} p^2 \mathrm{d}p \tag{18.66}$$

Now we convert the electron momentum spectrum into an electron energy spectrum

$$p = E/c \qquad\qquad \mathrm{d}p = \mathrm{d}E/c \tag{18.67}$$

Therefore,

$$4\pi p^2 n(p)\mathrm{d}p = N(E)\mathrm{d}E$$

$$n(p) = \frac{c^3}{4\pi} \frac{N(E)}{E^2}$$

$$\chi_v = -\frac{c^2}{8\pi v^2} \int_0^\infty j(v, E) \frac{\mathrm{d}}{\mathrm{d}E} \left(\frac{N(E)}{E^2}\right) E^2 \mathrm{d}E \tag{18.68}$$

For a power-law distribution of electron energies, $N(E) = \kappa E^{-p}$, we find

$$\chi_v = \frac{(p+2)\kappa c^2}{8\pi v^2} \int_0^\infty j(v, E) E^{-(p+1)} \mathrm{d}E \tag{18.69}$$

We now have to insert the expression for $j(v, E)$ into equation (18.69) and make the same reductions as in the previous subsections. Thus,

$$j(v, E) = 2\pi j(\omega, E) = \frac{\sqrt{3}e^3 B \sin\alpha}{4\pi\epsilon_0 c m_\mathrm{e}} F(x)$$

$$\chi_v = \frac{\sqrt{3}e^3 B c\kappa \sin\alpha}{32\pi^2 \epsilon_0 m_\mathrm{e} v^2}(p+2) \int_0^\infty F(x) E^{-(p+1)} \mathrm{d}E$$

We now use one of our standard results (18.38) to perform the integral. Setting

$$x = v/v_\mathrm{c} = v/\left(\tfrac{3}{2}\gamma^2 v_\mathrm{g} \sin\alpha\right) = A/E^2$$

We find that

$$\int_0^\infty F(x) E^{-(p+1)} \mathrm{d}E = \frac{1}{(p+2)} \left(\frac{A}{2}\right)^{-p/2} \Gamma\left(\frac{3p+22}{12}\right) \Gamma\left(\frac{3p+2}{12}\right)$$

Thus, the expression for the absorption coefficient becomes

$$\chi_v = \frac{\sqrt{3}e^3 B\kappa c \sin\alpha}{32\pi^2 \epsilon_0 m_\mathrm{e} v^2} \left(\frac{3eB \sin\alpha}{2\pi m_\mathrm{e}^3 c^4 v}\right)^{p/2} \Gamma\left(\frac{3p+22}{12}\right) \Gamma\left(\frac{3p+2}{12}\right)$$

$$= \frac{\sqrt{3}e^3 \kappa c}{32\pi^2 \epsilon_0 m_\mathrm{e}} \left(\frac{3e}{2\pi m_\mathrm{e}^3 c^4}\right)^{p/2} \Gamma\left(\frac{3p+22}{12}\right) \Gamma\left(\frac{3p+2}{12}\right) (B \sin\alpha)^{(p+2)/2} v^{-(p+4)/2}$$

$$\tag{18.70}$$

For a randomly oriented magnetic field, we average over the angle α. Following the same procedure as before, the probability distribution for the angle α is $\frac{1}{2} \sin\alpha$ $\mathrm{d}\alpha$, and hence we have to evaluate the following integral:

$$\int_0^\infty \frac{1}{2} \sin\alpha \sin^{(p+2)/2} \mathrm{d}\alpha = \frac{\sqrt{\pi}\Gamma\left(\frac{p+6}{4}\right)}{2\Gamma\left(\frac{p+8}{4}\right)}$$

Therefore, the absorption coefficient for synchrotron radiation in a randomly oriented magnetic field is

$$\chi_v = \frac{\sqrt{3\pi}e^3\kappa B^{(p+2)/2}c}{64\pi^2\epsilon_0 m_e}\left(\frac{3e}{2\pi m_e^3 c^4}\right)^{p/2}\frac{\Gamma\left(\frac{3p+22}{12}\right)\Gamma\left(\frac{3p+2}{12}\right)\Gamma\left(\frac{p+6}{4}\right)}{\Gamma\left(\frac{p+8}{4}\right)}v^{-(p+4)/2}$$

(18.71)

Now, to work out the emission spectrum from, say, a slab of thickness l, we write down

$$\frac{dI_v}{dx} = -\chi_v I_v + \frac{J(v)}{4\pi}$$

The solution is

$$I_v = \frac{J(v)}{4\pi\chi_v}[1 - \exp(-\chi_v l)]$$

(18.72)

If the source is optically thin, $\chi_v l \ll 1$, we obtain

$$I_v = \frac{J(v)l}{4\pi}$$

If the source is optically thick, $\chi_v l \gg 1$, we find

$$I_v = \frac{J(v)}{4\pi\chi_v}$$

The quantity $J(v)/4\pi\chi_v$ is often referred to as the *source function*. Substituting the expression (18.71) for the absorption coefficient $\chi(v)$ and the expression (18.49) for J_v, we obtain

$$I_v = (\text{constant})\ \frac{m_e v^{5/2}}{v_g^{1/2}}$$

where the constant is a number of order unity which involves numerous gamma functions. This is the same dependence as was found from our physical arguments above (expression (18.61)).

In a more complete analysis, we would work out separately the absorption coefficients in the two separate polarisations (see, for example, Ginzburg and Syrovatskii (1969) and references therein). It turns out that the absorption coefficients for the two polarisations are different. In the optically thick region, the radiation is expected to be polarised, but with the electric vector of the emitted radiation being parallel rather than perpendicular to the magnetic field. The degree of polarisation is

$$\Pi = \left|\frac{I_\perp - I_\parallel}{I_\perp + I_\parallel}\right| = \frac{3}{6p + 13}$$

18.1.8   *Useful numerical results*

It is convenient to have at hand a set of numerical constants for the various relations derived in the preceding sections. First, the total energy loss rate by synchrotron radiation is

$$-\left(\frac{dE}{dt}\right) = 2\sigma_T c U_{\text{mag}}\gamma^2\left(\frac{v}{c}\right)^2\sin^2\theta$$

(18.6)

and can be written

$$-\left(\frac{dE}{dt}\right) = 1.587 \times 10^{-14} B^2 \gamma^2 \left(\frac{v}{c}\right)^2 \sin^2 \theta \quad \text{W}$$

where the units of magnetic flux density B are tesla and γ is the Lorentz factor, $\gamma = (1 - v^2/c^2)^{-1/2}$. When averaged over an isotropic distribution of pitch angles θ, the result is

$$-\left(\frac{dE}{dt}\right) = \frac{4}{3}\sigma_T c U_{\text{mag}} \gamma^2 \left(\frac{v}{c}\right)^2 \tag{18.7}$$

which can be written

$$-\left(\frac{dE}{dt}\right) = 1.058 \times 10^{-14} B^2 \gamma^2 \left(\frac{v}{c}\right)^2 \quad \text{W}$$

The emission spectrum of a single electron is

$$j(v) = 2\pi j(\omega) = \frac{\sqrt{3} e^3 B \sin \alpha}{4\pi\epsilon_0 c m_e} F\left(\frac{v}{v_c}\right) \tag{18.36}$$

which becomes

$$j(v) = 2.344 \times 10^{-25} B \sin \alpha \; F\left(\frac{v}{v_c}\right) \quad \text{W Hz}^{-1}$$

where again B is expressed in tesla and the function $F(v/v_c)$ is given in Table 18.1. The critical frequency, v_c, is given by

$$v_c = \frac{3}{2}\gamma^2 \frac{eB}{2\pi m_e} = 4.199 \times 10^{10} \gamma^2 B \quad \text{Hz}$$

where B is measured in tesla.

The radiation spectrum of a power-law electron energy distribution $N(E) = \kappa E^{-p}$ in the case of a random magnetic field is

$$J(v) = 2\pi J(\omega) = \frac{\sqrt{3} e^3 B \kappa}{4\pi\epsilon_0 c m_e}\left(\frac{3eB}{2\pi v m_e^3 c^4}\right)^{(p-1)/2} a(p)$$

where

$$a(p) = \frac{\sqrt{\pi}}{2}\frac{\Gamma\left(\frac{p}{4} + \frac{19}{12}\right)\Gamma\left(\frac{p}{4} - \frac{1}{12}\right)\Gamma\left(\frac{p}{4} + \frac{5}{4}\right)}{(p+1)\Gamma\left(\frac{p}{4} + \frac{7}{4}\right)} \tag{18.49}$$

In SI units, this becomes

$$J(v) = 2.344 \times 10^{-25} a(p) B^{(p+1)/2} \kappa \left(\frac{1.253 \times 10^{37}}{v}\right)^{(p-1)/2} \quad \text{W m}^{-3}\text{Hz}^{-1} \tag{18.73}$$

The constant $a(p)$ depends upon the energy spectral index p, and appropriate values of $a(p)$ are given in Table 18.2. This relation is only useful for those who wish to write the energy of the particles in joules, that is, the energy spectrum $N(E)$ represents the number density of electrons per joule. This is highly non-standard, and so I also quote the result if the energies of the electrons are measured in giga-electron-volts and the units of $N(E)$ are particles per cubic metre per giga-electron-volt:

$$J(v) = 2.344 \times 10^{-25} a(p) B^{(p+1)/2} \kappa' \left(\frac{3.217 \times 10^{17}}{v}\right)^{(p-1)/2} \quad \text{W m}^{-3}\text{Hz}^{-1}$$

Table 18.2   *Constants for use with the synchrotron radiation formulae*

p	a(p)	b(p)	p	a(p)	b(p)
1	2.056	0.397	3.5	0.217	0.230
1.5	0.909	0.314	4	0.186	0.236
2	0.529	0.269	4.5	0.167	0.248
2.5	0.359	0.244	5	0.157	0.268
3	0.269	0.233			

Finally, it is convenient to have an expression for the absorption coefficient of synchrotron radiation. Following the same procedure as above, the absorption coefficient for a random magnetic field

$$\chi_v = \frac{\sqrt{3}e^3 c}{8\pi^2 \epsilon_0 m_e} \kappa B^{(p+2)/2} \left(\frac{3e}{2\pi m_e^3 c^4}\right)^{p/2} b(p) v^{-(p+4)/2}$$

$$b(p) = \frac{\sqrt{\pi}}{8} \frac{\Gamma\left(\frac{3p+22}{12}\right) \Gamma\left(\frac{3p+2}{12}\right) \Gamma\left(\frac{p+6}{4}\right)}{\Gamma\left(\frac{p+8}{4}\right)}$$

(18.74)

becomes, in SI units,

$$\chi_v = 3.354 \times 10^{-9} \kappa B^{(p+2)/2} (3.54 \times 10^{18})^p b(p) v^{-(p+4)/2} \quad \text{m}^{-1}$$

where the constant $b(p)$ depends upon the exponent p, as listed in Table 18.2. In this version, the energies of the particles are expressed in joules. If, instead, the $N(E)$ is expressed particles per cubic metre per giga-electron-volt, the expression becomes

$$\chi_v = 20.9 \ \kappa' B^{(p+2)/2} (5.67 \times 10^9)^p b(p) v^{-(p+4)/2} \quad \text{m}^{-1}$$

18.2   The radio emission of the Galaxy

It is very fortunate that we have the immediate possibility of testing the theory of synchrotron radiation in its astrophysical context by studying the intensity and spectrum of the Galactic radio emission. Continuum radio emission from the Galaxy was the first radio astronomical discovery, and was detected in the pioneering observations of Karl Jansky in 1933. Fig. 1.8(*a*) is a recent radio map of the whole sky at a frequency of 408 MHz, and is the result of many years of observation by a team led by Dr Glyn Haslam of the Max-Planck-Institut für Radioastronomie at Bonn. It can be seen that the radio map shows a very clear 'radio disc', similar, in general terms, to the optical disc of the Galaxy (Fig. 1.8(*d*)). In addition, there are various 'loops' which extend out of the Galactic plane, the most prominent being the feature known as the North Polar Spur, which originates at $l = 30°$ and extends toward the Galactic North Pole.

By the early 1940s, it was clear that the spectrum of the Galactic radio emission could not be accounted for by a black-body spectrum or by the spectrum of

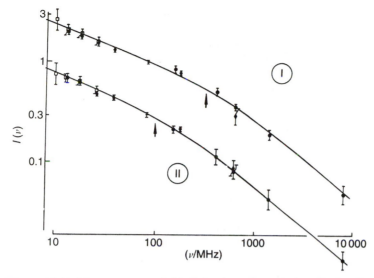

Figure 18.13. The spectrum of the Galactic radio emission. Region I corresponds to the anticentre direction at high galactic latitudes, and region II corresponds to the interarm region. (From A.S. Webster (1971). *Cosmic ray electrons, and Galactic radio emission.* Ph.D. dissertation; and A.S. Webster (1974). *Mon. Not. R. Astron. Soc.*, **166**, 355.)

thermal bremsstrahlung from hot gas clouds. There the matter rested until the 1950s, when the first detailed radio maps of the Galaxy were made and when, in particular, it was discovered that the radiation is highly polarised. It was soon realised that these properties of the 'non-thermal' Galactic emission are the signature of synchrotron radiation, the early history of which is described by Ginzburg and Syrovatskii (1965).

The determination of the Galactic radio spectrum is a particularly difficult observational problem because the Galactic radio emission extends over the whole sky, and so, even in directions far away from the direction in which the telescope is pointing, some radiation creeps into the receiver through far-out sidelobes of the telescope beam. Thus, the polar diagram of the radio telescope must be known very accurately. The best observations of the background spectrum are made with geometrically scaled aerials so that the reception pattern is identical at different wavelengths.

The spectra of the Galactic radio emission in the direction of the North Galactic Pole and in the anticentre direction are shown in Fig. 18.13. At frequencies less than about 200 MHz, the spectrum can be described by a power law of the form $I(\nu) \propto \nu^{-0.4}$; at frequencies greater than about 400 MHz, the spectrum steepens, the spectral index being about 0.8–0.9 (Webster (1971); see caption to Fig. 18.13 for reference). This spectrum can be compared with the predicted spectrum if the energy spectrum of cosmic ray electrons observed at the top of the atmosphere is assumed to be representative of the local interstellar medium as a whole. There are problems in making this comparison, both with regard to the observations of

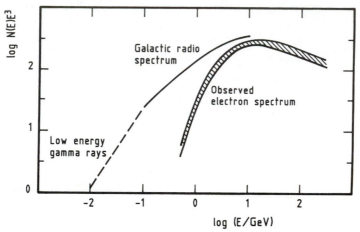

Figure 18.14. The spectrum of relativistic electrons in the local interstellar medium (see also Fig. 9.3). The observed electron spectrum is indicated by the hatched area. The spectrum deduced from the spectrum of the Galactic radio emission is indicated by a solid line. The low energy spectrum deduced from observations of low energy γ-rays is shown as a dashed line. The difference between the observed electron spectrum and that inferred from the Galactic radio emission provides a measure of the solar modulation of the flux of cosmic ray electrons. The units of $N(E)E^3$ are particles m^{-2} s^{-1} GeV^2. (From W. Webber (1983). In *Composition and origin of cosmic rays*, ed. M.M. Shapiro, p. 83. Dordrecht: D. Reidel Publishing Co.)

the cosmic ray electron spectrum and to the radio spectrum and emissivity of the interstellar medium.

   Taking the electron spectrum first, a summary of observations of the differential energy spectrum is shown in Fig. 18.14 (and Fig. 9.3 after Webber (1983)). The problem in making these observations is to ensure that the true interstellar flux of high energy electrons is observed, free from the contaminating effect of high energy secondary electrons created by nuclear collisions in the upper layers of the atmosphere (see Fig. 5.8). All the best observations lie within the shaded area shown in Fig. 18.14. It is apparent that the electron spectrum is strongly influenced by the effects of solar modulation at energies $E < 10$ GeV, and therefore the precise shape of the interstellar electron spectrum is uncertain at these energies. This is consistent with the effects of solar modulation observed in the spectrum of cosmic ray protons, since 10-GeV electrons have the same rigidities as 10-GeV protons. At higher energies, the electron spectrum can be well represented by a power law, and we adopt a differential intensity spectrum of the form

$$dN = 700\, E^{-3.3}dE \quad \text{particles m}^{-2}\text{s}^{-1}\text{sr}^{-1} \qquad (18.75)$$

where the energy E is measured in giga-electron-volts. This spectrum can be converted into a particle number density by writing

$$dn = \frac{4\pi\, dN}{c} = 2.9 \times 10^{-5}E^{-3.3}dE \quad \text{particles m}^{-3}\text{s}^{-1} \qquad (18.76)$$

Let us assume that this spectrum is representative of that of ultrarelativistic electrons in local interstellar space.

Electrons of energy $E = \gamma m_e c^2$ radiate most of their energy at a frequency $v \approx 28\gamma^2 B$ GHz, where B is measured in tesla. Let us adopt an average local magnetic field strength $B = 3 \times 10^{-10} x$ T. Then, 10-GeV electrons radiate most of their energy at a frequency $v \approx 3.2x$ GHz. Unfortunately, it can be seen that the frequency range over which the electron energy spectrum is free of the effects of solar modulation is just outside the range over which the Galactic radio spectrum has been accurately measured. This is just bad luck, and so the best we can do is to find out if the predicted spectrum matches smoothly onto the observed Galactic radio spectrum, since, as we have emphasised, the synchrotron radiation spectrum of a power-law distribution of electron energies is expected to be broad-band and smooth.

The next problem is to work out the local synchrotron emissivity of the interstellar medium. There are two alternatives. One approach is to estimate the local thickness of the Galactic disc of radio emission. Here, there are uncertainties about the exact thickness of the radio disc in our vicinity in the Galaxy. The intensity of the Galactic radiation in the direction of the Galactic Pole at a frequency of 10 MHz is 10^{-20} W m^2 sr^{-1} Hz^{-1} (see, for example, Webber (1983)). If the half-thickness of the disc is taken to be 1 kpc, the corresponding volume emissivity is 4.2×10^{-39} W m^{-3} Hz^{-1}. A second method is to make observations at very low radio frequencies, at which regions of ionised hydrogen of large angular size become optically thick because of thermal bremsstrahlung absorption. Then the radio emission in the direction of the opaque cloud must originate in the interstellar medium between the cloud and the Earth. Caswell (1976) has analysed his 10-MHz map of the Galactic radio emission in the direction of such clouds to measure the mean emissivity of the Galaxy locally and finds an average brightness temperature of $T_b = 240$ K pc^{-1} at 10 MHz, which corresponds to an intensity of

$$I(10 \text{ MHz}) = \frac{2kT_b}{\lambda^2} \quad \text{W m}^{-2}\text{Hz}^{-1}\text{sr}^{-1}\text{pc}^{-1}$$

and to a volume emissivity of 3×10^{-39} W m^{-3} Hz^{-1}. We will adopt this second value, and the Galactic radio spectrum has been normalised to it in Fig. 18.15.

We can now combine the synchrotron radiation formula (18.49) with the electron energy spectrum (18.76) so that $\kappa' = 2.9 \times 10^{-5}$ particles m^{-3} Gev$^{-(1-p)}$ and $p = 3.3$, for which $a(p) = 0.238$ (see Table 18.2). In Fig. 18.15, the predicted spectrum has been evaluated for magnetic field strengths $B = 0.15, 0.3$ and 0.6 nT, that is, $x = 0.5, 1$ and 2. The frequency at which 10-GeV electrons radiate is shown on each of these predictions. We can note the following points.

1 It can be seen that the predicted spectrum of the radio emission joins smoothly onto the observed spectrum of the Galactic radio emission, provided it is assumed that the magnetic field strength is high: $B = 6 \times 10^{-10}$ T. The most encouraging aspect of this analysis is that, with this assumption, it is entirely plausible that the local energy spectrum of

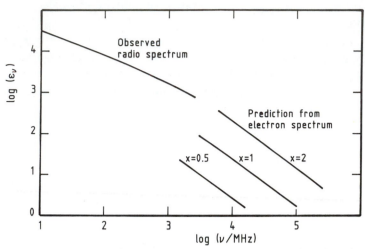

Figure 18.15. Comparison of the observed radio emissivity of the interstellar medium with that expected from the local electron energy spectrum for different values of the magnetic field strength. The radio emissivity is shown in relative units. The adopted radio emissivity at 10 MHz is 3×10^{-39} W m^{-2} Hz^{-1}.

the electrons is representative of the energy spectrum of the electrons in local interstellar space.

2 On the other hand, the mean value of the magnetic field strength required to achieve this agreement is larger than the typical values assumed for the average interstellar magnetic field as derived from pulsar rotation measures – these are found to lie in the range $(1.5–3) \times 10^{-10}$ T. There are various possible explanations for this discrepancy. First of all, it might be that the Earth is located within a region of low relativistic electron density relative to the general interstellar medium. Secondly, it will be noted that the intensity of the Galactic radio emission depends upon magnetic field strength as $B^{(p+1)/2} \propto B^{2.14}$, and hence, if the electron density is uniform, the intensity of emission along the line of sight is weighted as $\int B^{2.14} \, dl$, whereas the magnetic field strength derived from pulsar rotation measures is weighted as $\int B_{\parallel} N_e dl / \int N_e dl$. Thus, the intensity of synchrotron radiation gives greater weight to regions of high magnetic field. In either case, it is encouraging that the observed intensity is within a factor of about four of what might be reasonably expected, and thus one can assume with some degree of confidence that the Galactic radio emission is indeed synchrotron radiation.

3 Assuming one or both of the explanations discussed in (2) is correct, we can now use the observed Galactic radio spectrum to infer the energy spectrum of electrons in local interstellar space, free of the effects of solar modulation. This comparison is shown in Fig. 18.14. It is clear that the overall spectrum between 0.1 and 100 GeV must be gently curved, the spectral index p being about two between 0.1 and 1 GeV. In addition,

the comparison of the electron spectrum as measured in local interstellar space, and the spectrum measured in the vicinity of the Earth provides one of the best direct measures of the solar modulation of cosmic rays as a function of the energy (or rigidity).

18.3   The low energy γ-ray emission from the Galaxy

The arguments given in Section 18.2 enable us to determine the energy spectrum of relativistic electrons in the interstellar medium with energies between about 100 MeV and 1000 GeV. At the upper end of this range, the limitation is simply the fact that ultra high energy electrons are very rare, and long exposures are needed to detect a significant number of them. At the low energy end, the spectrum is inferred from the Galactic radio spectrum, and we have used observations down to only 10 MHz, which corresponds to an electron energy of about 500 MeV if the typical Galactic magnetic field strength is 3×10^{-10} T. The problem in extending this argument to lower radio frequencies is that the Galactic radio spectrum shows a rapid cut-off at frequencies below 10 MHz due to bremsstrahlung absorption in the interstellar medium. To extract the unabsorbed spectrum, the distribution of absorbing gas in the interstellar medium has to be known. The electron spectrum shown in Fig. 18.14 includes corrections for interstellar bremsstrahlung absorption below 10 MHz. An alternative is to attempt to demodulate the observed spectrum of energetic electrons but, as can be seen from Fig. 18.14, the modulation is large and the procedure is strongly dependent upon the modulation model.

One interesting possibility for investigating the electron energy spectrum in the range 10–100 MeV is to make use of observations of the Galactic γ-ray emission in the energy range 30–100 MeV. We will take up the story of the Galactic γ-ray emission in more detail in connection with the distribution of high energy protons and nuclei in the interstellar medium in Section 20.2. The dominant contribution at energies greater than 100 MeV is expected to be γ-rays produced in the decay of neutral pions, π^0, created in collisions of high energy protons and nuclei with protons and nuclei of the interstellar gas. The spectrum of π^0 decay γ-rays cuts off abruptly below about 100 MeV, and yet the SAS-II and COS-B observations have shown that there is intense Galactic γ-radiation in the energy range 30–100 MeV, much more than would be expected by an extrapolation of the high energy γ-ray flux to these energies using a π^0 decay spectrum (Fig. 18.16). One of the strong possibilities is that this excess is the bremsstrahlung of electrons with energies in the range 30–100 MeV. Let us look at this process in a little more detail.

The expression for the bremsstrahlung of a relativistic electron was derived in Section 3.6. For illustrative purposes, we will use the expression (3.68) with the modifications which come out of the more complete treatments (3.70) for the radiation spectrum of an ultrarelativistic electron interacting with cold interstellar material:

$$I(\omega)\mathrm{d}\omega = \frac{Z(Z+1.3)e^6 N}{16\pi^3 \epsilon_0^3 c^3 m_e^2 v} \ln\left(\frac{183}{Z^{1/3}}\right) \mathrm{d}\omega \qquad \hbar\omega < E \qquad (18.77)$$

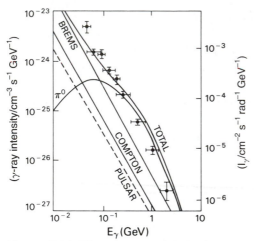

Figure 18.16. The spectrum of Galactic γ-ray emission from observations made by the SAS-II and COS-B satellites. The intensity is measured in the direction of the Galactic centre. (D.A. Kniffen and C.E. Fichtel (1981). *Astrophys. J.*, **250**, 389; P. Ramana Murthy and A.W. Wolfendale (1986). *Gamma ray astronomy*, p. 141, Cambridge: Cambridge University Press.)

In this approximation, we can assume that there is a sharp cut-off to the emission spectrum at energy $h\nu = E$. The important point is that the intensity spectrum of an individual electron is flat up to this cut-off energy. γ-ray astronomers normally work in units of photons per unit area per unit energy interval, and so we have to divide the intensity by the energy of each photon, ϵ. Putting in the values of the physical constants and setting $v = c$, we can write

$$N(\epsilon)d\epsilon = \frac{Z(Z + 1.3)e^6 N}{16\pi^3 \epsilon_0^3 m_e^2 c^4 \hbar} \ln\left(\frac{183}{Z^{1/3}}\right) \frac{d\epsilon}{\epsilon}$$

$$= \alpha N \frac{d\epsilon}{\epsilon}$$

where

$$\alpha = 3.62 \times 10^{-22} Z(Z + 1.3)(1 - 0.192 \ln Z^{1/3}) \quad \text{m}^3 \text{ s}^{-1} \tag{18.78}$$

In these expressions, the γ-ray photon flux is expressed in units of photons per second per mega-electron-volt and the energy ϵ is measured in mega-electron-volts. We have to make some assumption about the chemical composition of the interstellar gas, and we adopt the conventional assumption that it consists of 90% hydrogen and 10% helium by number. With these assumptions, the constant α has the value $\alpha = 9.33 \times 10^{-22}$ m^3 s^{-1}. Stecker (1975) has performed a more detailed analysis, and finds the constant $\alpha = 1.03 \times 10^{-21}$ m^3 s^{-1}. For our present purposes, we will adopt the value $\alpha = 10^{-21}$ m^3 s^{-1}. To work out the total intensity at a photon energy ϵ, we have to sum over all particles which can contribute at this energy. Only particles with energies $E \geq \epsilon$ are important, and therefore, adopting again a power-law spectrum $N_e(E) = \kappa E^{-p}$, the total intensity of radiation is

$$I_\gamma(\epsilon) = \int_\epsilon^\infty \frac{\alpha N}{\epsilon} \kappa E^{-p} \mathrm{d}E = \frac{\alpha N \kappa \epsilon^{-p}}{(p-1)} \tag{18.79}$$

This expression can also be written in terms of the total number of particles per unit volume with energies greater than ϵ in which case $I_\gamma(\epsilon) = (\alpha N/\epsilon) N_e(\geq \epsilon)$. The point of interest is that the spectrum of the γ-ray emission is of power-law form and that the spectral index of the photon intensity spectrum is the same as that of the electron energy spectrum.

Let us therefore put some figures into the above expression for the γ-ray emissivity of the interstellar medium to find out if it is likely that the 30–100 MeV γ-ray emission could be due to the bremsstrahlung of high energy electrons. The SAS-II and COS-B experiments provide integral measures of the intensity spectrum of the Galactic disc in units of photons per second per square metre per radian per mega-electron-volt in the direction of the Galactic centre. In a simple calculation, we can predict the γ-ray luminosity in the direction of the Galactic centre if we assume that the γ-rays originate from a disc of thickness 200 pc and radius 8 kpc. This disc is chosen to mimic crudely the distribution of neutral and molecular hydrogen in the direction of the Galactic centre. Within this disc, we assume that the number density of neutral material is 10^6 m^{-3}. With these assumptions, we can work out the average relativistic electron number density in the energy range 30–100 MeV. Performing this calculation, we find that the average energy spectrum of the electrons is $N(E) = 2.5 \times 10^3 E^{-1.6}$ particles m^3 GeV^{-1}, where the energy of the electrons E is measured in giga-electron-volts. Inspection of Fig. 18.14 shows that this is about four times the value found for the local electron energy spectrum at 100 MeV. In view of the many crude approximations involved, the agreement is encouraging, and suggests that it is entirely plausible that the low energy γ-ray emission of the Galaxy is due to relativistic bremsstrahlung. For illustrative purposes, a possible low energy extension of the relativistic electron spectrum is shown in Fig. 18.14, which joins smoothly onto the spectrum already derived. Much more exact calculations carried out by Kniffen and Fichtel (1981) confirm our order of magnitude estimates. It can be seen that, if this interpretation is correct, the relativistic electron spectrum turns over even further at these low energies. We will study the origin of this form of electron energy spectrum in Chapter 19.

An exact calculation would require much more effort, which is beyond our present needs. We should, however, note some of the complications which need to be considered carefully. For example, we cannot be certain how much of the 30–100 MeV γ-ray emission in the direction of the Galactic centre is due to discrete γ-ray sources. In addition, there may well be a contribution from inverse Compton scattering of optical photons, as is illustrated in Fig. 18.16. We should also note our grossly oversimplified model for the γ-ray emission of the central region of our Galaxy – we should take account of the fact that the gas is clumped and that the energy density of relativistic electrons may vary from place to place in the Galaxy.

18.4   The distribution of high energy electrons and magnetic field in the Galaxy

The Galactic radio emission provides direct evidence that relativistic electrons are present throughout the Galaxy and that the cosmic ray electrons which we observe at the top of the atmosphere are not peculiar to our local neighbourhood. From radio maps such as that shown in Fig. 1.8(*a*), it is possible to define the large-scale features of the distribution of relativistic electrons and magnetic field in the Galaxy. It should be recalled that, in interpreting such maps, we are located in the plane of the Galaxy itself and have a rather close up view of it.

From Fig. 1.8(*a*), it can be seen that our Galaxy possesses a central radio disc. The emissivity of the disc is rather uniform and has a radius of about 8 kpc and a half-thickness of about 700–1000 pc. The radius of the disc can be estimated from the fact that there is a somewhat rapid drop in intensity along the Galactic plane at about $\pm 60°$ in galactic longitude on either side of the Galactic centre. The estimate of thickness of the disc comes from the variation in intensity with galactic latitude. These are important facts – they tell us that the magnetic field in the Galaxy must fill a region at least about 1 kpc in thickness, and therefore any high energy particles released in the plane of the Galaxy have a long way to travel to escape from the disc.

What about spiral arm features? The feature at $l = 90°$ probably corresponds to emission from the local arm as we look along it. It can be seen that the half-width of the distribution in galactic latitude broadens in the anticentre direction, $l = 180°$. This would be consistent with the radio emission being associated with the local spiral arm, which should occupy a much larger angle on the sky because we are located on its inner edge. In the interarm regions, the radio emissivity appears to be weaker, which might suggest that the magnetic field is weaker or that there are fewer high energy electrons in these regions. There are now many beautiful radio maps of nearby spiral galaxies, and these reveal a diversity of behaviour. In many cases, the bulk of the radio emission originates in the spiral arm regions. Fig. 18.17 shows the distribution of radio brightness in the nearby spiral galaxies M31 and M51. This emission has a radio spectrum similar to that of our own Galaxy and provides convincing evidence that there is enhanced radio emission from the vicinity of spiral arms.

The Galactic radio emission is highly polarised, and, once account is taken of Faraday rotation, the intrinsic polarisation vectors provide information about the structure of the local magnetic field. The rotation measures of the polarised Galactic radiation are smaller than those of extragalactic sources and pulsars in the same region of sky, which suggests that most of the polarised emission originates quite close to the Sun, $r < 500$ pc. Therefore, the polarisation of the Galactic radio emission does not necessarily provide information about the large-scale magnetic field distribution in the Galaxy, although the observations are consistent with the general picture described in Section 16.4.

Finally, there is the question of whether or not there is a radio halo about our Galaxy. This is an important question because, if it does exist, the volume

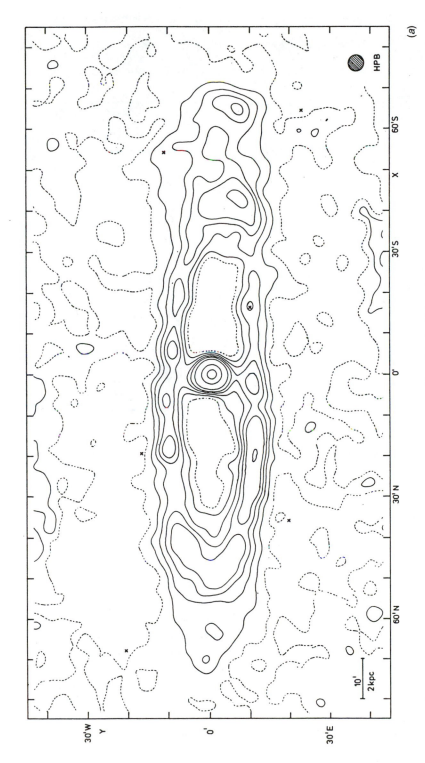

Figure 18.17 (*a*). The distribution of synchrotron radio emission from nearby normal galaxies. In the Andromeda galaxy (M31, NGC 224), the synchrotron radio emission originates from a ring of emission with no central concentration of the radio emission, as in our Galaxy. There is a weak diffuse source in the central regions. (From J.E.V. Bystedt, E. Brinks, A.G. de Bruyn, F.P. Israel, P.B.W. Schwering, W.W. Shane and R.A.M. Walterbos (1984). *Astron. Astrophys. Suppl.*, **56**, 245.)

(*a*)

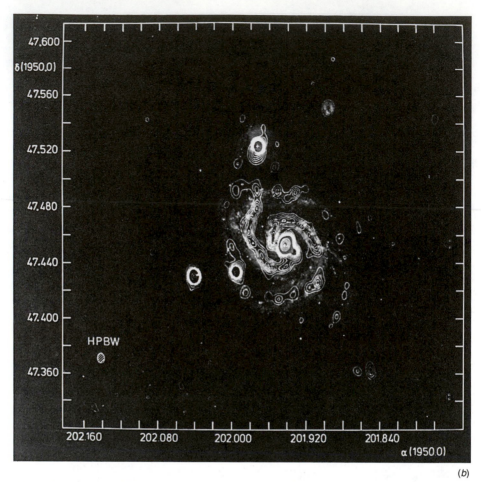

(b)

Figure 18.17 (*b*). In M51, the radio emission is strongly correlated with the spiral arms, but is displaced to the trailing edge of the spiral and is most clearly associated with the dust lanes. (From A. Segalovitz (1977). *Astron. Astrophys.*, **54**, 703.)

in which there is a relatively strong magnetic field is much larger than that of the disc and consequently provides a larger volume within which the high energy particles can be confined. The problem is that it is rather difficult to unravel the emission from a more or less spherical radio distribution about the Galactic centre in the presence of the strong emission from the Galactic plane. There is now evidence that our Galaxy does possess a rather weak radio halo, and that its magnetic field strength is about five times weaker than that of the disc. The precise dimensions of the radio halo are not well-defined, but the observations would be consistent with a flattened spheroidal distribution of radiation with radius in the plane of about 10 kpc and radius perpendicular to it of about 3–4 kpc (Webster (1978)). Similar haloes have been detected about some nearby edge-on spiral galaxies, a very beautiful example being the case of NGC 891

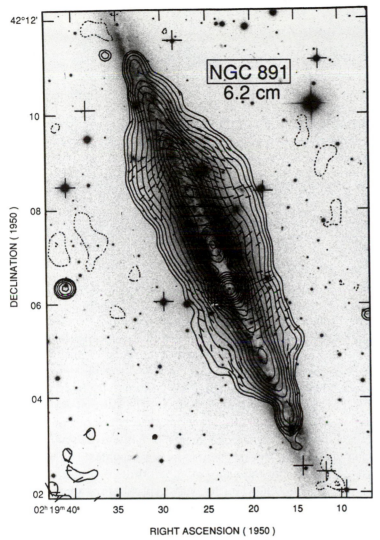

Figure 18.18. The distribution of synchrotron radio emission at a wavelength of 6 cm in the edge-on spiral galaxy NGC 891, showing the diffuse halo emission. (R.J. Allen and S. Sukumer (1991). *The interstellar disk-halo connection in galaxies*, ed. H. Bloemen. Dordrecht: Kluwer Academic Publishers.)

observed by R.J. Allen and S. Sukumer with the Very Large Array (Fig. 18.18). The sizes of these haloes are similar to those inferred to be present around our own Galaxy. We will have much more to say about haloes in our discussion of the lifetimes of high energy protons and nuclei in the Galaxy.

19

The origin of the electron energy spectrum in our Galaxy

19.1　Introduction

In Chapter 18, we showed that we can make a convincing case that the high energy electrons which are observed at the top of the atmosphere represent a sample of the high energy electrons present throughout the interstellar medium and which are responsible for the diffuse Galactic synchrotron radio emission. Our task in this chapter is to interpret these observations in terms of the propagation of these particles from their sources through the interstellar medium and the energetics of possible energy sources within the Galaxy. The key diagnostic tools are aging processes, which can result in features in the energy spectra of the electrons and estimates of the energy requirements of sources of synchrotron radiation. In this chapter, we will develop these tools in the context of the origin of the Galactic radio emission and the study of supernovae as sources of high energy electrons. These tools are, however, of very general applicability to the whole of high energy astrophysics. We will use them repeatedly in our discussion of the physics of radio sources and active galactic nuclei.

19.2　Energy loss processes for high energy electrons

High energy electrons are subject to a number of energy loss processes as they propagate from their sources through the interstellar medium. The loss processes cause distortions of the injection energy spectra of the particles from their sources and thus potentially provide information about the life histories of the high energy electrons. These loss mechanisms involve interactions with matter, magnetic fields and radiation. First, we convert the various loss mechanisms which we have studied into useful forms, which can be incorporated into the equations for propagation under interstellar conditions. We will find that it is the dependence of these loss processes upon the energy of the electrons which is most important, and so we rewrite the formulae, emphasising this aspect.

Ionisation losses If we put the figures into the formulae (3.1) and (3.2) quoted in Section 3.2, for the case of atomic hydrogen, we find

$$-\left(\frac{\mathrm{d}E}{\mathrm{d}t}\right)_{\mathrm{i}} = 7.64 \times 10^{-15} N(3\ln\gamma + 19.8) \quad \mathrm{eV\ s^{-1}} \tag{19.1}$$

where $\gamma = (1 - v^2/c^2)^{-1/2}$ is the Lorentz factor of the electron and N is the number density of hydrogen atoms in particles per cubic metre. Thus, an electron of energy $E(\mathrm{eV})$ loses all its energy in a time

$$\tau = \frac{E}{(\mathrm{d}E/\mathrm{d}t)_{\mathrm{i}}} = \frac{E(\mathrm{eV})}{7.64 \times 10^{-15} N(3\ln\gamma + 19.8)}$$

Thus, for example, an electron of energy 3 GeV, for which $\gamma = 6000$, loses all its energy in $3 \times 10^{14}/N$ years, so that, for a typical interstellar density of $N = 10^6$ m^{-3}, the lifetime of the electron is about 3×10^8 years. Because the dependence upon energy is only logarithmic, to a good approximation the energy loss of the electron amounts to about $10^{-5} N$ eV year^{-1}.

Bremsstrahlung We can use the results presented in Section 3.6 to derive suitable forms for the energy loss rate by bremsstrahlung for ultrarelativistic electrons. Following Heitler (1954), the energy loss rate can be written

$$-\frac{1}{E}\left(\frac{\mathrm{d}E}{\mathrm{d}t}\right)_{\mathrm{brems}} = 4NZ^2 r_{\mathrm{e}}^2 \alpha c \bar{g} \tag{19.2}$$

where r_{e} is the classical electron radius, α is the fine structure constant and $\bar{g}$ is a Gaunt factor. When the nuclei are unscreened, as in a fully ionised plasma,

$$\bar{g} = \ln(2\gamma) - \frac{1}{3} = \ln\gamma + 0.36$$

and so the loss rate of the electron in a fully ionised hydrogen plasma can be written

$$-\frac{1}{E}\left(\frac{\mathrm{d}E}{\mathrm{d}t}\right)_{\mathrm{brems}} = 7.0 \times 10^{-23} N(\ln\gamma + 0.36) \quad \mathrm{s^{-1}} \tag{19.3}$$

In the case of total screening,

$$\bar{g} = \ln(183 Z^{-\frac{1}{3}}) - \frac{1}{18}$$

and so, for neutral hydrogen, the loss rate associated with interactions between the ultrarelativistic electron and the nuclei of the hydrogen atoms is

$$-\frac{1}{E}\left(\frac{\mathrm{d}E}{\mathrm{d}t}\right)_{\mathrm{brems}} = 3.66 \times 10^{-22} N \quad \mathrm{s^{-1}} \tag{19.4}$$

If account is taken of the bremsstrahlung due to interactions with the bound electrons in the atoms as well, Z^2 should be replaced by $Z(Z + 1.3)$ in equation (19.2). Notice that, for a wide range of electron energies, say, $100 \leq \gamma \leq 10^5$, the expressions (19.3) and (19.4) are the same within about a factor of two, reflecting the fact that the differences in collision parameters occur within the logarithm. Notice that, in both cases, the energy loss rate is proportional to the energy of the electron.

Adiabatic losses The energy loss rate of an ultrarelativistic gas due to expansion of the volume within which the particles are contained was discussed in Section 11.3.2, and can be written in the form

$$-\left(\frac{dE}{dt}\right)_{ad} = \frac{1}{3}(\nabla \cdot \mathbf{v})E \qquad (11.27)$$

In this process, the relativistic gas does work adiabatically in the expansion, and consequently it loses internal energy. As an example, in the case of a uniformly expanding sphere, the velocity distribution inside the sphere is $v = v_0(r/R)$, where v_0 is the velocity of expansion of the outer radius R of the sphere. Then, as shown in Section 11.3.2, $\nabla \cdot \mathbf{v} = 3(v_0/R)$ and so

$$-\left(\frac{dE}{dt}\right)_{ad} = \left(\frac{v_0}{R}\right)E = \left(\frac{1}{R}\frac{dR}{dt}\right)E$$

or

$$-\frac{1}{E}\left(\frac{dE}{dt}\right)_{ad} = \left(\frac{1}{R}\frac{dR}{dt}\right) \qquad (19.5)$$

The condition that adiabatic losses be important is that the timescale of the expansion, $[(1/R)(dR/dt)]^{-1}$, should be of the order of the time the particles have been within the emitting volume. Evidently, if the particles have always been in the source during its expansion, adiabatic losses are always important. Notice, however, that the loss rate at any given time depends upon the instantaneous dynamics of expansion of the source through $(R^{-1}\dot{R})$.

This loss mechanism is important in the escape of particles from supernova remnants. When the particles exert a pressure which does work, adiabatic losses are important in the interstellar medium in general.

Synchrotron radiation The total energy loss rate by synchrotron radiation is given by the expression (18.6), which, in the ultrarelativistic limit $v \to c$, becomes

$$-\left(\frac{dE}{dt}\right)_{synch} = 2\sigma_T c\gamma^2 U_{mag} \sin^2\theta \qquad (19.6)$$

or

$$-\left(\frac{dE}{dt}\right)_{synch} = 1.587 \times 10^{-14}\gamma^2 B_\perp^2 \quad W$$

$$= 9.9 \times 10^4 \gamma^2 B_\perp^2 \quad \text{eV s}^{-1}$$

where $B_\perp = B\sin\theta$ is measured in tesla. If we average over an isotropic pitch angle distribution, we replace the factor of 2 in expression (19.6) by $\frac{4}{3}$, and then

$$-\left(\frac{dE}{dt}\right)_{synch} = \frac{4}{3}\sigma_T c\gamma^2 U_{mag} = 6.6 \times 10^4 \gamma^2 B^2 \quad \text{eV s}^{-1} \qquad (19.7)$$

Inverse Compton scattering From the analysis of Section 4.3.3, we can write the loss rate of an ultrarelativistic electron by inverse Compton scattering in a radiation field of energy density U_{rad} as

$$-\left(\frac{dE}{dt}\right)_{IC} = \frac{4}{3}\sigma_T c\gamma^2 U_{rad} \qquad (19.8)$$

Notice the remarkable similarity of the result (19.8) and the expression (19.7) for the energy loss rate by synchrotron radiation. The reason for this is that the energy loss rate by radiation depends upon the electric field which accelerates the electron in its rest frame, and the electron does not care what the origin of that field is. In the case of synchrotron radiation, the electric field is the $(\mathbf{v} \times \mathbf{B})$ field due to motion of the electron at almost the speed of light through the magnetic field, whereas, in the case of inverse Compton scattering, it is the sum of the electric fields of the electromagnetic waves incident upon the electron. Notice that, in the latter case, it is the sum of the squares of the electric field strengths which appear in the formulae for incoherent radiation, and so the energies of the waves just add linearly (see Section 4.3.1). Another way of expressing this similarity between the loss processes is to consider synchrotron radiation to be the scattering of 'virtual photons' observed by the electron as it gyrates about the magnetic field (see, for example, Jackson (1975), chapter 15).

Inverse Compton scattering is important whenever the high energy electron propagates through a radiation field. In the present instance, we are interested in the radiation fields in our own Galaxy. At a typical point in the Galaxy, there is an average energy density of optical photons due to the light of all the stars which amounts to about $U_{\text{rad}} \approx 6 \times 10^5$ eV m^{-3} (see, for example, Toller (1990)). In addition, the cosmic microwave background radiation fills the whole Universe (see Section 13.1.1), and its energy density is 2.65×10^5 eV m^{-3} if we adopt $T_{\text{r}} = 2.736$ K.

It is instructive to compare the importance of synchrotron and inverse Compton losses for electrons of the same energy in our Galaxy. Comparing equations (19.7) and (19.8), we see that

$$\frac{(\mathrm{d}E/\mathrm{d}t)_{\text{IC}}}{(\mathrm{d}E/\mathrm{d}t)_{\text{synch}}} = \frac{U_{\text{rad}}}{U_{\text{mag}}} \tag{19.9}$$

Putting in representative figures, $B = 3 \times 10^{-10}$ T and $U_{\text{rad}} = 6 \times 10^5$ eV m^{-3}, we find $U_{\text{rad}}/U_{\text{mag}} = 3$. Thus, inverse Compton losses as well as synchrotron losses are likely to be important for high energy electrons in our Galaxy. Furthermore, suppose the electrons were to escape from the Galaxy into intergalactic space. Then, the energy loss due to synchrotron radiation is likely to be very small because we can set low limits to the intergalactic magnetic field strength. The cosmic microwave background radiation is, however, omnipresent, and the electrons cannot escape from it. Therefore, they must lose all their energy by inverse Compton scattering of the photons of the cosmic microwave background.

Let us use the loss rate for inverse Compton scattering to work out the maximum lifetime of high energy electrons, anywhere in the Universe. The lifetime of an electron is

$$\tau = \frac{E}{\mathrm{d}E/\mathrm{d}t} = \frac{E}{\frac{4}{3}\sigma_{\text{T}}c\gamma^2 U_{\text{CBR}}} = \frac{2.3 \times 10^{12}}{\gamma} \quad \text{years} \tag{19.10}$$

taking U_{CBR} to be 2.65×10^5 eV m^{-3}. Thus, for 100-GeV electrons, which are observed at the top of the atmosphere, $\tau \le 1.15 \times 10^7$ years.

What are the energies of the photons scattered by the high energy electrons? We have shown that the typical energies of the high energy electrons which radiate in the radio waveband have $\gamma = 10^3$–10^4 (Section 18.2), and so scattering of the photons of the cosmic microwave background generates X-rays since

$$v \approx \gamma^2 v_0 \approx 10^6 \times 10^{11} \text{ Hz} = 10^{17} \text{ Hz} \quad ; \quad \epsilon = 0.4 \text{ keV}$$

The scattering of optical photons produces a flux of γ-rays with

$$v \approx \gamma^2 v_0 \approx 10^6 \times 10^{15} \text{ Hz} = 10^{21} \text{ Hz} \quad ; \quad \epsilon = 4 \text{ MeV}$$

Thus, part of the Galactic γ-ray emission in the 1–100 MeV energy band may be due to inverse Compton scattering (Fig. 18.16). Inverse Compton scattering is thus likely to be important whenever relativistic electrons propagate through regions containing large energy densities of radiation. For example, in the central regions of quasars, this mechanism may be one of the most important drains upon the energies of relativistic particles. An interesting and important situation may arise in some quasars in which the radiation density of synchrotron radiation is so high that inverse Compton scattering by the electrons which emitted the synchrotron radiation itself may be the dominant loss mechanism. This has been referred to as the *inverse Compton catastrophe* (see Volume 3).

19.3   The diffusion-loss equation for high energy electrons

We now work out how these energy loss processes influence the spectrum of high energy electrons as they propagate from their sources through the interstellar medium. It is already clear from our discussion of Chapter 11 that the propagation of high energy particles in magnetic fields is a complicated business, and to make progress we have to make some fairly sweeping approximations about the dynamics of the particles. We will study more detailed models for the propagation of high energy particles in the next chapter, but, for the moment, it is useful to study what is probably the simplest description of the dynamics of the particles, that is, that they diffuse through the interstellar medium. The rationale behind this picture is that there is abundant evidence of irregular structure of the interplanetary medium, and therefore it is reasonable to suppose that there will be irregularities in the magnetic field in the interstellar medium as well. These will result in large deviations in the particles' trajectories, as described in Sections 11.3 and 11.4 for the case of the interplanetary magnetic field. It is therefore reasonable to suppose that the particles can be considered to diffuse from their sources through the interstellar medium.

So, for simplicity and illustration, we adopt a scalar diffusion coefficient, D, to describe the motion of the electrons. We now need to find the partial differential equation which describes the energy spectrum and particle density at different points in the interstellar medium in the presence of continuous energy losses, and with the continuous supply of fresh particles from sources. This equation is known as the *diffusion-loss equation* for high energy electrons. We give two derivations, one somewhat pedestrian, the other rather more elegant.

19.3.1 *The simple approach*

Consider an elementary volume of space, dV, into which electrons are injected at a rate $Q(E,t)\,dV$. The electrons within dV are subject to energy gains and losses, and we write this

$$-\left(\frac{dE}{dt}\right) = b(E) \tag{19.11}$$

so that, if $b(E)$ is positive, the particles lose energy. Consider first the change in the energy spectrum of electrons due to the energy losses $b(E)$ in the absence of injection of electrons. At time t, the number of particles in the energy range E to $E + \Delta E$ is $N(E)\Delta E$. At a later time $t + \Delta t$, these particles have been replaced by those that had energies in the range E' to $E' + \Delta E'$ at time t, where

$$E' = E + b(E)\Delta t$$

$$E' + \Delta E' = (E + \Delta E) + b(E + \Delta E)\Delta t$$

Performing a Taylor expansion for small ΔE and subtracting, we find

$$\Delta E' = \Delta E + \frac{db(E)}{dE}\Delta E\Delta t$$

Therefore the change in $N(E)\Delta E$ in the time interval Δt is

$$\Delta N(E)\Delta E = -N(E,t)\Delta E + N[E + b(E)\Delta t, t]\Delta E'$$

Performing another Taylor expansion for small $b(E)\Delta t$ and substituting for $\Delta E'$, we obtain

$$\Delta N(E)\Delta E = \frac{dN(E)}{dE}b(E)\Delta E\Delta t + N(E)\frac{db(E)}{dE}\Delta E\Delta t$$

that is,

$$\frac{dN(E)}{dt} = \frac{d}{dE}[b(E)N(E)] \tag{19.12}$$

This equation describes the temporal evolution of the electron spectrum in the elementary volume dV subject only to energy gains and losses. We may now add other terms to this transfer equation. Thus, particles are continuously injected at a rate $Q(E,t)$ per unit volume, and hence

$$\frac{dN(E)}{dt} = \frac{d}{dE}[b(E)N(E)] + Q(E,t)$$

Particles enter and leave the volume dV by diffusion, and this process depends upon the gradient of particle density, $N(E)$. Adopting a scalar diffusion coefficient, D, we can write, in the normal way,

$$\frac{dN(E)}{dt} = \frac{d}{dE}[b(E)N(E)] + Q(E,t) + D\nabla^2 N(E) \tag{19.13}$$

This is the equation known as the *diffusion-loss equation* for relativistic electrons.

19.3.2 *The elegant approach*

An alternative and neater approach is to introduce a *coordinate space*, in which we plot energy along the ordinate and spatial coordinates along the abscissa; we can easily generalise to three spatial coordinates (Fig. 19.1). The ϕ's are fluxes of particles through different surfaces in the coordinate space. If we

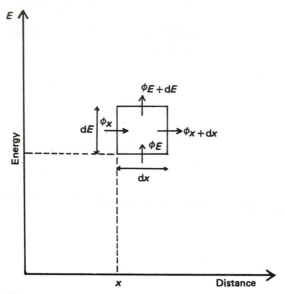

Figure 19.1. A coordinate-space diagram of energy against spatial coordinates used in deriving the diffusion-loss equation.

consider the little rectangle, particles move in the x direction by diffusion and in the y direction by energy gains or losses. The number of particles in the distance increment dx and energy increment E to $E + dE$ is $N(E, x, t)\, dEdx$. Therefore, the rate of change of particle density in this little box in coordinate space is

$$\frac{d}{dt} N(E, x, t)\, dEdx = [\phi_x(E, x, t) - \phi_{x+dx}(E, x + dx, t)]dE$$
$$+ [\phi_E(E, x, t) - \phi_{E+dE}(E + dE, x, t)]dx$$
$$+ Q(E, x, t)dEdx$$

where $Q(E, x, t)$ is the rate of production of electrons per unit volume of coordinate space. Now, performing a Taylor expansion and simplifying the notation, we find

$$\frac{dN}{dt} = -\frac{\partial \phi_x}{\partial x} - \frac{\partial \phi_E}{\partial E} + Q$$

ϕ_x is the flux of particles through the energy interval dE at the point x in space, and hence, by definition,

$$\phi_x = -D\frac{\partial N}{\partial x}$$
$$\frac{dN}{dt} = D\frac{\partial^2 N}{\partial x^2} - \frac{\partial \phi_E}{\partial E}$$

We can clearly generalise this to

$$\frac{dN}{dt} = D\nabla^2 N - \frac{\partial \phi_E}{\partial E} + Q$$

ϕ_E is the flux of particles through dx which have energy E at some time in interval dt. If $-dE/dt = b(E)$ is the loss rate of particles of energy E, then the number passing through E in unit time is

$$N(E)\frac{\mathrm{d}E}{\mathrm{d}t} = \phi_E = -b(E)N(E)$$

Therefore we obtain

$$\frac{\mathrm{d}N}{\mathrm{d}t} = D\nabla^2 N + \frac{\partial}{\partial E}[b(E)N(E)] + Q(E)$$

as before.

Notice that we could add other terms to this equation if we wished. For example, in Chapter 20 we will use a diffusion-loss equation of exactly the same form for nuclei, and then we have to include terms describing spallation gains and losses, catastrophic loss of particles, radioactive decay and so on.

19.3.3 *Solutions relevant to high energy electrons*

If we wished, we could now set about solving the diffusion-loss equation for a given distribution of sources and boundary conditions. This is done in, for example, the classic text by Ginzburg and Syrovatskii, *The origin of cosmic rays* (1964) and subsequent papers by Ginzburg, Dogel and Syrovatskii (see, for example, Bulanov, Dogel and Syrovatskii (1976)). We can, however, obtain some useful results by inspection of a few special steady-state solutions.

Suppose, first of all, that there is an infinite, uniform distribution of sources, each injecting high energy electrons with an injection spectrum $Q(E) = \kappa E^{-p}$. Then, diffusion is not important, and the diffusion-loss equation reduces to

$$\frac{\mathrm{d}}{\mathrm{d}E}[b(E)N(E)] = -Q(E)$$

$$\int \mathrm{d}[b(E)N(E)] = -\int Q(E)\mathrm{d}E$$

We assume $N(E) \to 0$ as $E \to \infty$, and hence on integrating we find

$$N(E) = \frac{\kappa E^{-(p-1)}}{(p-1)b(E)} \tag{19.14}$$

We now write down $b(E)$ for high energy electrons under interstellar conditions:

$$b(E) = -\left(\frac{\mathrm{d}E}{\mathrm{d}t}\right) = A_1\left(\ln\frac{E}{m_e c^2} + 19.8\right) + A_2 E + A_3 E^2 \tag{19.15}$$

The first term on the right-hand side, A_1, describes ionisation losses and depends only weakly upon energy; the second term, A_2, represents bremsstrahlung losses, and adiabatic losses if they are relevant; and the last term, A_3, describes inverse Compton and synchrotron losses. This analysis enables us to understand the effect of continuous energy losses upon the initial spectrum of high energy electrons. Thus, from expression (19.14),

1 if ionisation losses dominate, $N(E) \propto E^{-(p-1)}$, that is, the energy spectrum is flatter by one power of E;
2 if bremsstrahlung or adiabatic losses dominate, $N(E) \propto E^{-p}$, that is, the spectrum is unchanged;
3 if inverse Compton or synchrotron losses dominate, $N(E) \propto E^{-(p+1)}$, that is, the spectrum is steeper by one power of E.

These are also the equilibrium spectra expected whenever the continuous injection of electrons takes place over a timescale longer than the lifetimes of the individual electrons involved. For example, if we inject electrons continuously with a spectrum E^{-p} into a source component for a time t and synchrotron losses are the only important loss process, an electron of energy E_s loses all its energy in a time t such that $-(dE/dt)t = E_s$. From expression (19.7), we find that $E_s = 0.125/B^2 t$ eV, where B is measured in tesla and t in years. The electron energy spectrum in the source is different for electrons with energies greater and less than E_s; for lower energies, the electrons do not lose a significant fraction of their energy, and therefore the spectrum is the same as the injection spectrum, $N(E) \propto E^{-p}$. For energies greater than E_s, the particles have lifetimes less than t, and we only observe those produced during the previous synchrotron lifetime, τ_s, of the particles of energy E, that is, $\tau_s \propto 1/E$. Therefore the spectrum of the electrons is one power of E steeper, $N(E) \propto E^{-(p+1)}$, in agreement with the analysis proceeding from the steady-state solution of the diffusion-loss equation.

There are two useful analytic solutions for the electron energy distribution under continuous energy losses due to synchrotron radiation and inverse Compton scattering. In the first case, it is assumed that there is continuous injection of electrons with a power-law energy spectrum $Q(E) = \kappa E^{-p}$ for a time t_0. If we write the loss rate of the electrons in the form $b(E) = aE^2$, the energy spectrum after time t_0 has the form

$$N(E) = \begin{cases} \dfrac{\kappa E^{-(p+1)}}{a(p-1)}[1-(1-aEt)^{p-1}] & \text{if} \quad aEt_0 \leq 1 \\[2mm] \dfrac{\kappa E^{-(p+1)}}{a(p-1)} & \text{if} \quad aEt_0 > 1 \end{cases} \tag{19.16}$$

This form of spectrum is shown in Fig. 19.2(a), and agrees with the physical arguments given in the last paragraph.

A second useful case is that of the injection of electrons with a power-law energy spectrum at $t = 0$ with no subsequent injection of electrons. We can then write $Q(E) = \kappa E^{-p}\delta(t)$. It is straightforward to show that the solution of the diffusion-loss equation (19.13), ignoring the diffusion term, is

$$N(E) = \kappa E^{-p}(1-aEt)^{p-2}$$

Thus, after time t, there are no electrons with energies greater than $(at)^{-1}$. Notice that, if $p > 2$, the spectrum steepens smoothly to zero at $E = (at)^{-1}$; if $p < 2$, there is a cusp in the energy spectrum at $E = (at)^{-1}$. The number of electrons, however, remains finite and constant. These spectra are illustrated in Figs 19.2(b) and (c).

19.3.4   *Interpretation of the high energy electron energy spectrum in the local interstellar medium*

In Sections 18.2 and 18.3, we discussed the energy spectrum of high energy electrons in the local interstellar medium as derived from direct observations of the electrons themselves, from interpretation of the spectrum of the diffuse Galactic

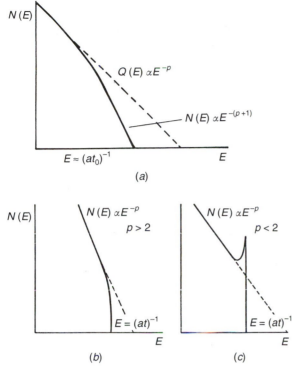

Figure 19.2. (a) A solution of the diffusion-loss equation for steady-state injection of electrons with a power-law energy spectrum $Q(E) \propto E^{-p}$ in the presence of energy losses of the form $dE/dt = -aE^2$. (b) The time evolution of a power-law energy distribution injected at $t = 0$ with no subsequent injection of electrons. In this case $p > 2$. (c) As in case (b), but with $p < 2$.

radio emission and from the low energy γ-ray emission from the general direction of the Galactic centre. We made the case that the electron energy spectrum in the local interstellar medium has the form shown in Fig. 19.3. Interpreting the energy spectrum by power laws of the form $N(E) \propto E^{-p}$, it can be seen that the spectrum steepens with increasing energy from $p \approx 1.6$ at about 30 MeV, to 2.8 between 1 and 10 GeV and to 3.3 at the highest energies, 10–100 GeV.

We can use the tools developed in Section 19.2.3 to interpret this spectrum. In a complete solution, we would have to take account of the diffusion of the electrons from their sources in the plane of the Galaxy to intergalactic space. This is a very major undertaking, and it is much simpler to work in an approximation in which the electrons only remain a certain time $\tau(E)$ within what is called the 'Galactic confinement volume' before escaping into intergalactic space. The diffusion term can then be replaced by a term of the form $N(E)/\tau(E)$. The form of $\tau(E)$ can be chosen to mimic diffusion if a Gaussian distribution of escape times is used or a 'leaky box' if an exponential distribution of escape times is adopted (see Sections 20.2 and 20.4). Also, if the escape time depends upon energy, simple

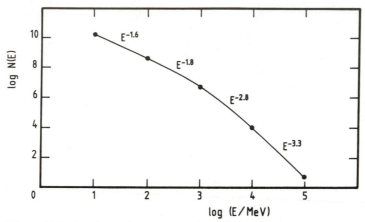

Figure 19.3. A schematic representation of the electron energy spectrum in the local interstellar medium from the data discussed in Sections 18.2 and 18.3. This spectrum has been subject to energy losses at high and low energies during propagation of the electrons through the interstellar medium. The units on the ordinate are relative units (see Fig. 18.14 for physical units).

solutions for the equilibrium energy spectrum can be found, particularly if the energy dependence is chosen to be of power-law form (see Section 20.2.2).

The simplest way of interpreting the observations is to study first those energy ranges in which the timescale for energy losses is less than the escape time $\tau(E)$. We have to anticipate our story somewhat by noting that the observed abundances of ^{10}Be nuclei in the cosmic rays suggest that cosmic ray nuclei have escape times from the Galaxy of about $(1-3) \times 10^7$ years (see Section 20.4.2). If the electrons have similar escape times, the timescales for energy losses are less than this value for the lowest energy electrons due to ionisation losses and for those with the highest energies by a combination of synchrotron and inverse Compton losses. Taking the ionisation loss rate to be $10^{-5}N$ eV year^{-1} and $N = 10^6$ m^{-3}, we find a lifetime of 3×10^7 years for 300-MeV electrons; taking $B = 6 \times 10^{-10}$ T, as suggested by the analysis of Section 18.2, we find the lifetime of 10-GeV electrons to be about 3×10^7 years due to synchrotron losses. Thus, if electrons are continuously injected uniformly into the interstellar medium, the electron spectra in these spectral regions should reach a steady-state under losses, that is, $dN/dt = 0$. From the analysis of Section 19.2.3, we find that, in the low energy regions in which ionisation losses dominate, $N(E) \propto E^{-(p-1)}$, and so the injection spectrum would be $Q(E) \propto E^{-2.6}$. In the high energy region, in which synchrotron losses dominate, $N(E) \propto E^{-(p+1)}$, and hence the injection spectrum is $Q(E) \propto E^{-2.3}$. These values are not too different, suggesting that the injection electron spectrum might be quite close to $E^{-2.5}$ throughout the energy range 100 MeV to 100 GeV.

Breaks in the spectra are expected at the energies at which the lifetimes of the electrons are equal to their escape times. Thus, the break in the radio

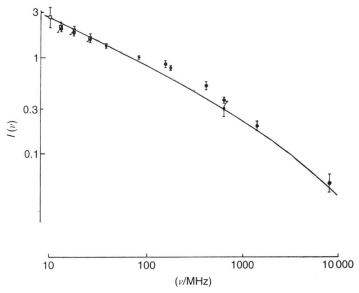

Figure 19.4. The solution of the diffusion-loss equation for a point at the centre of a spherical distribution of sources converted into a predicted radio spectrum and compared with observations of the Galactic radio spectrum. This is the sharpest break that can be produced by any simple diffusion model. (From A.S. Webster (1970). *Astrophys. Lett.*, **5**, 189.)

spectrum at 1–2 GHz (see Fig. 18.13) can be associated with 10-GeV electrons radiating in a magnetic field of strength 6×10^{-10} T. Likewise, for the low energy electrons, the break would reach to $E \approx 300$ MeV, corresponding to $v \sim 2$ MHz if $N = 10^3$ m^{-3}. There are many reasons why the breaks in the spectra of the electrons and their radio spectra would be smoothed out over a wide range of energies and frequencies about these critical values. As an example of the smoothing of these features, A.S. Webster (1970; for full reference, see the caption to Fig. 19.4) has solved the diffusion-loss equation for a distribution of sources in the plane of the Galaxy. The break in the radio spectrum at about 1 GHz is attributed to the combination of inverse Compton and synchrotron losses. Fig. 19.4 shows how the break is smoothed out over roughly a factor of ten in frequency. It is a general feature of the diffusion models that it is difficult to produce sharp features in the observed radio spectra. These calculations illustrate the point that, only in the intermediate range of energies at $E \sim 1$ GeV, is the spectrum expected to reflect the injection spectrum of the electrons.

For energies at which the loss time is of the same order as the synchrotron loss time, we might expect to see variations in the radio spectral index of the Galaxy in different directions. This is a notoriously difficult observation because whole sky surveys have been made with telescopes of very different angular resolutions and polar diagrams. Lawson *et al.* (1987) have made an attempt to do this for whole sky radio surveys in the frequency range 150 MHz to 5 GHz. They find little evidence for spectral variations at low radio frequencies, but significant fluctuations are found at high frequencies. This would be consistent with the idea that significant variations occur at frequencies at which the loss time is of the same order as or less than the escape time. By the same logic, there may well be significant fluctuations in the very low frequency spectrum of the Galaxy when the ionisation losses dominate.

Ideally, one would like to carry out similar analyses for the spectra of extragalactic systems, but the range of energies accessible to analysis is normally quite limited. We will find, however, that the 'universal' spectral index $p = 2.5$ is often found in very different astrophysical environments. This is a problem which we will have to tackle in considering models for the acceleration of high energy electrons.

19.4   Supernova remnants as sources of high energy electrons

The radio properties of supernova remnants provide convincing evidence that they are sources of the high energy electrons. In this section, we investigate some of their properties and ask whether or not they could be responsible for the flux of high energy electrons which we observe in the local interstellar medium. First of all, we review the evolution of supernova remnants (see also Section 15.2).

19.4.1   *The evolution of supernova remnants*

Supernova remnants come in two varieties. The most common are the *shell-like* remnants in which the radio, optical and X-ray emission originates from an expanding shell. In contrast, the *filled-centre* or *Crab-like* remnants possess a central radio source in the form of a young pulsar, and the brightness distribution is a maximum in the centre. Many of the observed features of *shell-like supernova remnants* can be accounted for in terms of general considerations of the expansion of a very hot gas cloud into the diffuse interstellar medium. The dynamical evolution can be divided into four stages.

(i) The energy liberated in the collapse of the central regions of the star is deposited in the outer layers, which are heated to a high temperature, and ejected with velocity about $(10–20) \times 10^3$ km s^{-1}. In a uniformly expanding sphere, the radial velocity of the expanding gas is proportional to the radial distance from the centre of the sphere, which remains undecelerated as long as the mass of interstellar gas swept up in the expansion is very much less than the mass of the ejected gas. The density and pressure distributions are largely dependent upon the initial conditions. If the temperature inside the sphere is uniform, it decreases adiabatically as the sphere expands according to the usual law $T \propto R^{-3(\gamma-1)}$, where γ is the ratio of specific heats. Because the expansion velocity is highly supersonic, a shock front forms in the interstellar gas and moves out ahead of the expanding sphere.

It is useful to introduce some of the terminology used to describe the physics of this phase since it will recur in many different circumstances. In the simplest picture of the supersonic expansion of a sphere of hot gas, there is an abrupt discontinuity between the expanding gas and the swept-up material. The situation resembles the case of the supersonic piston analysed in Section 10.6.2 (see Fig. 10.10). The expanding sphere of hot gas is represented by the piston, and the interface between the piston and the shocked gas immediately in front of it is called a *contact discontinuity*. In that example, it was shown that the shock wave runs ahead of the contact discontinuity, and that the region between the piston and the shock is heated to a high temperature. In the limit of very strong shocks, meaning that the Mach number, M_1, is very much greater than unity, the ratio of densities on either side of the shock wave is $\rho_2/\rho_1 = (\gamma+1)/(\gamma-1) = 4$ if the ratio of specific heats $\gamma = \frac{5}{3}$ (expression (10.31)). The temperature of the shocked gas is very high. In the limit of strong shocks, the ratio of temperatures behind and in front of the shock wave is $T_2/T_1 = 2\gamma(\gamma - 1)M_1^2/(\gamma + 1)^2 = \frac{5}{16}M_1^2$ if $\gamma = \frac{5}{3}$, where M_1 is the Mach number of the shock wave with respect to the sound speed of the unshocked gas (expression (10.32)). Since supernova explosions are highly supersonic with respect to the sound speed in the interstellar gas, it is evident that the shocked gas must be heated to very high temperatures, and therefore it is not at all surprising that young supernova remnants are intense X-ray emitters (see Fig. 7.14).

(ii) When the swept-up mass becomes greater than the ejected mass, the dynamics are described by the adiabatic blast-wave similarity solution of Taylor (1950) and Sedov (1959). During this stage, the overall dynamics are determined by the total mass of expanding gas, which is almost entirely swept-up interstellar gas, and the energy released in the initial explosion. This is a classic example, in which the dynamics can be derived by dimensional analysis. The only variables which can enter the solution are the energy of the explosion and the density of the ambient medium which provides all the swept-up mass. The dynamical variables are the radius and time. The dimensions of E/ρ_0 are $\mathrm{L}^5\mathrm{T}^{-2}$, and, consequently, we can find a dimensionless quantity $(E/\rho_0)t^2/r^5$ which describes the dynamics of the expansion, that is, $r \propto (E/\rho_0)^{1/5}t^{2/5}$. Taylor's studies of atomic bomb explosions have shown that they obey this relation closely.

When the deceleration of the expanding sphere becomes significant, there are important changes in the simple physical picture described in (i). The most important is that the outer shells of the expanding sphere are decelerated first, and so the material inside the sphere begins to catch up with the material in the outer layers. Therefore, the matter density begins to increase at the boundary of the expanding sphere. As the deceleration continues, the flow of gas into the outer layers becomes supersonic relative to the sound speed inside the sphere itself, and so a shock wave forms on the inner edge of the compressed outer layers. This behaviour is illustrated in Fig. 19.5. The formation of this shock wave has the effect of heating strongly the matter in the outer shells. The net result is that, although the material inside the sphere cooled during the adiabatic phase, the gas is reheated by the conversion of a large fraction of the kinetic energy of expansion back into heat. The internal shock wave propagates back through the expanding gas towards the origin, and in the process heats up all the ejected gas. Whereas, in phase (i), the kinetic energy of expansion of the gas was being communicated to the swept-up interstellar gas, in the deceleration phase the kinetic energy is also fed back into the ejected gas itself. This heated gas is a strong soft X-ray emitter, entirely consistent with the X-ray images of supernova remnants (Fig. 7.14). As we will discuss in Section 19.6, this structure is unstable to Rayleigh–Taylor instabilities, and this is likely to be responsible for the irregular structures seen in the shells of young supernova remnants.

In the case of the young supernova remnant Cassiopeia A, it is estimated that the mass ratio is about unity, and so it is intermediate between phases (i) and (ii). In the case of Tycho's supernova, there is evidence that it is already in phase (ii) and is expanding according to the Sedov relation $r \propto t^{2/5}$. The distribution of radio and X-ray emission is also consistent with what is expected in the Sedov blast-wave solution.

(iii) As the remnant continues to expand, the temperature in the region behind the shock front, which now contains most of the expanding mass, drops below 10^6K, and cooling by line emission of heavy ions becomes important. The resulting compression to preserve pressure balance at the shock front increases and the shell forms a dense 'snowplough'. Older remnants, such as the Cygnus Loop, which is about 50 000 years old, are associated with stage (iii), and the optical line emission of oxygen and sulphur observed from the filaments of the shell are associated with cooling gas.

(iv) The expansion eventually becomes subsonic, $v \leq 20$ km s^{-1}, and the supernova remnant loses its identity; it is dispersed by random motions in the interstellar medium.

The evolution of supernovae with central energy sources, such as the Crab Nebula, is somewhat different dynamically. The major difference is that the Nebula is constantly receiving a supply of energy either in the form of relativistic particles or electromagnetic energy from the central pulsar.

19.4.2   *Radio observations of supernova remnants*

Radio emission is observed from both young and old supernova remnants, the young ones being particularly intense emitters. Four examples of the radio structures of these remnants are shown in Fig. 19.6. In the case of Cassiopaeia A (Fig. 19.6(a)), the emission results from a spherical shell with a great deal of fine-scale structure. In the case of the Crab Nebula (Fig. 19.6(b)), the radio image looks remarkably like the optical image (Fig. 15.3), the enhanced radiation from the filaments presumably being due to compression of the gas and magnetic field in these regions. Tycho's supernova (Fig. 19.6(c)) is also shell-like, but has a very sharp, almost circular, boundary. In the case of the old supernova remnant the Cygnus Loop (Fig. 19.6(d)), there is again a close correlation between regions of enhanced radio emissivity and the presence of optical emission lines, again suggestive of regions where high compression of gas and radio emitting electrons has taken place.

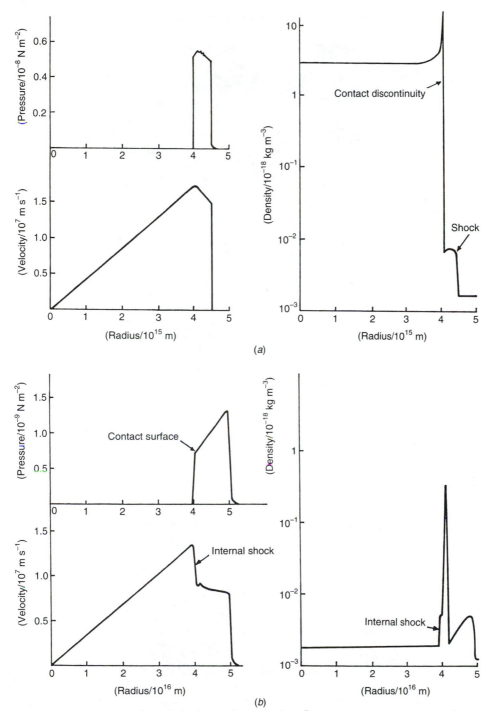

Figure 19.5. The evolution of the internal properties of a young supernova remnant from the time when (a) the mass ratio is very much less than unity to (b) the Sedov phase, when the swept-up mass dominates the dynamics. Features of importance are the formation of the dense shell of material just inside the contact discontinuity and the role of the reverse inner shock in heating up the material inside the expanding sphere. (After S. F. Gull (1975). *Mon. Not. R. Astron. Soc.*, **171**, 263.)

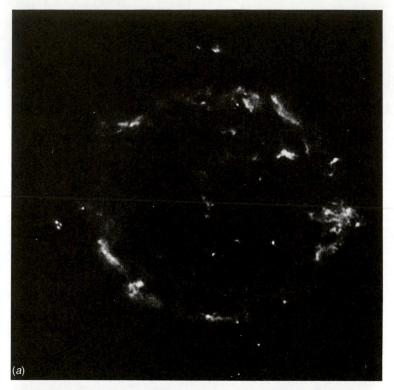

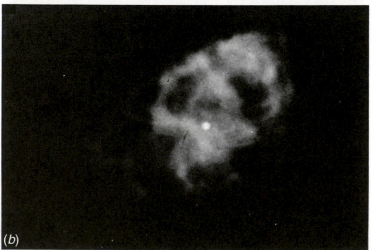

Figure 19.6 (*a*), (*b*). 'Radio photographs' of supernova remnants: (*a*) Cassiopeia A (from A.R. Bell (1977). *Mon. Not. R. Astron. Soc.*, **179**, 574, reprocessed by S.F. Gull); (*b*) the Crab Nebula (from E. Swinbank and G.G. Pooley (1979). *Mon. Not. R. Astron. Soc.*, **186**, 776).

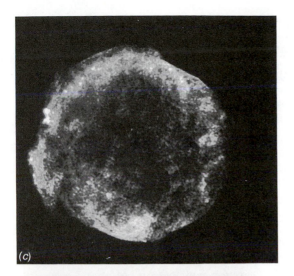

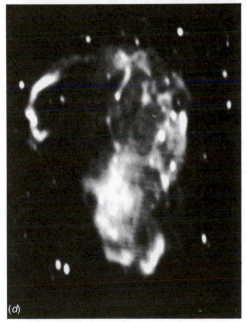

Figure 19.6. (*c*), (*d*). 'Radio photographs' of supernova remnants: (*c*) Tycho's supernova (from S.M. Tan and S.F. Gull (1985). *Mon. Not. R. Astron. Soc.*, **216**, 949); and (*d*) the Cygnus Loop (from D.A. Green (1984). *Mon. Not. R. Astron. Soc.*, **211**, 433).

The radio emission of both young and old remnants is highly polarised, and the radio spectra are of power-law form. Whereas the radio spectra of young shell-like supernova remnants have spectral indices $\alpha > 0.5$, the filled-centre remnants have $\alpha \approx 0.0 - 0.3$. By analogy with the Galactic radio emission, it is identified with the synchrotron radiation of ultrarelativistic electrons gyrating in the magnetic field in the shell of the remnant. In young supernova remnants, the intensity of the radio emission is so high that the energy densities of relativistic electrons and magnetic fields greatly exceed the values which could be obtained by compressing the Galactic flux of high energy electrons and magnetic field in a strong shock. Therefore, the relativistic particles and fields must originate within the remnant itself. On the other hand, it is likely that the radio emission of some old remnants can result from compression of the interstellar magnetic field and relativistic electron fluxes. High compressions are possible in the shells of old remnants because of the rapid cooling in the shocked region, when the temperature falls below about 10^6 K. The radio spectra of some of the radio filaments in old remnants like the Cygnus Loop have spectral indices $\alpha \sim 0.5$, which would be consistent with the compression of the Galactic flux of high energy electrons.

It is of importance to know how much relativistic particle energy there is present in supernova remnants, and estimates of this can be found from what is known as the *minimum energy requirements* for synchrotron radiation.

19.5   The minimum energy requirements for synchrotron radiation

Suppose a source has luminosity L_ν at frequency ν and its volume is V. The spectrum of the radiation is of power-law form, $L_\nu \propto \nu^{-\alpha}$, and the emission mechanism is assumed to be synchrotron radiation. The following arguments can be applied to the synchrotron radiation emitted by the source at any frequency, be it radio, optical or X-ray wavelengths. The radio luminosity can be related to the energy spectrum of the ultrarelativistic electrons and the magnetic field B present in the source through the formula (18.73) for synchrotron radiation:

$$L_\nu = A(\alpha)V\kappa B^{1+\alpha}\nu^{-\alpha} \qquad (19.17)$$

where the electron energy spectrum per unit volume is $N(E)\,\mathrm{d}E = \kappa E^{-p}\,\mathrm{d}E, p = 2\alpha + 1$ and $A(\alpha)$ is a constant. From the expression (18.73), the constant $A(\alpha)$ is

$$A(\alpha) = 2.344 \times 10^{-25}(1.253 \times 10^{37})^\alpha a(2\alpha + 1)$$

where the units are standard SI units and $a(2\alpha+1)$ is given in Table 18.2. Writing the energy density in relativistic electrons as ϵ_e, the total energy present in the source responsible for the radio emission is

$$W_{\text{total}} = V\epsilon_e + V\frac{B^2}{2\mu_0}$$

$$= V\int \kappa E N(E)\mathrm{d}E + V\frac{B^2}{2\mu_0} \qquad (19.18)$$

From equation (19.17), it can be seen that the luminosity of the source, L_ν, determines only the product $V\kappa B^{1+\alpha}$. If V is assumed to be known, the radio luminosity may either be produced by a large flux of relativistic electrons in a weak magnetic field, or vice versa. There is no way of deciding which combination of ϵ_e and B is appropriate from observations of L_ν. Between the extremes of dominant magnetic field and dominant particle energy, however, there is a minimum total energy requirement.

Before proceeding to that calculation, we should consider the problem of how much energy is also present in the form of relativistic protons, which presumably must also be present in the radio source. There are, unfortunately, very few sources for which estimates of both the electron and proton fluxes are known. On the one hand, in our own Galaxy, there seems to be about 100 times as much energy in relativistic protons as there is in electrons, whereas, in the Crab Nebula, the energy in relativistic protons cannot be much greater than the energy in the electrons from dynamical arguments. Therefore, to take account of the protons, it is customary to assume that they have energy β times that of the electrons, that is,

$$\epsilon_{\text{protons}} = \beta\epsilon_e$$

$$\epsilon_{\text{total}} = (1+\beta)\epsilon_e = \eta\epsilon_e \tag{19.19}$$

We therefore write

$$W_{\text{total}} = \eta V \int_{E_{\min}}^{E_{\max}} \kappa E N(E)\mathrm{d}E + V\frac{B^2}{2\mu_0} \tag{19.20}$$

The energy requirements as expressed in the expression (19.20) depend upon the unknown quantities κ and B, but they are related through equation (19.17) for the observed luminosity of the source, L_v. We also require the relation between the frequency of emission of an ultrarelativistic electron of energy $E = \gamma m_e c^2 \gg m_e c^2$ in a magnetic field of strength B. We use the results of Section 18.1.4 that the maximum intensity of synchrotron radiation occurs at a frequency

$$v = v_{\max} = 0.29v_c = 0.29\frac{3}{2}\gamma^2 v_g$$

$$= 0.29 \times 4.199 \times 10^{10}\gamma^2 B = CE^2 B \tag{19.21}$$

where v_g is the non-relativistic gyrofrequency and $C = 1.22 \times 10^{10}/(m_e c^2)^2$. Therefore, the relevant range of electron energies in the integral (19.20) is related to the range of observable radio frequencies through

$$E_{\max} = \left(\frac{v_{\max}}{CB}\right)^{1/2} \quad ; \quad E_{\min} = \left(\frac{v_{\min}}{CB}\right)^{1/2} \tag{19.22}$$

$v_{\max}$ and $v_{\min}$ are the maximum and minimum frequencies for which the radio spectrum is known, or the range of frequencies relevant to the problem at hand. Then

$$W_{\text{particles}} = \eta V \int_{E_{\min}}^{E_{\max}} E\kappa E^{-p}\mathrm{d}E$$

$$= \frac{\eta V \kappa}{(p-2)}(CB)^{(x-2)/2}\left[v_{\min}^{(2-p)/2} - v_{\max}^{(2-p)/2}\right]$$

Substituting for κ in terms of L_v and B from equation (19.17),

$$W_{\text{particles}} = \frac{\eta V}{(p-2)}\left[\frac{L_v}{A(v)VB^{1+\alpha}v^{-\alpha}}\right](CB)^{(p-2)/2}[v_{\min}^{(2-p)/2} - v_{\max}^{(2-p)/2}]$$

Preserving only the essential variables,

$$W_{\text{particles}} = G(\alpha)\eta L_v B^{-3/2} \tag{19.23}$$

where $G(\alpha)$ is a constant which depends weakly on α, v_{max} and v_{min} if $\alpha \approx 0.75$. From the above analysis,

$$G(\alpha) = \frac{1}{a(p)(p-2)}[v_{min}^{-(p-2)/2} - v_{max}^{-(p-2)/2}]v^{(p-1)/2}$$

$$\times \frac{(7.4126 \times 10^{-19})^{-(p-2)}}{2.344 \times 10^{-25}}(1.253 \times 10^{37})^{-(p-1)/2}$$

Therefore,

$$W_{total} = G(\alpha)\eta L_v B^{-3/2} + V\frac{B^2}{2\mu_0} \tag{19.24}$$

The variations of the energies in particles and the magnetic field are shown in Fig. 19.7 as a function of B. There is a minimum total energy which can be found by minimising equation (19.24) with respect to B:

$$B_{min} = \left[\frac{3\mu_0}{2}\frac{G(\alpha)\eta L_v}{V}\right]^{2/7} \tag{19.25}$$

This magnetic field strength, B_{min}, corresponds to approximate equality of the energies in the relativistic particles and magnetic field. Substituting B_{min} in equation (19.23), we find

$$W_{mag} = V\frac{B_{min}^2}{2\mu_0} = \frac{3}{4}W_{particles} \tag{19.26}$$

Thus, the condition for minimum energy requirements corresponds closely to the condition that there are equal energies in the relativistic particles and the magnetic field. This condition is often referred to as *equipartition*. The minimum total energy is

$$W_{total}(min) = \frac{7}{6\mu_0}V^{3/7}\left[\frac{3\mu_0}{2}G(\alpha)\eta L_v\right]^{4/7} \tag{19.27}$$

The expressions (19.25) and (19.27) are the answers we have been seeking. These are the magnetic field strength and minimum total energy needed to account for the observed radio luminosity of the source. These results are frequently used in the study of the synchrotron radiation from radio, optical and X-ray sources but their limitations should be appreciated.

1 There is no physical justification for the source components being close to equipartition. It might be that the particle and magnetic field energies in the source components tend towards equipartition, but there is no proof that this must be so. For example, it has been conjectured that the magnetic field in the source components may be stretched and tangled by motions in the plasma, so there might be rough equipartition between the magnetic energy density and the energy density in turbulent motions. The turbulent motions might also be responsible for accelerating the high energy particles, and these particles might come into equipartition with the turbulent energy density if the acceleration mechanism were very efficient. In this way, it is possible that there might be a physical justification for the source components being close to equipartition, but this is really no more than a conjecture.

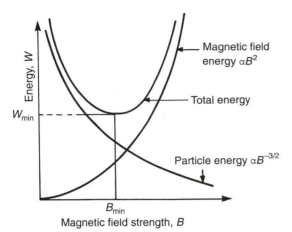

Figure 19.7. Illustrating the origin of the minimum energy requirements of a source of synchrotron radiation as a function of magnetic field strength B.

2 The amount of energy present in the source component is sensitive to the value of η, that is, the amount of energy present in the form of protons and nuclei.

3 The total amount of energy in relativistic particles is dependent upon the upper and lower energy limits assumed to the energy spectrum of the particles. It can be seen that, if $\alpha = 0.75$, we need only consider the dependence upon ν_{min}, which is quite weak, $W_{min} \propto \nu_{min}^{-0.25}$. However, there might be large fluxes of low energy relativistic electrons present in the source components with a quite different energy spectrum, and we would have no way of knowing that they are present from the radio observations.

4 Even more important is the fact that the energy requirements depend upon the volume of the source. The calculation has been carried out assuming that the particles and magnetic field fill the source volume uniformly. In fact, the emitting regions might occupy only a small fraction of the apparent volume of the source, for example, if the synchrotron emission originated in filaments or sub-components within the overall volume V. In this case, the volume which should be used in the expressions (19.25) and (19.27) should be smaller than V. Often, a *filling factor f* is used to describe the fraction of the volume occupied by radio emitting material. Clearly, the energy requirements are reduced if f is small.

5 On the other hand, we can obtain a firm lower limit to the *energy density* within the source components since

$$U_{min} = \frac{W_{total}(min)}{V} = \frac{7}{6\mu_0} V^{-4/7} \left[\frac{3\mu_0}{2} G(\alpha)\eta L_\nu \right]^{4/7} \tag{19.28}$$

For dynamical purposes, the energy density is more important than the total energy since it is directly related to the pressure within the source components $p = (\gamma - 1)U$, where γ is the ratio of specific heats. In the case of an ultrarelativistic gas, $\gamma = \frac{4}{3}$, and so $p = \frac{1}{3}U$ as usual. As we will find, one of the important problems of these sources is to understand how this energy density can be confined within the source components.

For these reasons, the values of the magnetic field strength and minimum energy which come out of these arguments should be considered only order of magnitude estimates. Obviously, if the source components depart radically from the equipartition values, the energy requirements are increased, and this can pose problems for some of the most luminous sources.

It is often cumbersome to have to go through the procedure of working out $G(\alpha)$ to estimate the minimum energy requirements and magnetic field strengths. A simplified calculation can be performed in the following way. If we assume that the spectral index $\alpha = 0.75$, we can neglect the upper limit, ν_{max}, in comparison with ν_{min} in evaluating $G(\alpha)$. Then, if we know the luminosity, $L(\nu)$, at a certain frequency, ν, we obtain a lower limit to the energy requirements if we set $\nu = \nu_{min}$. Making these simplifications, we find that the minimum energy requirement is

$$W_{min} \approx 3.0 \times 10^6 \, \eta^{4/7} V^{3/7} \nu^{2/7} L_\nu^{4/7} \quad \text{J} \tag{19.29}$$

where the volume of the source, V, is measured in cubic metres, the luminosity $L(\nu)$ in watts per hertz and the frequency ν in hertz. In the same units, the minimum magnetic field strength is

$$B_{min} = 1.8 \left(\frac{\eta L_\nu}{V} \right)^{2/7} \nu^{1/7} \quad \text{T} \tag{19.30}$$

This line of reasoning was very important in the late 1950s, when the energy requirements of extragalactic radio sources were first estimated by Burbidge (1956). A good example was provided by the radio source Cygnus A, the brightest extragalactic radio source in the northern sky. At that time, it was thought that the source consisted of two components roughly 100 kpc in diameter. The source had luminosity roughly 8×10^{28} W Hz^{-1} at 178 MHz. Inserting these values into the expression (19.29), we find that the minimum total energy is $2 \times 10^{52} \eta^{4/7}$ J, which corresponds to the rest mass energy of $3 \times 10^5 \eta^{4/7} M_\odot$ of matter. The realisation of the enormous energy demands of extragalactic radio sources in the form of relativistic particles and magnetic fields was one of the most important problems which stimulated the very rapid growth of high energy astrophysics in the 1960s. Evidently, a very considerable amount of mass has to be converted into relativistic particle energy and ejected from the nucleus of the galaxy into enormous lobes well outside the body of the galaxy. We will take up this story in much more detail in Volume 3.

19.5.1   *Supernova remnants as sources of high energy electrons*

Let us estimate the minimum energy in the form of relativistic particles and magnetic field present in the young supernova remnant Cassiopeia A (Fig. 19.6(a)). Cassiopaeia A is the brightest radio source in the northern sky and has flux density 2720 Jy $= 2.72 \times 10^{-23}$ W m^{-2} Hz^{-1} at a frequency of 1 GHz.

(For the definition of the jansky, Jy, see Appendix A2.) The bulk of the radiation originates within a roughly spherical volume of angular diameter 4 arcmin. For our present purposes, it is sufficient to assume that the radio source is a uniform sphere. Adopting a distance of 2.8 kpc, the radio luminosity of the source at 1 GHz is $L_{1GHz} = 2.6 \times 10^{18}$ W Hz^{-1} and its volume $V = 5.3 \times 10^{50}$ m^3. The minimum energy requirement is therefore $W_{min} = 2 \times 10^{41} \eta^{4/7}$ J, and the equipartition magnetic field strength $B = 10 \eta^{2/7}$ nT. This is a large amount of energy. We recall that the rest mass energy of one solar mass of material is $M_\odot c^2 = 2 \times 10^{47}$ J. Another interesting comparison is with the kinetic energy ejected in the explosion. There are optical filaments more or less coincident with the northern half of the radio shell, and, by taking photographic plates at different epochs, the radial motions of the filaments from the centre of the explosion can be found. Velocities and masses of the optical filaments can be found from spectroscopic analyses of the intense emission lines they emit. It is estimated that the total kinetic energy in optical filaments is about 2×10^{44} J. When we recall that the figure of $2 \times 10^{41} \eta^{4/7}$ J is the minimum energy to produce the radio emission, it can be seen that there must exist a highly efficient mechanism for converting the gravitational energy of collapse into high energy particle and magnetic field energy. We can therefore be confident that supernova remnants are very powerful sources of high energy electrons. Notice that if $\eta \approx 100$, a value appropriate to protons and electrons observed at the top of the atmosphere, the minimum energy requirements would be increased by a factor $\eta^{4/7}$, that is, about an order of magnitude.

Let us carry out some simple estimates to find out if, in principle, supernova remnants such as Cassiopeia A can account for the energy density of high energy particles which we observed in our vicinity in the interstellar medium. There is considerable flexibility in the numbers. The mean energy density in high energy particles is determined by the following quantities: the average time interval between Galactic supernovae, t_{SN}; the volume within which the high energy particles are confined, V; the characteristic escape time of particles from the confinement volume, t_c; and the average energy release in high energy particles per supernova, E_0. Thus, the local energy density of high energy particles in a steady state is expected to be

$$\epsilon_{CR} = \frac{t_c}{t_{SN}} \frac{E_0}{V} \tag{19.31}$$

For the purpose of illustration, let us suppose that the lifetime of the high energy particles in the disc of the Galaxy is 10^7 years (see section 19.3.4). If the thickness of the disc is assumed to be 700 pc and its radius to be 10 kpc, its volume is 6×10^{60} m^3. The local energy density of high energy particles is roughly 10^6 eV m^{-3} (Section 9.3), and so, adopting a supernova rate of one per 30 years, the average energy release per supernova must be $E_0 \approx 2 \times 10^{42}$ J. This should be compared with the above estimate for Cassiopaeia A of $2 \times 10^{41} \eta^{4/7}$ J.

The conclusion of this analysis is that, energetically, it is quite feasible to account for the total energy density of high energy particles in terms of their origin in supernova remnants. We will return to this subject when we consider the confinement volume of high energy protons and nuclei, but the basic conclusion will remain unaltered.

19.5.2   *The adiabatic loss problem and supernova remnants*

Before we are lulled into a false sense of security, we should note that it is not immediately obvious that we can make good use of the high energy particles present in supernova remnants because of what is known as the *adiabatic loss problem* which pervades much of the astrophysics of clouds of relativistic plasma. The concern is that, if high energy particles are accelerated in a supernova explosion, they lose all their energy adiabatically during phases (i) and (ii) of the evolution described in Section 19.4.1 before they are released into the interstellar medium.

Suppose the relativistic electrons were accelerated very soon after the catastrophic event which gave rise to the supernova. Then, the relativistic gas exerts a pressure on its surroundings and consequently suffers adiabatic losses, as described in Section 19.2. For a relativistic particle, the energy decreases with increasing radius as $E \propto r^{-1}$. This also applies to the total relativistic particle energy, and thus, if the total energy is W_0 at radius r_0, when the remnant expands to radius r, the internal energy of the relativistic gas is only $(r_0/r)W_0$. The energy of the particles has gone into the kinetic energy of expansion of the relativistic gas. There are two ways of looking at this result. We can say that, unless we find a mechanism for regenerating the energy in the relativisic gas, we will not obtain adequate energy from sources such as Cassiopaeia A, by the time they have expanded to phases (iii) and (iv), to account for the local energy density of high energy particles. Alternatively, we can say that, if r_0 were too small, the energy requirements in relativistic particles would become very large indeed in the early stages of expansion.

There are also consequences for the radio emission of the remnant. If the expansion were purely adiabatic and the magnetic field strength decreased as $B \propto r^{-2}$, as is expected if magnetic flux freezing is applicable (Section 10.5), the radio luminosity should decrease rapidly as the remnant expands. From equation (19.13), we can work out how the electron energy spectrum changes with radius in an adiabatic expansion. In this case, the diffusion-loss equation reduces to

$$\frac{\mathrm{d}N(E)}{\mathrm{d}t} = \frac{\partial}{\partial E}[b(E)N(E)] \tag{19.32}$$

where $b(E) = (1/r)(\mathrm{d}r/\mathrm{d}t)E$. Note that $N(E)$ now refers to all the particles in the remnant rather than the number per unit volume. Therefore, during the expansion, we can write $N(E) = V\kappa(r)E^{-p}$, since, as we have shown in Section 19.3.3, the spectral index does not change under adiabatic losses. If $N(E)$ were the number per unit volume, we would have to add the term $-N\nabla \cdot \mathbf{v}$ to the right-hand-side of equation (19.32). Therefore, assuming $N(E) = KE^{-p}$,

$$\frac{\mathrm{d}N(E)}{\mathrm{d}t} = \frac{\partial}{\partial E}\left(\frac{1}{r}\frac{\mathrm{d}r}{\mathrm{d}t}KE^{-(p-1)}\right)$$

$$\frac{\mathrm{d}N(E)}{N(E)} = -(p-1)\frac{\mathrm{d}r}{r}$$

$$\frac{N(E,r)}{N(E,r_0)} = \left(\frac{r}{r_0}\right)^{-(p-1)}$$

that is,

$$\frac{K(r)}{K(r_0)} = \left(\frac{r}{r_0}\right)^{-(p-1)} \tag{19.33}$$

We can now work out how the synchrotron radio luminosity varies with radius because

$$I_v = A(\alpha)K(R)B^{(1+\alpha)}v^{-\alpha}$$

and hence

$$I_v(r) \propto r^{-(p-1)}r^{-2(1+\alpha)}$$

$p = 2\alpha + 1$, and therefore

$$I_v(r) \propto r^{-2(2\alpha+1)} = r^{-2p} \tag{19.34}$$

Whilst it is observed that Cassiopaeia A is decreasing in radio luminosity at roughly the rate predicted by this expression, it cannot be extrapolated back to the earliest phases of expansion because young supernovae in external galaxies would then be very intense radio sources, which is contrary to observation. Furthermore, the magnetic field strength in Cassiopaeia A at radius $r \approx 5 \times 10^{16}$ m is $\approx 10^{-8}$ T, which, if compressed to the dimensions of a giant star which has radius $r \approx 10^{11}$ m, would far exceed conceivable field strengths.

All these arguments suggest that both the relativistic particles and the magnetic field cannot have been created in the initial explosion of the supernova. Rather, the acceleration of the high energy particles and the generation of strong magnetic fields must take place during the expansion phase of the supernova. Fortunately, there exist plausible physical mechanisms by which this can be achieved.

The origin of the strong magnetic fields in young supernova remnants has been elegantly explained by Gull (1975), who has studied the dynamics of young supernova shells during phases (i) and (ii) of the scheme outlined in Section 19.4.1. As explained in that section, the deceleration of the expanding sphere causes the matter to accumulate in a dense shell just inside the contact discontinuity between the sphere and the interstellar medium (see Fig. 19.5). The force of deceleration acting upon this dense shell is in pressure balance with the much less dense shocked interstellar gas, and this results in a *Rayleigh–Taylor instability*. This is the instability which occurs when a heavy fluid is supported by a light fluid in a gravitational field (Fig. 19.8). In these instabilities, magnetic field amplification is possible as a 'seed' magnetic field, however small initially, is stretched and sheared in this turbulent unstable zone. Calculations of the amount of energy which may be channelled into the magnetic field show that field strengths of up to 10^{-7} T can be obtained for a supernova such as Cassiopaeia A. The instability stretches the field in the radial direction, and this is in agreement with the observed magnetic field configuration in young shell-like supernovae. The subsequent evolution of

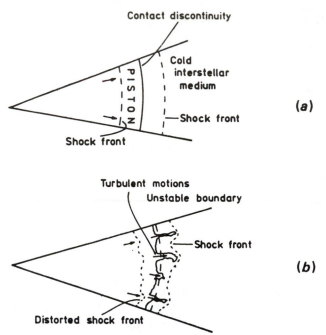

Figure 19.8. (*a*) Illustrating the structure of a supersonically expanding sphere of hot gas. The sphere of gas acts as a piston, and ahead of it runs a shock front which heats and compresses the cold interstellar gas. As the sphere expands into the interstellar gas, it is decelerated, and matter inside the hot sphere beings to pile up at the interface (or contact discontinuity) between the shocked interstellar gas and the interior of the sphere. An internal shock front forms as matter piles up at the interface, and the temperature of the gas inside the sphere cools. (*b*) Because of the piling up of matter at the interface and the deceleration of the shell, there is an effective force of gravity acting on the dense shell of gas, which is supported by the much more rarefied shocked interstellar gas. This results in a Raleigh–Taylor instability as illustrated. (From S. F. Gull (1975). *Mon. Not. R. Astron. Soc.*, **171**, 263.)

these instabilities may lead to the formation of transient optical filaments as observed in young remnants.

There remains the problem of accelerating the high energy particles which are present in the shells of supernova remnants. The breakthrough came in the late 1970s when four independent papers appeared which showed how high energy particles can be accelerated efficiently in strong shock waves. These papers by Axford, Leer and Skadron (1977), Krymsky (1977), Bell (1978) and Blandford and Ostriker (1978) were the seeds of a spectacular development of the understanding of the processes of particle acceleration in shock waves in as diverse astrophysical situations as the Earth's magnetospheric boundary, supernova shock waves and extragalactic radio sources. This is key topic for the whole of high energy astrophysics, and we return to it in detail in Chapter 21.

The important point of principle is that it is possible to overcome the problem of adiabatic losses by using the kinetic energy of expansion of the expanding remnant. Thus, contrary to what might be expected, when the sphere expands into the interstellar medium, the internal energy of the expanding sphere is not lost, but is stored as the kinetic energy of the expanding material. The great lesson of the radio observations of supernova remnants is that there exist physical mechanisms for retrieving this energy and converting it into 'useful' forms, by which we mean high energy particles and magnetic fields.

20

The origin of high energy protons and nuclei

20.1 γ-ray observations of the Galaxy

Just as the Galactic radio emission outlines the distribution of high energy electrons and magnetic fields in the Galaxy, so the distribution of γ-radiation can provide information about high energy protons and the overall distribution of interstellar gas. As described in Section 5.4, in collisions between high energy particles and protons and nuclei of atoms and molecules of the interstellar gas, pions of all charges, π^+, π^0 and π^-, are produced. The positive and negative pions decay into positive and negative muons, which, in turn, decay into positrons and electrons with relativistic energies (see Fig. 5.11). The latter may make a contribution to the low energy electron spectrum, and the predicted presence of positrons provides a direct test of the importance of the pion production mechanism in interstellar space. The neutral pions decay almost instantly into two γ-rays. In proton–proton collisions, the cross-section for the production of a pair of high energy γ-rays is roughly the geometric size of the proton, $\sigma_\gamma \approx 10^{-30}$ m^2. The spectrum of γ-rays produced in such collisions is shown in Fig. 20.1. The characteristic signature of this process is that the spectrum of γ-rays has a broad maximum at about 70 MeV. Knowing the physical conditions in the interstellar gas, the γ-ray production rates by various mechanisms can be worked out. Stecker (1977) has carried out these calculations assuming an interstellar gas density of 10^6 particles m^{-3} and a local energy density of starlight of 4.4×10^5 eV m^{-3}. The principal emission mechanisms are the π^0 decay process described above, bremsstrahlung of high energy electrons (see Section 18.3) and inverse Compton scattering of starlight by the high energy electrons responsible for the Galactic radio emission (see Section 19.2). The results of these calculations are shown in Fig. 20.2, from which it is apparent that, at energies greater than about 100 MeV, the π^0 decay mechanism is by far the most important production mechanism for γ-rays.

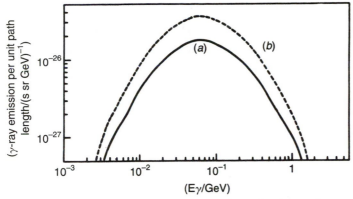

Figure 20.1. The γ-ray spectrum resulting from the decay of neutral pions, π^0, created in collisions involving (a) proton–proton interactions and (b) proton–proton, proton–α, α–proton and α–α interactions. The cosmic abundances of hydrogen and helium are assumed. (From F. W. Stecker (1975). In *The origin of cosmic rays*, eds J. L. Osborne and A. W. Wolfendale, p. 267. Nato Advanced Study Institute Series, Dordrecht: D. Reidel.)

20.1.1 γ-ray observations of the disc of the Galaxy

Intense γ-radiation from the Galaxy was first detected by the OSO-III satellite in 1967. Since then, spark chamber detectors have been flown on the SAS-II and COS-B satellites, and these observations have resulted in maps of the Galactic distribution of γ-ray emission with an angular resolution of about 3° (for details of the experiments, see Section 7.4). Clues to the emission process are provided by the spectrum of the γ-ray emission (Fig. 18.16). At energies greater than 100 MeV, the spectrum is steep, but there is a definite flattening at about 50–100 MeV. This result strongly suggests that most of the γ-radiation at energies $h\nu > 100$ MeV results from the π^0 decay process, and, as such, provides direct information about the distribution of high energy particles, principally the proton component, and of interstellar matter. At lower energies, $h\nu < 100$ MeV, the γ-ray spectrum of the Galaxy does not flatten to the extent predicted by a pure π^0 decay spectrum, suggesting that either the inverse Compton or bremsstrahlung processes make a significant contribution to the total γ-ray intensity. We argued in Section 18.3 that it is entirely plausible that this radiation is the bremsstrahlung of high energy electrons with energies $30 \leq E \leq 100$ MeV.

The COS-B satellite has produced a detailed map of the Galactic high energy γ-ray emission (Fig. 1.8(f)). Various cuts through that map are shown in Fig. 20.3, illustrating the intensity distribution along and across the Galactic plane. Some important features of these observations are as follows.

1 There is an intense ridge of γ-radiation in the direction of the Galactic centre extending between Galactic longitudes 310° and 45°. This contrasts with the much lower intensity observed in the anticentre direction.

2 It can be seen from Fig. 20.3(a) that there are two components present in the emission from the Galaxy. There is a very narrow component which

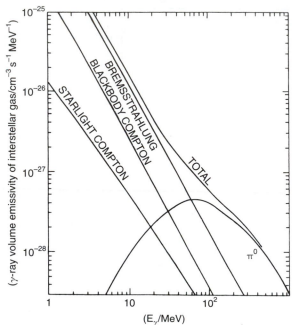

Figure 20.2. γ-ray production rates by various emission mechanisms for typical interstellar conditions. The gas density is assumed to be 10^6 particles m^{-3} and the energy density of starlight is assumed to be 4.4×10^5 eV m^{-3}. (From F. W. Stecker (1977). In *The structure and content of the Galaxy and Galactic gamma-rays*, ed. C. E. Fichtel, p. 315. Greenbelt: Goddard Space Flight Center Publications.)

is barely resolved by the beam of the COS-B telescope in the direction of the Galactic centre and a rather broader component with half-width about 7°. In the anticentre direction, the width of the γ-ray disc is significantly broader than the resolution of the telescope, and is about 7° in half-width. The COS-B observers interpret these observations as indicating that there is γ-ray emission from a disc similar to that seen at radio wavelengths, as well as emission from more local features such as the local spiral arm and the major local concentration of young massive stars, known as Gould's belt.

3   There is considerable fine structure in the distribution of γ-ray emissivity, some of which is certainly associated with discrete γ-ray sources. Strong signals are observed from the direction of the Crab and Vela supernova remnants, and pulsed γ-ray emission with the periods of their respective pulsars has been detected. In fact, at energies greater than 400 MeV, it appears that most of the γ-ray emission from the Crab Nebula is in the pulsed component. Significant pulsed signals were also detected from the pulsars PSR 1818-04 and PSR 1747-46, although they do not stand out prominently as discrete sources in Fig. 1.8(f). The COS-B observers detected a total of 29 γ-ray sources, most of them lying within

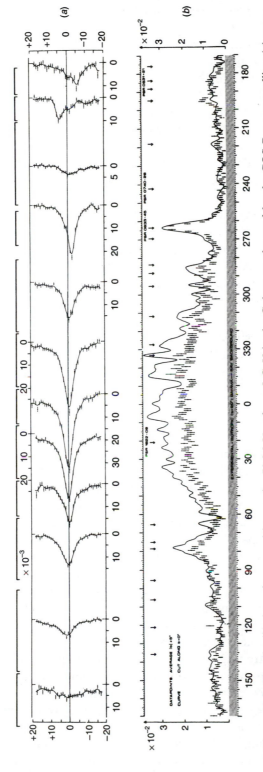

Figure 20.3. The distribution of γ-rays with energies 70 MeV $\leq h\nu \leq$ 5 GeV in the Galaxy as observed by the COS-B γ-ray satellite: (*a*) cuts across the Galactic plane at different galactic longitudes; (*b*) the γ-ray distribution along the Galactic plane. The solid curve represents the intensity along $b = 0$, and the data points are average intensities within galactic latitudes $|b| \leq 5°$. (Courtesy of K. Bennett and the European Space Agency.)

the Galactic plane. An exception to this is the quasar 3C 273, the first of these objects to be detected as a γ-ray source. More recent observations from high flying balloons and the SIGMA satellite have shown that the active galaxies NGC 4151, Centaurus A and NGC 1275 are also γ-ray sources. Among the first results of the GRO mission has been a remarkable observation of the quasar 3C 279 by the EGRET instrument. It was not detected by the SAS-2 and COS-B telescopes, although it was within their fields of view. In June 1991, 3C 279 outshone 3C 273 as the brightest extragalactic high energy γ-ray source known. It is unlikely to be coincidental that this radio quasar is highly variable in intensity and is currently going through a phase of intense radio activity. It is also one of the best studied superluminal radio sources (see Volume 3). At the time of writing (April 1993), 24 active galactic nuclei have been detected by the GRO, 20 of them being quasars and 4 BL-Lac objects. They are all core-dominated, flat-spectrum radio sources and half of them are superluminal sources. Such objects must make a substantial contribution to the γ-ray background radiation (Padovani *et al.* (1993)).

The status of γ-ray astronomy and the understanding of the physics of γ-ray sources prior to the epoch of the Gamma-Ray Observatory are comprehensively surveyed in the monograph *Gamma-ray astronomy* by Ramana Murthy and Wolfendale (1993).

20.1.2 *Interpretation of the γ-ray emission of the Galaxy*

To determine the γ-ray luminosity of the Galaxy, it is necessary to convert the distributions shown in Figs 1.8(f) and 20.3 into a plot of γ-ray emissivity per unit volume as a function of Galactic coordinates. This is not an unambiguous procedure, but, using estimates by Stecker, who analysed the SAS-II data, the γ-ray luminosity of the Galaxy is roughly 10^{32} W at γ-ray energies $hv \geq 100$ MeV.

In interpreting the spatial distribution of γ-radiation shown in Fig. 20.3, it is revealing to compare it with other tracers of the distribution of high energy particles and interstellar gas. The most striking comparisons are with the distributions of radio emissivity (Fig. 1.8(a)) and of molecular hydrogen as derived from surveys of Galactic carbon monoxide (CO) emission. As discussed in Section 17.2.2, molecular hydrogen is the dominant form of interstellar gas at distances less than 10 kpc from the Galactic centre, the maximum density occurring at radii about 5–6 kpc from the centre (Fig. 17.2). Mean densities of about 1–2×10^6 m^{-3} are inferred from the CO observations. These distributions are remarkably similar to the distribution of γ-radiation. There is also some evidence that the Galactic distributions of HII regions and supernova remnants have maxima at distances of about 6–8 kpc from the Galactic centre.

Let us perform a simple calculation to work out the γ-ray luminosity of the Galaxy due to the π^0 decay process. The probability, P_{coll}, that a high energy proton undergoes an inelastic collision per second with a nucleus of the interstellar gas is

$$P_{\text{coll}} = \sigma_{\text{pp}} N c$$

where σ_{pp} is the proton–proton inelastic cross-section, which is about 2.5×10^{-30} m^2, N is the number density of the interstellar gas and it is assumed that the proton travels at the speed of light. Now, roughly one-third of the pions produced in each collision are neutral pions which decay into two γ-rays with mean energy 180 MeV. Therefore, if we suppose the disc of the Galaxy has volume V and is filled uniformly with interstellar gas and high energy particles, the number of collisions per second of high energy protons of energy E is $V N_{\text{CR}}(E)\sigma_{\text{pp}} N c$, where $N_{\text{CR}}(E)$ is the number density of high energy particles with energy E. Therefore, the total energy liberated as γ-rays is roughly

$$L_\gamma = \frac{1}{3}\sigma_{\text{pp}} N c \sum N_{\text{CR}}(E)E = \frac{1}{3}P_{\text{coll}}\epsilon_{\text{CR}} V$$

where ϵ_{CR} is the local energy density of high energy particles. Taking the disc to have half-thickness 200 pc and radius 8 kpc, we find $V \approx 2 \times 10^{60}$ m^3. Therefore, adopting $N = 10^6$ m^{-3} and $\epsilon_{\text{CR}} \approx 10^6$ eV m^{-3}, we find $L_\gamma \approx 10^{32}$ W, in good agreement with the total observed γ-ray luminosity of the Galaxy. Obviously, this argument needs to be made much more precise by performing a proper integral over the energy spectrum of the high energy particles and the π^0 decay spectrum shown in Fig. 20.1.

It is generally agreed that, if it is assumed that the local flux of high energy particles permeates the inner regions of the Galaxy, the γ-radiation at energies $h\nu \geq 100$ MeV can be accounted for by the π^0 decay process involving high energy interactions with the nuclei of the atoms of molecular hydrogen. There is, thus, every reason to regard the high energy γ-radiation as a tracer of the distribution of high energy protons and nuclei and of interstellar matter in the Galaxy. This is convincing evidence that high energy protons and nuclei fill up the region of the Galactic plane with a distribution similar to that of the electrons. It also leads to all sorts of exciting possibilities – the ability to observe high energy particles in distant parts of the Galaxy and how they diffuse away from their sources, the proton–electron ratio in supernova shells and so on.

Since we can account for the radio emission and the γ-ray emission of the Galaxy if we assume that roughly the local energy densities of high energy protons and electrons are present throughout the plane of the Galaxy, this suggests that the ratio of the energy densities of high energy protons to electrons throughout the Galactic plane is roughly the value observed in local interstellar space, that is, $\eta \sim 100$.

20.2 The origin of the light elements

We can now investigate the origin of the light elements in the cosmic rays. The elemental abundances in the cosmic rays were compared with the Solar System abundances in Tables 9.1 and 9.2 and in Fig. 9.4. There are three obvious differences which need to be to explained, in each case the cosmic rays having much greater elemental abundances as compared with the Solar System abundances. These are:

1 the light elements lithium, beryllium and boron;
2 the ratio of ^{3}He to ^{4}He;
3 the elements just lighter than iron.

We have already noted that light elements can be produced as fragmentation products of the interactions of high energy particles with cold matter, the process known as *spallation*. What we need in order to carry out these calculations is a table of fragmentation probabilities which describes all the ways in which heavy nuclei are split up when they make inelastic collisions with hydrogen or helium nuclei in the interstellar gas. The basic data needed for this analysis were discussed and presented in Section 5.2 and Tables 5.1(*a*) and (*b*). We will now use these data to analyse the three abundance anomalies listed above.

20.2.1 *The transfer equation for light nuclei*

We can modify the diffusion-loss equation we have already developed for the propagation of high energy electrons through the interstellar gas (equation (19.13)) for the case of the propagation of nuclei in the presence of fragmentation gains and loss of a particular species i:

$$\frac{\partial N_i}{\partial t} = D\nabla^2 N_i + \frac{\partial}{\partial E}[b(E)N_i] + Q_i - \frac{N_i}{\tau_i} + \sum_{j>i} \frac{P_{ji}}{\tau_j} N_j \qquad (20.1)$$

The first four terms of this equation are already familiar. N_i is the number density of nuclei of species i and is a function of energy, that is, we should really write $N_i(E)$; the term $D\nabla^2 N_i$ is the usual diffusion term; $\partial[b(E)N_i]/\partial E$ takes account of the effect of energy gains and losses upon the energy spectrum of the particles (see Sections 19.3.1 and 19.3.2); and Q_i is the rate of injection of particles of species i from sources per unit volume. The last two terms describe the effects of spallation gains and losses, where τ_i and τ_j are the spallation lifetimes of particles of species i and j. It is the spallation of all species with $j > i$ which give contributions to N_i, as indicated by the sum in the last term of equation (20.1). P_{ji} is the probability that, in an inelastic collision involving the destruction of the nucleus j, the species i is created. This is the form of equation which was presented by Ginzburg and Syrovatskii (1964), who also included a term to take account of the stochastic acceleration of the particles (see Section 21.3), but we will neglect that for our present analysis. The very brave can set about solving the full diffusion-loss equation, but that is a major undertaking. Instead, let us consider the simplest case and see how it has to be modified to obtain agreement with observation.

First of all, let us neglect diffusion and energy losses. We also note that the species we are interested in have very low cosmic abundances, and so let us suppose that there is no injection of these particles from the sources of high energy particles, $Q_i = 0$. It is traditional to write the transfer equation in terms of the number of kilogrammes per square metre traversed by the particle, just as was described in Section 2.4.2 for the case of ionisation losses. As in that case, we write the path length through the interstellar gas in the form $\xi = \rho x = \rho v t$, where v is the velocity of the particle and ρ is the gas density. We can therefore write the transfer equation for light nuclei in the form

$$\frac{\mathrm{d}N_i(\xi)}{\mathrm{d}\xi} = -\frac{N_i(\xi)}{\xi_i} + \sum_{j>i}\frac{P_{ji}}{\xi_j}N_j(\xi) \tag{20.2}$$

where ξ_i and ξ_j are now the mean free paths for inelastic collisions, that is, spallation, expressed in kilogrammes per square metre. This is a much more manageable transfer equation, which can be solved to determine the number of particles of species i after the sample has traversed ξ kg m^{-2} of the interstellar gas. Notice the essential assumption made in adopting this transfer equation: it is assumed that *all* the particles traverse the same amount of material between 0 and ξ, that is, there is a one-to-one relation between path length, ξ, and species produced. This model is often referred to as a *slab model* and is obviously an oversimplification, but we can improve our answer afterwards.

To follow the evolution of the chemical abundances of all species, we have to write down an equation like (20.2) for every isotope of every element in the periodic table, so that, even without taking account of diffusion and other complications such as ionisation losses and so on, there is a fairly complicated matrix to be inverted.

Let us display the essential physics by studying the evolution of the abundances of broad groups of particles. We tackle first the problem of the origin of the light elements, lithium, berylium and boron in the cosmic rays – we will refer to these as the *light*, or L, group of elements. From the table of abundances of the cosmic rays (Table 9.1), we note that carbon, nitrogen and oxygen are far more abundant than all the other heavy elements, and therefore it is sensible to investigate first the spallation of this group of elements, which we will refer to as the *medium*, or M, group of elements, since they are bound to be the progenitors of the light group of elements.

Initially, there are no particles in the L group at $\xi = 0$, and then we obtain two simple differential equations which describe how the abundances of the L and M groups of elements change with path length:

$$\frac{\mathrm{d}N_{\mathrm{M}}(\xi)}{\mathrm{d}\xi} = -\frac{N_{\mathrm{M}}(\xi)}{\xi_{\mathrm{M}}} \tag{20.3}$$

$$\frac{\mathrm{d}N_{\mathrm{L}}(\xi)}{\mathrm{d}\xi} = -\frac{N_{\mathrm{L}}(\xi)}{\xi_{\mathrm{L}}} + \frac{P_{\mathrm{ML}}}{\xi_{\mathrm{M}}}N_{\mathrm{M}}(\xi) \tag{20.4}$$

These can be easily solved to give the abundances of the L elements after they have passed through a path length ξ kg m^{-2}. Equation (20.3) can be immediately integrated to give

$$N_{\mathrm{M}}(\xi) = N_{\mathrm{M}}(0)\exp(-\xi/\xi_{\mathrm{M}})$$

Table 20.1. *Abundance-weighted probabilities of creating lithium, beryllium and boron by the spallation of carbon, nitrogen and oxygen nuclei*

	Abundance-weighted probability of formation (millibarns)	Measured abundance relative to [Si] = 100
Lithium	24	136
Beryllium	16.4	67
Boron	35	233

Multiplying equation (20.4) by an integrating factor $\exp(\xi/\xi_L)$, we obtain

$$\frac{d}{d\xi}[\exp(\xi/\xi_L)N_L(\xi)] = \frac{P_{ML}}{\xi_M}\exp\left(\frac{\xi}{\xi_L} - \frac{\xi}{\xi_M}\right)N_M(0)$$

The solution of this equation is

$$\frac{N_L(\xi)}{N_M(\xi)} = \frac{P_{ML}\xi_L}{(\xi_L - \xi_M)}\left[\exp\left(\frac{\xi}{\xi_M} - \frac{\xi}{\xi_L}\right) - 1\right] \tag{20.5}$$

In our simplified treatment, we now need suitable average values of ξ_L, ξ_M and P_{ML} for the L and M group of elements, and we can find these from Table 5.1. For illustrative purposes, we adopt a weighted average total inelastic cross-section for the M elements of 280 millibarns and a weighted average fragmentation probability P_{ML} of 0.28. We also need the average total inelastic cross-section for destruction of the L group of elements, which we take to be 200 millibarns. Notice that, following our prescription in Section 5.1, the total inelastic cross-sections are roughly proportional to the geometric cross-sections of the nuclei. For collisions with hydrogen nuclei, the corresponding values of the mean free paths are $\xi_M = 60$ kg m^{-2} and $\xi_L = 84$ kg m^{-2}.

The data presented in Table 9.1 indicate that the observed ratio of the abundances of the L to the M elements in the cosmic rays is $N_L(\xi)/N_M(\xi) = 0.25$. We can therefore insert these values into the expression (20.5) and find the typical path length through which the M elements would have to pass in order to create the observed abundance ratio of the L to M elements. We find that $\xi = 48$ kg m^{-2}, of the same order of magnitude as the mean free path of the M elements, which is hardly surprising.

There is another encouraging piece of evidence which we can derive from the table of partial cross-sections (Table 5.1) and the observed abundances of the light elements. We can work out the abundance-weighted cross-sections for the production of lithium, beryllium and boron from carbon, nitrogen and oxygen and compare these with the observed abundance ratios of these light elements. This comparison is shown in Table 20.1. It can be seen that the predicted relative probabilities of creating lithium, beryllium and boron are in roughly the same proportions as the observed abundances.

Exactly the same type of calculation can be performed for the production of ^{3}He from ^{4}He by spallation of ^{4}He in the interstellar gas. In this case the important interactions are

$$^4\text{He} + \text{p} \rightarrow {}^3\text{He} + \text{p} + \text{n}$$

$$^4\text{He} + \text{p} \rightarrow {}^3\text{H} + \text{p} + \text{p} \rightarrow {}^3\text{He} + \text{e}^- + \text{p} + \text{p}$$

There is a neat little nest of interactions involving H, D, T, ^{3}He and ^{4}He. This analysis also gives a best value of about 50 kg m^{-2}, suggesting that the helium nuclei have passed through roughly the same amount of material as the L group of elements (Shapiro (1991)).

This leaves only the spallation products of iron. We can see immediately that there is a problem because the inelastic cross-section for the iron nuclei, $\sigma_{\text{Fe}} = 764$ millibarns, is more than twice those of the M nuclei and therefore the iron must be much more severely depleted by spallation. This is particularly worrying because, even for the C, N, O elements, the amount of material traversed was of the order of one interaction mean free path. In this case, the ratio of spallation products to remaining iron should be

$$\frac{[\text{products}]}{[\text{primaries}]} = \frac{1 - \exp(-\xi/\xi_{\text{Fe}})}{\exp(-\xi/\xi_{\text{Fe}})} \tag{20.6}$$

Taking $\xi = 50$ kg m^{-2} and $\xi_{\text{Fe}} = 22$ kg m^{-2}, we find [products]/[primaries] $= 8.7$.

We can understand the implication of this result from the table of abundances of cosmic rays (Table 9.1). This table shows that many of the products are just lighter than the parent nucleus, and inspection of the table of partial spallation cross-sections (Table 5.1) suggests that about one-third of the total cross-section would result in nuclei such as Mn, Cr, V. We would therefore expect the abundances of these elements in total to be significantly greater than that of iron. This is not the case. Another way of demonstrating the discrepancy is to compare the ratio of abundances of the elements between chlorine and vanadium with that of iron, [Cl→V]/[Fe], since most of the products of spallation of iron fall in this range of atomic numbers. The observed value is 1.5, which is significantly smaller than 8.7. These discrepancies are confirmed by more detailed calculations. Notice that, although it is unusual to argue about factors of two in astrophysics, these ratios are all *relative* values, which depend upon physical parameters which are known with some precision.

Fortunately, there is an obvious way of resolving this problem. We have assumed that the high energy particles all pass through the same amount of matter in reaching the Earth, that is, we have not put in any of the physics to describe the fact that we actually observe high energy particles with a wide range of different path lengths. We can understand how to develop models for the distributions of path length from the diffusion-loss equation (20.1), from which we remove the source and energy loss terms:

$$\frac{\partial N}{\partial t} = D\nabla^2 N - \frac{N}{\tau_e} \tag{20.7}$$

where τ_e is now a characteristic escape time which describes how long the particles remain within the system. If the particles diffused from their sources

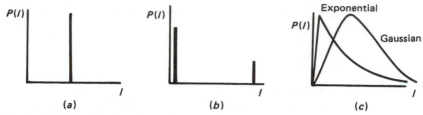

Figure 20.4. Models for the path length distribution for the propagation of high energy particles in the interstellar gas: (*a*) a 'slab' model, in which all high energy particles traverse the same path length of interstellar gas; (*b*) a model in which high energy particles arriving at the Earth traverse different path lengths of interstellar gas so that a wider range of spallation products can be explained; (*c*) a Gaussian distribution of path lengths expected in diffusion models and an exponential distribution with a low energy cut-off.

to the observer, that is, $\tau_e = \infty$ in equation (20.7), we would expect a *Gaussian distribution* of path lengths between the source and the observer. If, alternatively, we adopted a model in which the particles remain within the Galactic disc for some characteristic time, τ_e, and then escape from it, that is $D = 0$, we would obtain an *exponential* distribution of path lengths, since, if

$$\frac{\partial N}{\partial t} + \frac{N}{\tau_e} = 0, \quad \text{then} \quad N \propto \exp(-t/\tau_e) \quad \text{or} \quad N \propto \exp(-\xi/\xi_e) \tag{20.8}$$

in terms of path lengths. These correspond to different interpretations of the ways in which particles diffuse and escape from the Galaxy and to plausible physical pictures. For the moment, let us consider the problem from an empirical point of view; what distributions of mean free paths can result in the observed abundances of the cosmic rays?

The complete answer can only come from detailed studies using a full set of transfer equations, but we can give a simple illustration of how a distribution of path lengths helps. Suppose, for example, one-third of the matter traverses a path length of 100 kg m^{-2} and two-thirds traverses a negligible amount. Then, we may use equations (20.4) and (20.5) to work out the expected ratios of abundances of the L to M elements and the ratio of iron products to iron nuclei which survive. Evidently, the one-third which traverses 100 kg m^{-2} suffers a large amount of spallation. Inserting $\xi_M = 100$ kg m^{-2} into the expression (20.5), we find [L]/[M] = 0.6, whereas equation (20.6) indicates that essentially all the iron is converted into iron products. The other two-thirds survives unaltered. On adding together the products of these two different paths, we obtain

[L]/[M] = 0.25   [products of Fe]/[Fe] = 0.5

in good agreement with observation. What we have done is to modify the predictions of the slab model by considering a *distribution* of path lengths. Some models of the distribution of path lengths, including the above, are shown in Fig. 20.4.

Table 20.2   *Cosmic ray source abundances compared with the local Galactic abundances, both normalised to [Si] = 100*

Element	Cosmic ray source abundance (1990 update)	Local Galactic abundance
H	$8.9 \pm 2.2 \times 10^4$	$2.7 \pm 0.3 \times 10^6$
He	2.4×10^4	$2.6 \pm 0.7 \times 10^5$
C	431 ± 34	1260 ± 330
N	19 ± 9	225 ± 90
O	511 ± 20	2250 ± 560
F	< 2.5	0.09 ± 0.06
Ne	64 ± 8	325 ± 160
Na	6 ± 4	5.5 ± 1.0
Mg	106 ± 6	105 ± 3
Al	10 ± 4	8.4 ± 0.4
Si	100	100
P	< 2.5	0.9 ± 0.2
S	12.6 ± 2.0	43 ± 15
Cl	< 1.6	0.5 ± 0.3
Ar	1.8 ± 0.6	11 ± 5
K	< 1.9	0.3 ± 0.1
Ca	5.1 ± 0.9	6.2 ± 0.9
Sc	< 0.8	$3.5 \pm 0.5 \times 10^{-3}$
Ti	< 2.4	0.27 ± 0.04
V	< 1.1	0.026 ± 0.005
Cr	2.2 ± 0.6	1.3 ± 0.1
Mn	1.7 ± 1.7	0.8 ± 0.2
Fe	93 ± 6	88 ± 6
Co	0.32 ± 0.12	0.21 ± 0.03
Ni	5.1 ± 0.5	4.8 ± 0.6
Cu	0.06 ± 0.01	0.06 ± 0.03
Zn	0.07 ± 0.01	0.10 ± 0.02
Ga	$5.6 \pm 2.8 \times 10^{-3}$	$\sim 3.7 \times 10^{-3}$
Ge	$7.4 \pm 1.0 \times 10^{-3}$	$\sim 11.4 \times 10^{-3}$

From J.P. Wefel (1991). In *Cosmic rays, supernovae and the interstellar medium*, eds M.M. Shapiro, R. Silberberg and J.P. Wefel, p. 44. Dordrecht: Kluwer Academic Publishers.

There is now good agreement about the results of detailed calculations of the primary abundances of the high energy particles in their sources and about the types of path length distributions which can account for the overall chemical abundances of the cosmic rays arriving at the top of the atmosphere. Examples of the results of these computations are discussed by Simpson (1983) and Wefel (1991). Fig. 20.5, which is presented by Shapiro (1991), displays the typical results of these calculations. In the diagram, the inferred source isotopic abundances are indicated by the shaded histogram, and next to these are the abundances observed in the vicinity of the Earth, showing separately the amount of each isotope which is primordial (open histogram) and how much is produced by the spallation of heavier elements (black histogram). The source abundances and the cosmic abundances of the elements are compared in Table 20.2, which is taken from Wefel's review.

It is apparent from Fig. 20.5 that all the isotopes of the light elements in the cosmic rays, lithium, beryllium and boron, are secondary products, as are species such as ^{15}N, ^{17}O, ^{18}O, ^{19}F and ^{21}Ne. In addition, significant fractions of the observed abundances of the elements just lighter than iron are the products of the spallation of iron. On the other hand, substantial fractions of the common elements, carbon, oxygen, neon, magnesium and silicon, have survived unaffected by spallation between their sources and the Earth.

Inspection of the data in Table 20.2, which have been normalised to [Si] = 100, shows that there are significant differences between the source abundances of the cosmic rays and the local Galactic abundances. Wefel (1991) discusses the uncertainties in making this comparison, but the broad features of the table are clear. It is apparent that the elements up to about neon are underabundant as compared with the local Galactic abundances. This feature of the source abundances has been known for many years, and it is found that the underabundances are correlated with the atomic properties of the elements. In particular, if the species are ordered by their first ionisation potentials, there is a systematic trend in the underabundances (Fig. 20.6). It can be seen that the elements with first ionisation potentials greater than about 10 eV are underabundant by a factor of about five relative to the cosmic abundances. The hydrogen and helium points show an even greater discrepancy.

Exactly the same pattern of underabundances is found in the solar cosmic rays, as was discussed in Section 12.5.2. The standard interpretation of the correlation shown in Fig. 20.6 is that it provides information about the ionisation state of the region in which the particles were accelerated. Those elements with ionisation potentials less than about 10 eV would be ionised and the local Galactic composition is preserved after acceleration. For elements with higher ionisation potentials, there must be some suppression mechanism which reduces the element abundances by about a factor of five. Presumably, some other mechanism is required to account for the even lower abundances of hydrogen and helium, but they are rather special primordial elements, and it is perhaps not so surprising that they are anomalous. According to Wefel (1991), these data demonstrate that the cosmic ray source abundances can be derived from the local cosmic abundances of ordinary matter, provided the acceleration region is such that it allows selection by first ionisation potential or some related atomic property of the elements.

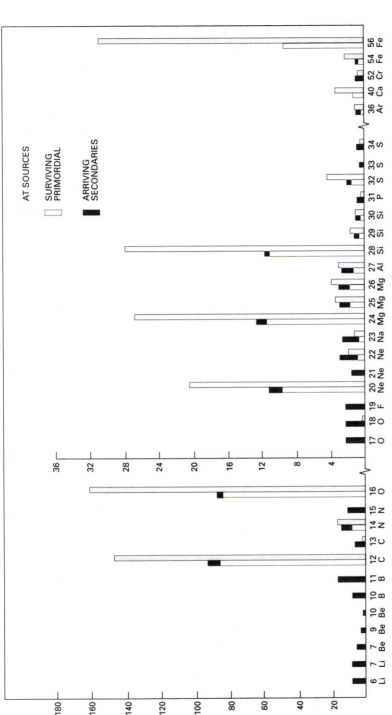

Figure 20.5. The relative isotopic abundances of the cosmic rays as observed near the Earth and as inferred to be present in their sources, once the effects of spallation between the sources and the Earth are taken into account. The abundances have been normalised to 100 for $^{12}C + ^{13}C$. The grey histograms show the inferred source abundances and the neighbouring histograms show the observed abundances, the open parts depicting the surviving primary elements and the black parts the amount produced by spallation. (From M. M. Shapiro (1991). In *Cosmic rays*, *supernovae and the interstellar medium*, eds M. M. Shapiro, R. Silberberg and J.P. Wefel, p. 14. Dordrecht: Kluwer Academic Publishers.)

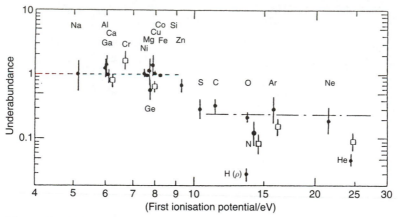

Figure 20.6. The ratio of the cosmic ray source abundances to the local Galactic abundances as a function of the first ionisation potential of the elements. Open squares show recent improved estimates of some of the abundance ratios. (After J.P. Wefel (1991). In *Cosmic rays, supernovae and the interstellar medium*, eds M.M. Shapiro, R. Silberberg and J.P. Wefel, p. 45. Dordrecht: Kluwer Academic Publishers.)

20.2.2   *Variations in the chemical composition of cosmic rays with energy*

As discussed in Section 9.2, we cannot leave this story here because the relative abundances of the cosmic rays vary with energy. In Fig. 9.2 we showed examples of these variations, and it was pointed out that some of these were associated with differences in the primary injection spectra and others with variations of the secondary to primary ratio as a function of energy. More recent data have been surveyed by Wefel (1991), who confirms the trends described in Section 9.2. Examples of some of these recent studies are shown in Fig. 20.7, which displays the variation with energy of the boron-to-carbon [B/C] and the chromium-to-iron [Cr/Fe] ratios.

The simplest interpretation of the energy dependence of the secondary to primary ratio is that the path length of the primary particles through the interstellar gas decreases with increasing energy. Let us assume that the energy dependence of the path length for escape from the Galaxy is of the form $\xi_e(E) = \xi_0(E/E_0)^{-\alpha}$, α being positive so that the path length of the particles for spallation decreases at high energies. We can construct a simple model to show how this idea works in practice. We begin by simplifying equation (20.4) for the number density of light elements, N_L, produced by spallation of the medium group of elements. We now need to include a term of the form $-N_L/\xi_e(E)$ on the right-hand side of equation (20.4), which represents the escape of particles from the system. Let us study the steady-state solution $dN_L/dt = 0$, which means that we assume that $N_M(\xi)$ reaches a steady value under the combination of spallation losses, escape from the system and injection from sources. We can therefore write the expression for the abundance of the light elements as

$$-\frac{N_L}{\xi_e(E)} + \frac{P_{ML}}{\xi_M}N_M(\xi) - \frac{N_L}{\xi_L} = 0$$

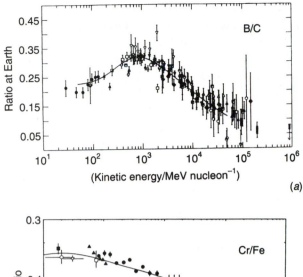

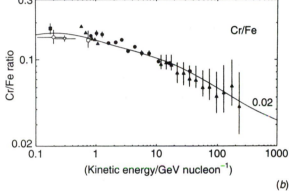

Figure 20.7. (*a*) The boron-to-carbon ratio as a function of energy. The curve shows the results of propagation calculations by Garcia-Munoz *et al.* (1987). (*b*) The chromium-to-iron ratios as a function of energy. The curve shows the results of full propagation calculations by Mewaldt and Webber (1990), which assumed that the source Cr/Fe ratio was 0.02. (Both diagrams from J.P. Wefel (1991). In *Cosmic rays, supernovae and the interstellar medium*, eds M.M. Shapiro, R. Silberberg and J.P. Wefel, p. 38. Dordrecht: Kluwer Academic Publishers.)

Rearranging this expression, we find

$$N_L = \frac{P_{ML} N_M(\xi)/\xi_M}{[1/\xi_e(E)] + (1/\xi_L)}$$

If we consider the high energy limit at which the escape path length is much less that the spallation path length, $\xi_E \ll \xi_L$, this expression reduces to

$$\frac{N_L(\xi)}{N_M(\xi)} = P_{ML} \frac{\xi_e(E)}{\xi_M}$$

Since P_{ML} and ξ_M are independent of energy, it can be seen that the energy dependence of the ratio of secondary to primary particles is directly related to the energy dependence of the escape path length, $\xi_e(E)$. From the analysis of Garcia-Munoz *et al.* (1987), dependence of the form $\xi_e(E) \propto E^{-0.6}$ can provide a good fit to the observations. This form of dependence is entirely consistent with the observations shown in Fig. 20.7 for the case of the boron-to-carbon ratio.

In the case of the chromium-to-iron ratio shown in Fig. 20.7(*b*), inspection of Fig. 20.6 shows that only part of the chromium is created by spallation and some of it is primordial. Therefore, because the escape path length decreases with energy, at a high enough energy there will only be a small contribution from spallation and the primordial chromium abundance should be observed. This is illustrated by the model calculations of Mewaldt and Webber (1990), which are shown in Fig. 20.7(*b*). These variations in the path length with energy provide information about the processes of diffusion of high energy particles in the interstellar medium, and we will discuss physical reasons for these variations in Section 20.4.

In addition to the variation of the secondary to primary ratios of the cosmic ray abundances with energy, there are differences in the energy spectra of what are inferred to be primary nuclei as well. Wefel (1991) gives an assessment of these differences. If the energy spectrum of the particles is taken to be of power-law form, $N(E)dE \propto E^{-x}dE$, the spectral index of high energy protons is $x = 2.73 \pm 0.09$ over the energy interval 100 to 10^5 MeV. In the high energy range, $E > 10^4$ MeV, the spectral index of helium nuclei is slightly steeper, $x = 2.87 \pm 0.13$. On the other hand, species such as carbon, oxygen and iron seem to have spectral indices closer to $x = 2.5$. Presumably, these spectral indices reflect differences in the processes of acceleration of the high energy particles.

An important clue to the sites of acceleration may be provided by isotopic anomalies in the cosmic rays. With the increasing mass resolution of cosmic ray telescopes (see Section 7.2), it is now possible to study isotopic as well as chemical abundances of the elements, and these prove to be much subtler probes of the sites of particle acceleration because the effects of chemical fractionation can be neglected. We will discuss the key significance of the isotopes of beryllium in the next section, but here we note only the most striking anomalies. The largest anomaly is associated with the ratio of the isotopes of neon. It is found that the isotopic ratio [^{22}Ne/^{20}Ne] is about four times its cosmic abundance. Smaller but significant isotopic excesses are also found for the abundance ratios [^{25,26}Mg/^{24}Mg] and [^{29,30}Si/^{28}Si]. These anomalies suggest that at least some of the cosmic rays have originated in neutron-rich environments. As an example of the type of information which can potentially be obtained from these studies, stars in which a similar isotope anomaly of neon is found are the *Wolf–Rayet stars* (see Section 14.5). These are massive stars which undergo substantial amounts of mass loss, resulting in abnormal chemical abundances in their atmospheres. It might be that these types of stars are prime sites for the ejection of particles into the regions in which acceleration of the cosmic rays takes place.

20.3   The confinement volume of cosmic rays and the Galactic halo

According to the calculations of Section 20.2.2, typically, high energy particles must traverse about 50 kg m^{-2} of matter. If they were to traverse more, the spallation reactions would be so drastic that all the primary species would be destroyed and incorrect relative abundances of cosmic ray nuclei would be produced. We can convert this path length into an escape time, τ_e, for the cosmic rays if we assume a value for the mean density, $\bar{\rho}$, of matter through which they have travelled, $\xi = \bar{\rho}c\tau$. A typical reference density for the interstellar gas is $N = 10^6$ m^{-3}, and so, assuming that the particles propagate at the speed of light, the upper limit to the time which the particles can remain within this gas density is $\tau_e \approx 3 \times 10^6$ years. If they also spend time in lower density regions, their total escape time can be longer without affecting the abundances of the cosmic rays.

These data provide direct evidence that the particles do not escape unimpeded from our Galaxy. If we take a typical dimension of, say, $1-10$ kpc for our Galaxy, then, if the high energy particles propagated freely at a velocity close to that of light, they would escape from the system in about 3×10^3 to 3×10^4 years, which is very much less than the time necessary to produce the spallation products. Therefore, the particles must travel by much more tortuous paths to escape from the Galaxy. It is for this reason that we refer to the *confinement volume* within which the cosmic rays propagate before they escape from our Galaxy. Why is this an important topic? Unlike all other information about high energy particles in cosmic environments, we have more clues about the dynamics of relativistic plasmas in our own Galaxy than in any other system. The ability to measure directly the chemical composition of the cosmic rays and their lifetimes within the local confinement volume, as well as the properties of the relativistic electrons, is unique to our Galaxy. It is therefore worth some effort to understand how much we can learn about how they propagate from their sources and the volume within which they are confined.

The high energy particles must either diffuse from their sources or be 'confined' within some volume before escaping into intergalactic space. The path length distribution for the production of the secondary elements is best described by an exponential function which is consistent with what is commonly known as the *leaky box model* of cosmic ray confinement. In this model, the high energy particles diffuse freely inside the confinement volume and they are reflected at its boundaries (Fig. 20.8). There is a certain probability that the particles escape from this volume at each encounter with the boundary, and this results in an exponential path length distribution:

$$\frac{\partial N}{\partial t} + \frac{N}{\tau_e} = 0$$

$$N \propto \exp(-t/\tau_e)$$

where τ_e is the characteristic escape time from the confinement volume. The question of interest is the size of the region within which the particles are confined before they escape into intergalactic space. It might be that the confinement

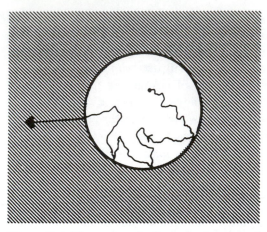

Figure 20.8. Illustrating the 'leaky box' model of the confinement of high energy particles.

volume is simply the disc of our Galaxy, which we can model by a flattened cylinder of radius about 10–15 kpc and thickness about 300–500 pc. Alternatively, the confinement volume may be taken to be the halo of our Galaxy. These models can be put on a much more exact quantitative basis by making assumptions about the diffusion coefficients in the disc and halo.

What is the evidence for a halo around our Galaxy? We have already mentioned in Section 18.4 that there is good evidence that there exists a radio halo about our Galaxy, although this remained a controversial subject for many years. We find Baldwin writing in his review of the problem in 1975: 'In the discussion so far, I have avoided the use of the phrase *radio halo*. It arouses antagonism in otherwise placid astronomers and many have sought to deny its existence.' The halo described in Section 18.4, which was inferred to be present around our Galaxy by Webster (1978), is consistent with all the existing data. It consists of a somewhat flattened ellipsoid with semi-major axis 10 kpc and semi-minor axis about 3–4 kpc perpendicular to the plane, similar to the haloes seen around other edge-on spiral galaxies such as NGC 891 (Fig. 18.18).

A second highly suggestive piece of evidence has come from ultraviolet observations of blue halo stars and B stars in the Magellanic clouds. The highly ionised lines of CIV and SiIV are observed as absorption troughs with velocity widths up to about 150 km s^{-1}. The gas responsible for these broad absorption troughs is associated with our own Galaxy. The lines of sight to these stars pass through the halo of our Galaxy. Savage and de Boer (1979) first interpreted these lines as originating in hot gas which forms a halo around our Galaxy. In Jenkins's survey of the use of these lines to probe the Galactic halo, he shows that the data are consistent with a somewhat flattened halo, similar in shape to that inferred by Webster for the radio halo (Jenkins (1987)). It is natural that any halo gas should be hot because, otherwise, it would be confined to the plane of the Galaxy. The temperatures and scale heights of the gas estimated from

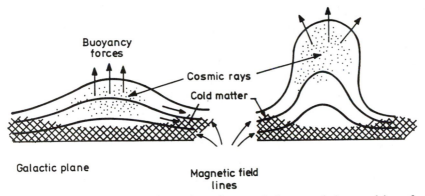

Figure 20.9. Illustrating the formation of magnetic loops and the expulsion of relativistic gas from the Galactic disc due to the instability described by Parker. These are often referred to as Parker's loops. (From E. N. Parker (1965). *Astrophys. J.*, **142**, 584.)

these observations are consistent with it being in hydrostatic equilibrium in the Galactic gravitational potential.

In addition, there are excellent theoretical reasons why there should be a halo of high energy particles around the Galaxy. First of all, Ginzburg has pointed out that there must be a high energy particle halo because, at the very least, high energy particles must be able to diffuse out of the Galactic plane, and this must produce an extended volume around the Galaxy containing a significant flux of high energy particles.

Secondly, in a classic analysis, Parker (1965) showed that the gaseous disc of our Galaxy is unstable to perturbations of the magnetic field lines which lie predominantly parallel to the Galactic plane. One of the key observations in these studies is the rough equality of the energy densities in high energy particles and in the Galactic magnetic field. The energy density of the Galactic magnetic field is $B^2/2\mu_0$, which, for $B = 3 \times 10^{-10}$ T, is 2×10^5 eV m^{-3}. The best estimate which we derived for the energy density in cosmic rays (Section 9.3) was about 10^6 eV m^{-3}. It can hardly be a coincidence that these values are of the same order of magnitude because of the close coupling between the relativistic gas and the interstellar magnetic field.

In Parker's analysis of the stability of the gaseous disc of the Galaxy, the unperturbed state consists of a magnetic field running parallel to the plane of the disc embedded in a cold gas, which is partially ionised, and the relativistic gas (see Fig. 20.9). The crucial point for understanding the stability of this disc is that the magnetic field is tied to the relativistic gas and to the cold gas. It is obvious why the magnetic field is coupled to the ionised component because of the considerations of flux freezing which we described in Section 10.5. However, the neutral component is also coupled by particle collisions and, thus, the magnetic field is also coupled to the cold gas, which contains all the inertia and which consequently 'holds down' the magnetic field.

Now, suppose there is a kink in the magnetic field lines which causes them to bulge out of the plane. Then, the cold matter, having a low temperature, falls down the potential gradient to the minimum value, whilst the relativistic gas 'inflates' the kink. In truth, what is happening is that the relativistic gas in the kink forms a 'bubble' of hot (in fact, relativistic) plasma, which rises buoyantly up the potential gradient and out of the Galactic plane. The natural tendency of a mixture of relativistic gas and magnetic field is to expand if unrestrained. This is the instability which Parker suggested is responsible for the formation of Galactic loops and, more important from our point of view, may be responsible for the formation of a halo of high energy particles.

The instability also suggests a physical reason for the local equality of the cosmic ray and the magnetic field energy densities. If $\epsilon_{CR} \ll \epsilon_B$, the pressure is dominated by the magnetic field, which remains firmly tied to the cold matter. If, however, the opposite inequality is true, $\epsilon_B \ll \epsilon_{CR}$, the local pressure is dominated by the relativistic gas, which forms a bubble, and the instability begins to grow, thus expelling the local excess pressure into the halo. Therefore, we expect that, in general, the high energy particle and magnetic field energy densities should be comparable. The dominant scale length for the instability is probably of the order of the scale height of the disc, that is, about 200–300 pc.

These theoretical arguments appear to me to be eminently plausible, and it seems inevitable that there should exist a high energy particle halo. Let us therefore review some of the evidence which bears upon the confinement volume for the high energy particles.

20.3.1   Cosmic ray clocks

One of the most elegant approaches to the question of the confinement timescale for the cosmic rays uses the fact that some of the species created in the spallation reactions are radioactive, and hence their abundances can be used to 'date' the samples of cosmic rays observed near the Earth. The most important of these is the radioactive isotope of beryllium, ^{10}Be, which has a characteristic lifetime $\tau_r = 3.9 \times 10^6$ years. If the particles are relativistic, the lifetime we measure is $\gamma\tau_r$, where γ is the Lorentz factor, $\gamma = (1 - v^2/c^2)^{-1/2}$. ^{10}Be is produced in significant quantities in the spallation of carbon and oxygen, the fraction of the total spallation cross-section for the production of ^{10}Be being about 10% of the total cross-section for the production of beryllium (see Table 5.1). The ^{10}Be then undergoes a β-decay into ^{10}B. Therefore, the relative abundances of the isotopes of Be and B can tell us whether or not all the ^{10}Be has decayed and consequently provide an estimate of the mean age of the cosmic rays observed at the Earth. Combining this age estimate with the path length of 50 kg m^{-2}, we can find the average particle density traversed by the cosmic rays during their journey to the Earth.

Formally, we have to introduce another term into the diffusion-loss equation (20.1) to describe the radioactive decay of species i with decay constant $1/\tau_r$, that is, a term of the form $-N_i(\xi)/\tau_r$. Evidently, if the confinement time of the high energy particles, or rather the typical time it takes them to reach the Earth from their sources, is less than τ_r, the ratio $[^{10}\text{Be}]/[^7\text{Be} + {}^9\text{Be} + {}^{10}\text{Be}]$ should correspond to the relative production rates of these species, which is about 10%. If the typical propagation time is much longer than τ_r, then this ratio should be very much less than 10%. Crudely, the isotopic abundance ratio should be of order τ_r/t of the production ratio if $t \gg \tau_r$, but, evidently, a much better calculation is needed which takes account of the destruction of ^{10}Be by spallation and of energy loss processes.

Let us give a simple derivation of the expected abundance of ^{10}Be in the 'leaky box' model, in which we characterise the Galaxy by a volume from which there is characteristic escape time, τ_e. We begin with the diffusion-loss equation (20.1) but make a number of modifications to it. First, to model the escape of particles from the Galaxy, we replace the diffusion term $D\nabla^2 N_i$ by $-N_i/\tau_e(i)$. Secondly, we assume that the system is in a steady state so that $\partial N_i/\partial t = 0$. Thirdly, we assume that all the beryllium isotopes are produced by spallation of the M elements. To simplify the notation, we write the production rate of the isotope i as

$$C_i = \sum_{j>i} \frac{P_{ij}}{\tau_j} N_j$$

With these simplifications, we can write the expression for the steady-state abundance of a non-radioactive isotope as

$$-\frac{N_i}{\tau_e(i)} + C_i - \frac{N_i}{\tau_{\text{spal}}(i)} = 0$$

$$N_i = \frac{C_i}{\left(1/\tau_e(i)\right) + \left(1/\tau_{\text{spal}}(i)\right)} \tag{20.9}$$

where $\tau_{\text{spal}}(i)$ is the timescale over which the isotope i is destroyed by inelastic collisions. If the isotope j is radioactive, we have to add another loss term to the transfer equation. We write the characteristic decay time of the radioactive nucleus as $\tau_r(j)$, and then we find

$$-\frac{N_j}{\tau_e(j)} + C_j - \frac{N_j}{\tau_{\text{spal}}(j)} - \frac{N_j}{\tau_r(j)} = 0$$

$$N_j = \frac{C_j}{\left(1/\tau_e(j)\right) + \left(1/\tau_r(j)\right) + \left(1/\tau_{\text{spal}}(j)\right)} \tag{20.10}$$

We can now write down the steady state ratio of the radioactive to non-radioactive species. Considering the specific case of the isotopes ^{10}Be and ^{7}Be, we expect

$$\frac{N(^{10}\text{Be})}{N(^7\text{Be})} = \frac{(1/\tau_e(^7\text{Be})) + (1/\tau_{\text{spal}}(^7\text{Be}))}{(1/\tau_e(^{10}\text{Be})) + (1/\tau_r(^{10}\text{Be})) + (1/\tau_{\text{spal}}(^{10}\text{Be}))} \frac{C(^{10}\text{Be})}{C(^7\text{Be})} \tag{20.11}$$

If we suppose that the timescale for the destruction of the beryllium isotopes is greater than their escape times, we obtain a simple expression for the escape of the particles in terms of the observed abundance ratio and the inferred source abundance ratio since we know the half-life of the ^{10}Be isotope:

$$\frac{N(^{10}\text{Be})}{N(^7\text{Be})} = \frac{(1/(\tau_e(^7\text{Be}))}{(1/(\tau_e(^{10}\text{Be})) + (1/(\tau_r(^{10}\text{Be}))} \frac{C(^{10}\text{Be})}{C(^7\text{Be})} \tag{20.12}$$

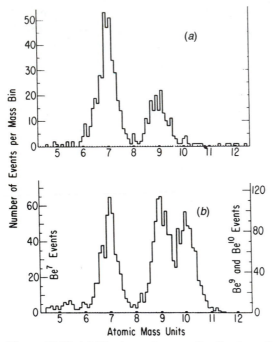

Figure 20.10. (*a*) The isotopic mass distribution of the isotopes of beryllium as observed by the cosmic ray telescopes on board IMP-7 and IMP-8. (*b*) Calibration of the expected distribution of beryllium isotopes in laboratory experiments showing the ability of the telescopes to distinguish the isotopes ^{9}Be and ^{10}Be. (From M. Garcia-Munoz, G.M. Mason and J.A. Simpson (1977). *Astrophys. J.*, **217**, 859.)

Fig. 20.10 shows the results of an experiment carried out by the Chicago group, in which it can be seen that the isotopes of beryllium have been clearly separated. According to Simpson and his colleagues (see Simpson (1983)), the ratio $[^{10}\text{Be}]/[^7\text{Be} +^9\text{Be} +^{10}\text{Be}] = 0.028$. If we insert this value into expression (20.12), we find an escape time of 10^7 years. When we combine this result with the fact that the path length through the interstellar gas for these particles amounts to about 50 kg m^{-2}, we can find the average interstellar density through which the particles have travelled. From these rough estimates, we find that the average interstellar gas density is about 3×10^5 m^{-3}.

In a much better calculation, we would use the full expression (20.11) including the destruction terms for the beryllium isotopes. Since this term depends upon the path length through the interstellar gas in kilogrammes per square metre, that is, $\zeta(\text{Be}) = \rho v \tau_{\text{spal}}(\text{Be})$, it is simplest to present the results in terms of the inferred particle number density. This procedure is shown in Fig. 20.11, in which the theoretical curves take account of ionisation energy losses as well as the loss processes described by the expression (20.9). Several determinations of the ratio of the beryllium isotopes in the cosmic rays are shown, all of them corresponding

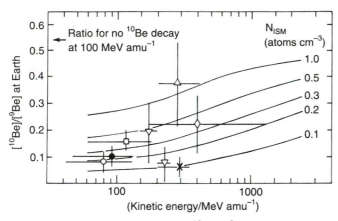

Figure 20.11. Comparison of the $[^{10}\mathrm{Be}]/[^9\mathrm{Be}]$ ratio measured at the Earth with propagation calculations assuming different densities of the interstellar gas. (From J. Simpson (1983). *Ann. Rev. Nucl. Particle Sci.*, **33**, 359.)

to values of the mean particle density of about 2×10^5 m^{-3} and to escape times τ_e close to 10^7 years.

How is one to interpret this result in terms of the propagation of particles in the disc and halo? First, we may say that, if the mean density of matter in the disc is 10^6 m^{-3}, then the particles must have spent most of their lifetime outside the disc, presumably forming a rather flattened halo about the high density disc. Secondly, can we dispose of models in which the confinement time might be much greater than 10^7 years? It might look as if we could, but in fact we cannot. If the sources of the cosmic rays were homogeneously distributed throughout the halo, then it would be true that we could exclude this model. This is not the case, however. The Solar System is located in a highly preferred position in the Galaxy in that it lies in the Galactic plane, where the sources of high energy particles are found. Thus, although a high energy particle might take, say, 10^8 years to diffuse from its source to the edge of the halo, this is not necessarily the typical time it takes a high energy particle to propagate from its source to the Solar System, because we are sitting preferentially close to the sources. Thus, there is not necessarily any inconsistency in assuming that high energy particles fill up a large spherical halo. All the ^{10}Be clock tells us is that the typical time it takes the high energy particles to get from their sources to the Earth is about 10^7 years.

All sorts of other complications could be introduced into the picture. Is the diffusion constant the same in the disc and halo? Are the sources inhomogeneously distributed in the disc? Are we really seeing a typical sample of Galactic high energy particles, or are we located by chance in a region which was recently engulfed by a supernova remnant?

The result is that we have learned something significant about the mean time that the high energy particles have taken to travel from their sources to the Earth but not a great deal about the high energy particle halo of our Galaxy. Since

Table 20.3. *Half-lives of some important radioactive isotopes created in spallation reactions*

Isotope	Half-life
^{26}Al	7.4×10^5 years
^{36}Cl	3.1×10^5 years
^{53}Mn	3.7×10^6 years
^{54}Mn	303 days
^{59}Ni	8.0×10^4 years

this test was first proposed, several other suitable radioactive spallation products have been suggested as possible clocks. Some of these are listed in Table 20.3 with their corresponding half-lives.

20.3.2   *The isotropy of cosmic rays*

If the high energy particles streamed freely out of the Galaxy, then their distribution on the sky would be highly anisotropic, most of the flux coming from the general direction of the central regions of the Galaxy. This would be in conflict with the very high degree of isotropy of cosmic rays with energies 10^{11}–10^{14} eV (see Fig. 9.7). Alternatively, we might expect some anisotropy in the arrival directions of cosmic rays if they streamed along the local spiral arm. Again, we can exclude this possibility.

The obvious interpretation of these observations is that the particles diffuse in much the same way as we discussed for the case of high energy particles in the interplanetary medium. We must assume that the high energy particles are effectively scattered, either by irregularities in the magnetic field, or, as we will show in a moment, by waves excited by the high energy particles themselves (Section 20.4).

Let us therefore assume for the moment that we can use an isotropic diffusion model as a possible description of the dynamics of particles in the disc of the Galaxy. In this case, the particles must be able to diffuse a distance roughly the half-thickness of the Galactic disc, about 300 pc, in about 10^7 years. Then, if τ_e is the typical escape time,

$$D\nabla^2 N - \frac{N}{\tau_e} = 0$$

If L is the scale of the system, then, to order of magnitude, the diffusion coefficient, D, can be found by writing

$$\frac{DN}{L^2} \sim \frac{N}{\tau_e} \quad ; \quad D \approx L^2/\tau_e = 3 \times 10^{23} \text{ m}^2 \text{ s}^{-1}$$

The diffusion coefficient, D, is related to the mean free path of the particles by $D = \frac{1}{3}\lambda v$, and hence we find $\lambda \approx 0.1$ pc. Authors such as Ginzburg and Syrovatskii (1964) interpret this mean free path as the typical scale of inhomogeneities in the interstellar medium, and, indeed, there is abundant evidence that there exist irregularities on this scale, associated with supernova shells, regions of ionised hydrogen and so on. Notice that this corresponds to a very much longer mean free path than was inferred to be appropriate for the propagation of high energy particles through the interplanetary medium. One has to suppose that, on fine scales, the interstellar medium is much more quiescent than the interplanetary medium or else the high energy particles would not be able to travel very far at all.

We can now work out how *isotropic* we would expect the flux of cosmic rays to be from the net streaming velocity, V, of the high energy particles due to diffusion:

$$D\frac{dN}{dx} = VN \quad ; \quad V \approx \frac{D}{L} = 10^{-4}c$$

Therefore, if we were sitting at the edge of the Galactic disc, we would expect a net streaming velocity of about $10^{-4}c$, which would be reflected in an anisotropy of the cosmic ray flux of the order of 10^{-4}. The exact relation between a measured cosmic ray anisotropy, $\delta = (I_{max} - I_{min})/(I_{max} + I_{min})$, and the streaming velocity of the particles is in fact $v = [\delta/(p+2)]c$, where p is the spectral index of the differential energy spectrum of the cosmic rays. This effect is due to the small Doppler shifts of the particle spectrum in the approaching and receding directions, which means that particles of energy E are observed with slightly different energies. This is known as the *Compton–Getting effect*. Taking the observed anisotropy to be $\delta = 6 \times 10^{-4}$ and $p = 2.5$, we find $v \approx 10^{-4}c$, consistent with the above analysis.

There are several reasons, however, why we cannot assume that the quoted upper limits to the anisotropy are immediately applicable to this version of the story. First, we are not sitting at the edge of the Galactic disc, where we would expect the flux of high energy particles to be the greatest. We are located quite close to the central plane of the Galaxy, and it is likely that the gradient of particle density is much smaller than it is towards the edge of the disc. Secondly, the diffusion model results in a Gaussian distribution of path lengths for the high energy particles, and we have argued above that the abundances of the elements can best be accounted for by an exponential distribution of path lengths, as is found in the leaky box model. According to Wdowczyk and Wolfendale (1989), however, the amplitudes and phases of the anisotropies seen in the cosmic rays with energies $E = 10^{11}$–10^{14} eV are the same, suggesting a general drift of the cosmic rays out of the Galaxy.

20.4 The scattering of high energy particles by Alfvén and hydromagnetic waves

The question we want to address is the following. Suppose a uniform magnetic field is embedded in a partially ionised plasma and a flux of high energy particles propagates along the magnetic field direction with its normal spiral motion, but at a high streaming velocity. What is the interaction between the stream of high energy particles and the magnetoactive plasma?

The results of these investigations are as follows: if the plasma is fully ionised, the high energy particles resonate with irregularities in the magnetic field and are scattered in pitch angle, exactly as described in Section 11.4 for the propagation of high energy particles through the Solar Wind. In the present case, we are interested in scattering by Alfvén and hydromagnetic waves which grow in amplitude under the influence of the streaming motions so that, even if there were no irregularities to begin with, they will be generated by the streaming of the high energy particles.

The full theory is somewhat complicated, and we make no attempt to do it justice here. Wentzel (1974) and Cesarsky (1980) provide excellent reviews of these areas of plasma physics and summarise the essential results. The following simplified arguments indicate the issues of interest.

The first part of the argument follows exactly that given in Section 11.4.2. If the perturbation has field strength B_1, pitch angle scattering results in changing the pitch angle of the particles by about 90° after a mean free path of about $\lambda_{SC} \approx r_g/\phi^2$, where $\phi = B_1/B_0$, and the corresponding diffusion coefficient is $D \approx v\lambda_{SC}$. This is clearly a mechanism which converts streaming motion into a random distribution of pitch angles after distance λ_{SC}.

The complications arise because (i) the waves with which the particles of the beam resonate are Alfvén and hydromagnetic waves, which are the characteristic low-frequency 'sound' waves found in a magnetised plasma – the circularly polarised hydromagnetic waves are particularly important because they can resonate with the spiral motion of the particles; and (ii) the strength of the perturbed component, B_1, is due to the streaming of the particles themselves. Physically, it is clear why this comes about. The forward momentum of the beam is transferred to the waves, which must grow as a result.

Let us show how the growth rate of the instability can be derived from the simple physical picture described above. We aim to find the *growth rate*, Γ, of the instability, which is defined by the equation for the exponential increase in the energy density, U, of the Alfvén waves, $U = U_0 \exp \Gamma t$. First, we convert the expression for the mean free path of the high energy proton into a timescale, τ_s, for scattering through 90°,

$$\tau_s = \frac{\lambda_{SC}}{v} = \left(\frac{r_g}{v}\right)\left(\frac{B_0}{B_1}\right)^2 \tag{20.13}$$

The energy density of the Alfvén or hydromagnetic waves is just the energy density in the perturbing magnetic field, B_1, $U_A = B_1^2/2\mu_0$, and the Alfvén speed is $v_A = B_0/(\mu_0\rho)^{-1/2}$, where $\rho = N_p m_p$ is the mass density of the fully ionised plasma. Making these substitutions, we find

$$\tau_s \approx \left(\frac{r_g}{v}\right)\left(\frac{v_A^2 N_p m_p}{U_A}\right) \tag{20.14}$$

We are seeking the rate of momentum transfer into the waves, and it is therefore simplest to work in terms of their *momentum density*. For all types of wave motion, the momentum density is given by $P_{wave} = U_{wave}/v$, where P_{wave} and U_{wave} are the energy and momentum densities, respectively, and v is the speed of the wave.

In the present case, the speed of the waves is the Alfvén speed, and so we find

$$\frac{\mathrm{d}P_{\mathrm{wave}}}{\mathrm{d}t} = \frac{\mathrm{d}}{\mathrm{d}t}\left(\frac{U_{\mathrm{wave}}}{v_A}\right) \tag{20.15}$$

This has to be equal to the rate at which momentum is lost from the streaming relativistic particles. We note that the momentum supplied to unit volume over the timescale τ_s is $EN(E)v/c^2$, where the E is the energy of the protons which are resonant with the wavelength of the Alfvén waves, that is, $r_g(E) \sim \lambda_A$, and $N(E)$ is their number density. Therefore, the equation for the growth rate of the momentum of the waves becomes

$$\frac{1}{v_A}\frac{\mathrm{d}U_{\mathrm{wave}}}{\mathrm{d}t} = \frac{EN(E)v}{\tau_s c^2} \tag{20.16}$$

Substituting for τ_s, we find

$$\frac{\mathrm{d}U_{\mathrm{wave}}}{\mathrm{d}t} = \frac{EN(E)v^2}{r_g v_A N_p m_p c^2}U_{\mathrm{wave}} \tag{20.17}$$

This simple analysis shows how the amplitude of the waves grows as a result of the scattering of the high energy particles by the magnetic fields in the waves. From the equation (20.17), we immediately find the growth rate of the waves:

$$\Gamma = \frac{EN(E)v^2}{r_g v_A N_p m_p c^2} \tag{20.18}$$

Now, we write the gyroradius of the protons in terms of the particle's velocity, v, and total energy, E, $r_g = (Ev/eBc^2)$, and so

$$\Gamma = \frac{eB}{m_p}\frac{N(E)}{N_p}\left(\frac{v}{v_A}\right) \tag{20.19}$$

But $\Omega_0 = (eB/m_p)$ is the non-relativistic angular gyrofrequency of the proton, and so we obtain the answer we have been seeking:

$$\Gamma = \Omega_0\frac{N(E)}{N_p}\left(\frac{v}{v_A}\right) \tag{20.20}$$

This result is of exactly the same form as that quoted by Cesarsky (1980) for the typical growth rate of the instability:

$$\Gamma(k) = \Omega_0\frac{N(\geq E)}{N_p}\left(-1 + \frac{|v|}{v_A}\right) \tag{20.21}$$

$N(\geq E)$ means all those particles with energies greater than or equal to that energy E which resonates with the wave. The result is that the instability develops until the streaming velocity of the high energy particles is reduced to the Alfvén velocity, $v_A = B_0/(\mu_0\rho)^{1/2}$.

If we take the density of the ionised component of the interstellar gas to be $N = 10^5$ m^{-3} and the magnetic field strength to be $B_0 = 3 \times 10^{-10}$ T, we find $v_A = 2 \times 10^4$ m s^{-1}. It therefore appears as if this mechanism provides an attractive solution to the problem of the isotropy of the cosmic rays, namely, that they generate Alfvén and hydromagnetic waves if they stream faster than the Alfvén velocity and so their net streaming velocity is $v \leq v_A \approx 10^{-4}c$.

The story is not as simple as this, however, because we have not taken into

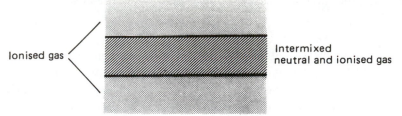

Figure 20.12. Illustrating the distribution of diffuse neutral and ionised gas in the plane of the Galaxy. It is possible to account for the isotropy of the cosmic rays in the modified model because the particles stream freely inside the volume of the disc.

account the damping of the Alfvén waves. The instability is only effective if the waves produced by it are not damped before they have time to grow to significant amplitude. The presence of neutral particles in the interstellar plasma can readily abstract energy from the Alfvén waves by neutral-ion collisions in a time that is short compared with the growth time. The significance of the neutral particles is that they provide a mechanism for removing kinetic energy from the waves, whereas ionised particles are simply constrained to oscillate with the waves. The damping rate for the waves is given by Kulsrud and Pierce (1969) for temperatures of 10^3 and 10^4 K, and they are $\Gamma^* = \Gamma_0 N_H = (3.3$ and $8.4) \times 10^{-9} N_H$ s^{-1}, where N_H is the number density of neutral hydrogen atoms.

At low cosmic ray energies, $E \sim 3$ GeV, the high energy particles are sufficiently numerous that the growth rate of the instability exceeds the damping rate, and so the streaming velocities are restricted to the Alfvén velocity in the neutral intercloud medium. It can be noted, however, that the typical timescale for the growth of the instability is a strong function of energy since it depends upon the number of high energy particles which can resonate with the waves, $N(\geq E)$. The timescale $\tau = \Gamma^{-1}$ for the growth of the instability varies as $E^{1.7}$, and so the high energy particles are not scattered as strongly as those of low energy. It is intriguing that the streaming instabilities automatically lead to an escape time from the Galaxy which is energy dependent, as is inferred must be the case from the energy dependence of the spallation products of common nuclei (see Section 20.2.2). According to Cesarsky, particles with energies $E > 100$ GeV should not be confined to the Galaxy.

The result is that, in regions where there is a large density of neutral material, the particles can diffuse rapidly, but, as soon as they encounter fully ionised plasmas, their streaming velocity drops to the Alfvén speed. This has suggested a model which provides a physical realisation for the leaky box picture of the propagation of high energy particles in the Galactic disc. Within the Galactic disc, the plasma consists of a mixture of ionised and neutral material, whereas outside the disc the gas is ionised (Fig. 20.12). The particles can stream freely inside the disc, but outside they are limited to the Alfvén velocity. Therefore, by continuity at the interface, we find that the rate of loss of particles from the disc of the Galaxy must be

$$N_{\text{int}}c = N_{\text{ext}}v_A$$

Thus, if the net outward streaming velocity from the disc is only v_A, the typical residence time inside the disc is $(L/c) \times (c/v_A) \approx 10^7$ years. This figure is similar to the figures derived above from the ^{10}Be cosmic ray clock. Skilling (1971) has shown that, adopting the observed values for the distribution of ionised and neutral matter in the disc, the mean free path of the particles should be proportional to $E^{-0.4}$, in reasonable agreement with the requirements of the observations.

Another aspect of this question which could be addressed observationally is how we expect the high energy particles to diffuse away from their sources. The flux of high energy particles produced in a supernova is liberated into the interstellar medium after about 10^5 years. Because of these streaming instabilities, we would expect the high energy particles to be confined to a 'bubble' which expands along the local magnetic field direction. These bubbles might be observable as diffuse γ-ray sources, because of the π^0 decay mechanism, at the sites of extinct supernovae. There is also the possibility that we might be located within one of these bubbles and thus obtain a biased view of what the typical properties of Galactic high energy particles really are.

We have only been able to give the briefest description of an important area of the physics of high energy particles under cosmic conditions. Cesarsky (1980) provides an excellent survey of the many variations upon this theme. We will use some of the ideas developed in this section when we discuss the acceleration of charged particles in the vicinity of strong shocks, in Chapter 21.

20.5   The origin of cosmic rays

We have implicitly assumed that the cosmic rays observed in the vicinity of the Earth originate within our Galaxy from sources such as supernovae. In the first edition of this book, I devoted considerable space to discussing whether or not we can rule out the hypothesis that the cosmic rays which we observe locally could, in fact, be universal, or fill up very much larger volumes of space than our own Galaxy. I attempted to survey all the evidence so that some of the many options could be excluded. I used the occasion to illustrate the diversity of approaches taken by different astrophysicists to astronomical problems, and I repeat some of these arguments, despite the fact that some of the issues appear to me to be somewhat dated as I write in 1993.

20.5.1   *The psychology of astronomers and astrophysicists*

The question of the confinement volume of cosmic rays is one which generated a great deal of strong feeling among certain astrophysicists, and it is worthwhile drawing a few morals about the nature of the astrophysical sciences from this debate. Let me first quote from an interchange in the literature between two eminent astrophysicists. First of all, V.L. Ginzburg (1970), one of the authors of the classic text *The origin of cosmic rays*, states:

> The author has already written several papers and reviews on the origin of cosmic rays and for quite some time now he has felt dissatisfied with having to discuss many times the same problems, even taking into account some new facts. ... In general within the limits of evolutionary cosmology and with the cosmological interpretation of the redshift for quasars, it seems that extragalactic models must be considered as very improbable. At the same time, steady-state cosmology and the local model for quasars are at present especially improbable. So there is every reason to ignore the extragalactic models.

In response, G.R. Burbidge and K. Brecher (1971) write:

> Objections to the idea that cosmic rays with high energy densities pervade much larger volumes of space than that occupied by our own Galaxy have been based on the claim that...

They then list four of the topics which we will discuss, and end:

> (5) The universal cosmic ray hypotheses are, to quote Ginzburg, 'discussed together with other unorthodox theories such as steady-state cosmology and the local theory of quasars (perhaps because the same misguided scientists are involved in each case). Since Ginzburg states that he is 'strongly opposed to these latter theories', it necessarily follows that they must clearly be incorrect and hence, using the technique of guilt by association, it follows that the universal cosmic ray theory is also incorrect!

I have intentionally selected the most polemical parts of the interchange to highlight the basic differences in outlook.

Part of the problem stems from the fact that astrophysics is an observational science dealing with phenomena on a scale which vastly exceeds laboratory scales. There might be some new laws of physics which only become significant on a very large scale, and we would have no way of knowing about them except through astronomical observations. This is part of the background against which Burbidge would defend his stance on topics such as the local theory of quasars and steady-state cosmology. Burbidge and other astronomers, such as H. Arp, have regularly sought out astronomical observations which might cast doubt upon the conventional structure of physics. The problem lies in deciding what constitutes convincing evidence that there really is something wrong with physics as we know it. Ginzburg adopts the opposite point of view that, within certain well-defined physical limitations, the laws of physics as we know them can be trusted.

A second important point about controversies of this type is what I have heard called *Redman's theorem*. The late Professor Redman was Professor of Astronomy at Cambridge University. He is reputed to have stated that 'Any competent theoretician can fit any given theory to any given set of facts.' Obviously, this was a jest at the expense of some of his theoretical colleagues, but it contains an important element of truth. A clever theoretician can very often invent or appeal to processes which would be difficult to detect observationally but which would render a critical test of a theory inconclusive. To put this comment another way, there are many theorists for whom the statement 'It is impossible that...' is an immediate challenge to invent ways of circumventing the conjecture. This is a very profitable way of generating new ideas and concepts, and many exciting pieces of astrophysics have originated in this way. There is, thus, a value judgement which has to be made about when a particular argument may be reasonably discarded as unreasonable.

Before we begin the discussion, let me lay my cards on the table and confess that I belong to the class of astrophysicists, like Ginzburg, who believe that, within certain well-defined physical limitations, the laws of physics as we know them can be trusted. I have not been convinced by the evidence produced by a number of astronomers, including Burbidge, that the laws of physics are at present inadequate to explain most of the properties of the objects we observe in the Universe. Very few astrophysicists give credance nowadays to the local theory of quasars or to steady-state cosmology, since conventional astrophysical and cosmological theories turn out to be remarkably successful in accounting for the basic physical requirements of quasars and the hot Big Bang model of the Universe. This does not mean that I do not have a healthy respect for the approach taken by Burbidge and his like-minded colleagues. The observational evidence must be looked at as objectively as possible, and it is exactly this approach of inquiry into our preconceived notions which might in the end produce the most important evidence of the need for new physics.

With this lengthy introduction, let us study whether or not we can exclude the hypothesis that the local flux of cosmic rays fills all space.

20.5.2   *Evidence on the confinement volume for cosmic rays*

We can rephrase the question in terms of the *confinement volume* for the cosmic rays, which we introduced in Section 20.3, in the sense that we assume that the particles are more or less uniformly distributed within this volume. Going up progressively in scale, this volume could be the disc of our Galaxy, the Galactic halo, the local supercluster or, in the extreme extragalactic hypothesis, the entire Universe.

High energy electrons Let us begin with a strong argument. The cosmic microwave background radiation permeates all space, and therefore all relativistic electrons are subject to inverse Compton losses by scattering of these low energy photons. Using the formula we derived in Section 19.2, we know that the lifetime of an electron against these losses is

$$\tau = \frac{2.3 \times 10^{12}}{\gamma} \text{ years}$$

where $\gamma = (1 - v^2/c^2)^{-1/2}$ is the Lorentz factor of the electron. Therefore, we can ask how far the electrons we observe near the Earth could have travelled in this time. For example, taking $E = 100$ GeV, we find a distance of 3.6 Mpc, if the electrons travel in straight lines. In fact, just like the cosmic ray protons and nuclei, the electrons must diffuse from their sources to the Earth and therefore their sources must be very nearby. We conclude that the electrons could not fill up a volume, say, the size of the local supercluster and must have a Galactic origin.

The chemical composition of cosmic rays and the mean path lengths of 50 kg m^{-2}
We have shown that, to obtain the observed chemical abundance of the cosmic rays through a combination of intrinsic (or primordial) source composition and spallation in the interstellar medium, the cosmic rays should pass through 50 kg m^{-2} of matter (Sections 20.2.2 and 20.3).

One possibility is that spallation in the *intergalactic medium* could significantly modify the chemical composition. Except within rich clusters of galaxies, there is little direct evidence for intergalactic gas. All the methods of investigating the interstellar gas which we described in Chapter 17 can be adapted for studying the intergalactic gas in the space between clusters, and all have given negative results. So, let us suppose, for argument's sake, that the intergalactic gas has the critical cosmological density $\rho_{crit} \approx 5 \times 10^{-27}$ kg m^{-3} (see Section 13.1.3) and that it is at such a temperature that we would not be able to detect it by any of the techniques described in Chapter 17. Then, to attain a path length of 50 kg m^{-2}, the distance travelled by the cosmic rays would be about 3×10^5 Mpc, which would take about 10^{12} years, exceeding the age of the Universe by a large factor. Therefore, there is no danger of exceeding the spallation limit of 50 kg m^{-2}, even if the cosmic rays travelled through a high density intergalactic gas. We conclude that there is no evidence against the extragalactic hypothesis because of excessive spallation in the intergalactic medium for the last 10^{10} years.

Spectra of cosmic ray sources Radio observations of supernova remnants and powerful extragalactic radio sources indicate that they are strong sources of high energy electrons. It turns out that the radio spectral indices of supernova remnants and extragalactic radio sources are remarkably similar, $\alpha \approx 0.75$. Since the radio emission is identified as the synchrotron radiation of these ultrarelativistic electrons, the corresponding electron energy spectra in the source regions must be

$$N(E) \propto E^{-2.5}$$

(see Section 18.1.5). These spectra are similar to those of the cosmic ray protons and nuclei. Thus, it seems that both Galactic and extragalactic sources could produce the observed energy spectra of cosmic ray protons and electrons, and these observations provide no evidence one way or the other on the question of their origin.

Cosmic ray clocks From the observations of the isotopes of beryllium described in Section 20.3, it was concluded that the local flux of cosmic rays must have originated within the last 10^7 years, suggesting that a considerable fraction of the cosmic rays which we observe must have originated within our Galaxy. This would seem to be difficult to reconcile with the extragalactic hypothesis.

This result still does not exclude the possibility, however, that a major fraction of the cosmic rays which we observe locally are indeed of extragalactic origin. To take an extreme example, suppose that only 25% of the local cosmic rays are generated in supernovae in our Galaxy and the remainder are extragalactic. Then, if the propagation time of the local cosmic rays is less than 10^6 years, the ^{10}Be which we observe locally would be produced with roughly its production ratio. This would be diluted by the extragalactic component, which would contain very little ^{10}Be, resulting in the abundances observed in Fig. 20.10.

A common argument against such a model is that it would be a remarkable coincidence if the Galactic and extragalactic components were to turn out to be of roughly the same magnitude at the Solar System despite the fact that their origins are totally different. It could also be objected to on the grounds that it is an inelegant solution. One may have some sympathy with this line of reasoning, but it must be recognised that they are not physical arguments.

γ-rays from π^0 decays The same process which is responsible for the γ-ray emission from the Galactic plane, that is, the decay of neutral pions, π^0, created in collisions between cosmic rays and interstellar matter, should also give rise to an intergalactic flux of γ-rays if the cosmic rays permeate all space. The observed γ-ray background radiation can thus be used to set limits to the universal flux of cosmic rays, if the mean intergalactic gas density is known. Let us perform a simple calculation similar to that used in working out the γ-ray luminosity of the Galaxy in Section 20.2, which illustrates the essential point.

The probability that a cosmic ray proton undergoes an inelastic collision with nuclei of the intergalactic gas (assumed to be hydrogen in this calculation) over a cosmological timescale is

$$P_{\text{coll}} = \sigma_{\text{pp}}\Omega_{\text{igg}}N_0 l$$

σ_{pp} is the proton–proton inelastic cross-section, which is about 2.5×10^{-30} m^2, $\Omega_{\text{igg}} = N/N_0$ is the density of the intergalactic gas, measured in units of the critical density, $N_{\text{crit}} \approx 3$ m^{-3}, and $l = c/H_0 = 2 \times 10^{26}$ m $= 6000$ Mpc is the cosmological distance scale. Substituting these numbers, we find $P_{\text{coll}} = 1.5 \times 10^{-3}\,\Omega_{\text{igg}}$. In an inelastic collision, roughly one-third of the pions are neutral pions which decay into two γ-rays with mean energy ≈ 180 MeV. Therefore, if ϵ_{ex} is the energy density of intergalactic cosmic rays, the expected energy density of γ-ray photons, ϵ_γ, is

$$\epsilon_\gamma = \frac{1}{3}P_{\text{coll}}\epsilon_{\text{ex}}$$

From the SAS-II observations, the upper limit to ϵ_γ in the relevant energy range is $\epsilon_\gamma \leq 10$ eV m^{-3}, and hence we find

$$\epsilon_{\text{ex}} \leq 3\epsilon_\gamma/P_{\text{coll}} = 2 \times 10^4\,\Omega_{\text{igg}}^{-1} \text{ eV m}^{-3}$$

Thus, if the intergalactic gas density is high, $\Omega_{\text{igg}} \sim 1$, $\epsilon_{\text{ex}} \leq 4 \times 10^4$ eV m^{-3}, which is an order of magnitude less than the local energy density of cosmic rays in the Galaxy. There is no evidence, however, that there is any intergalactic gas in the space between clusters of galaxies. If, for example, the density parameter, Ω_{igg}, were 0.1, there would be no contradiction with the postulate that the energy density of cosmic rays is everywhere the same as at the Earth. Indeed, at the present moment there is no direct observational reason to believe there is any intergalactic gas at all in the space between clusters, i.e. $\Omega_{igg} \ll 0.1$, and consequently this argument cannot be used to rule out the cosmological hypothesis.

There are, however, interesting variants on this theme. According to the universal hypothesis, there are about 10^6 eV m^{-3} of cosmic ray energy everywhere in the Universe. Consequently, if we look in directions in which we know that there is diffuse gas, we can evaluate the predicted γ-ray flux. Ginzburg (1970) has evaluated this flux for the gas present in our nearest neighbours in space, the Magellanic Clouds. He predicted a γ-ray flux at γ-ray energies $\hbar\omega > 100$ MeV of 3×10^{-3} photons m^{-2} s^{-1}. Fichtel *et al.* (1991) predicted a similar γ-ray flux density of 2.3×10^{-3} photons m^{-2}s^{-1}. This experiment has now been carried out by the Gamma-Ray Observatory (Sreekumar *et al.* (1993)). The Large Magellanic Cloud has been detected in γ-rays, and the flux density at energies greater than 100 MeV is $(1.9 \pm 0.4) \times 10^{-3}$ photons m^{-2}s^{-1}, remarkably close to the predicted value. The inference is that the energy density of cosmic rays in the Large Magellanic Cloud is similar to that in our own Galaxy. As the authors point out, this is the first time that the flux of high energy protons and nuclei has been measured in a galaxy other than our own. As a test of the Galactic and extragalactic hypotheses of the origin of cosmic rays, the results are inconclusive, being consistent with both hypotheses. At the time of writing (August, 1993), the test has been repeated for the Small Magellanic Cloud. The expected flux was 2.4×10^{-3} photons m^{-2}s^{-1}, and an upper limit of 0.5×10^{-3} photons m^{-2}s^{-1} has been obtained. Thus, it is probable that the bulk of the cosmic rays are of Galactic origin.

According to the extragalactic hypothesis, the cosmic ray flux should be constant throughout the Galaxy and its environs. The γ-ray observations of the Galaxy by the COS-B satellite can be used to determine the spatial distribution of the high energy electrons and protons throughout the Galaxy. As shown in Fig. 20.13, at low energies, $E < 100$ MeV, the γ-rays are primarily due to the inverse Compton scattering of starlight and the bremsstrahlung of high energy electrons. At high energies, $E > 300$ MeV, the γ-rays are primarily due to the decay of neutral pions created in collisions between the proton and nuclear components of the high energy particles and the cold interstellar gas. Wolfendale (1991) has analysed these data in order to derive the radial dependence of both the electron and the proton and nuclear components shown in Fig. 20.3. It can be seen that, according to his analysis, the inferred gradients of the flux of both the electron and the proton and nuclear components indicate a decrease in these fluxes with increasing distance from the centre of the Galaxy, suggesting that the cosmic rays originate within the Galaxy.

The isotropy of cosmic rays If the cosmic rays are universal, there is no problem in explaining why their distribution is so isotropic. The question is whether or not the observations are consistent with the Galactic hypothesis. Let us treat separately the low, high and very high energy cosmic rays.

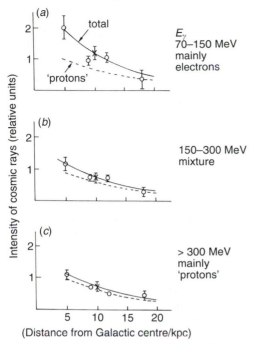

Figure 20.13. The gradient of the cosmic ray flux from analyses of the COS-B maps of the Galactic γ-ray emission. (*a*) At low energies, the γ-rays are mainly associated with the electron component of the cosmic rays and are due to a combination of inverse Compton scattering and bremsstrahlung (see Fig. 20.2). At high energies, (*c*), the γ-rays are mostly associated with the proton and nuclear component, and are due to the decay of neutral pions created in collisions between these particles and the cold interstellar gas. At intermediate energies, (*b*), the observed γ-ray gradient is due to the combination of both the electron and the proton and nuclear components. (From A.A. Wolfendale (1991). In *Cosmic rays, supernovae and the interstellar medium*, eds M.M. Shapiro, R. Silberberg and J.P. Wefel, p. 71. Dordrecht: Kluwer Academic Publishers.)

Low energies, $10^{11} \leq E \leq 10^{14}$ eV. Observations of the distribution of cosmic rays in this energy range show a small anisotropy at the level $\Delta I / I \approx 6 \times 10^{-4}$. Measurements of the direction of maximum intensity show that it is the same throughout this energy range and is of the same order of magnitude expected from the streaming arguments of Section 20.3. This anisotropy is consistent with a general drift of the cosmic rays out of the Galaxy. Therefore, these data cannot be used as evidence against the Galactic hypothesis, and indeed are consistent with outflow of cosmic rays from sources within the Galaxy.

High energies, $10^{14} \leq E \leq 10^{19}$ eV. In this energy range, the anisotropy of the cosmic ray flux increases, consistent with a picture in which the particles escape more easily from the Galaxy with increasing energy. Diffusion arguments involving the bulk streaming of the particles are no longer relevant because they are too rare to give rise to Alfvén waves. We are more concerned with the dynamics of individual particles and in particular with how much the particle's

Table 20.4   *The radii of curvature, r_g, of cosmic ray protons in a magnetic field of strength 3×10^{-10} T*

Rigidity, R (V)	Radius of curvature of a proton's trajectory
10^{15}	0.36 pc
10^{17}	36 pc
10^{19}	3.6 kpc
10^{21}	360 kpc

trajectory is influenced by a magnetic field. We showed in Section 11.1 that the gyroradius, r_g, of the particle's orbit is given by

$$r_g = \frac{\gamma m_0 v}{ze} \frac{\sin \theta}{B} = \left(\frac{pc}{ze}\right) \frac{\sin \theta}{Bc} = \frac{R \sin \theta}{Bc} \tag{11.6}$$

where R, the rigidity, is pc/ze and θ is the pitch angle of the particle's trajectory with respect to the magnetic field direction. For relativistic energies, the rigidity of a proton in volts is the same as its energy in electron-volts. Representative values of r_g for a Galactic magnetic field strength of 3×10^{-10} T are shown in Table 20.4. We recall that the thickness of the Galactic disc is about 300 pc and the radius of the halo is about 3–10 kpc. Thus, in disc models, the gyroradii of protons of energy 10^{18} eV are equal to the thickness of the disc.

There are several points to be made about these numbers. Fig. 9.14 shows that the overall spectrum of cosmic rays steepens at about 10^{15} eV. Ginzburg and Syrovatskii noted many years ago (1964) that the gyroradius of a proton of energy 10^{15} eV is roughly equal to the scale of the irregularities, which they inferred from the diffusion argument was responsible for the scattering of the cosmic rays, $\lambda \sim 0.1$ pc (see Section 20.3.2). It is natural to assume that, for energies higher than this, the diffusion approximation is no longer appropriate and that particles can escape more freely from the Galaxy.

This argument has been given rough quantitative form by Hillas (1984), who showed that the observed cosmic ray energy spectrum in the energy range $10^{14} < E < 10^{19}$ eV is consistent with an power-law injection energy spectrum of the form $Q(E) \propto E^{-2.47}$, and with the assumption that the anisotropy is due to the fact that the particles have shorter escape times from the Galaxy at higher energies. In a simple picture, the anisotropy is expected to be inversely proportional to the time, $T(E)$, during which the particles remain within the Galaxy. The expected anisotropy can therefore be compared with the deviations of the observed energy spectrum from the injection spectrum, since we expect $N(E) \propto T(E)Q(E)$. This comparison is illustrated in Fig. 9.7, in which the solid line shows the predicted anisotropy, A, as a function of energy derived from $T(E)$, $A^{-1} \propto T(E) \propto N(E)/Q(E) \propto E^{2.47}N(E)$.

Another way of presenting the evidence on the origin of these particles has

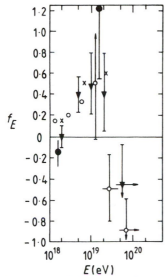

Figure 20.14. The dependence of the enhancement factor, f_E, upon the energy of the cosmic rays. An isotropic distribution of cosmic rays corresponds to $f_E = 0$. If the particles are concentrated towards the Galactic plane, $f_E > 0$. Values of f_E less than zero imply that the particles avoid the Galactic plane and are found preferentially at high galactic latitudes. (From A.A. Wolfendale and J. Wdowczyk (1991). In *Cosmic rays, supernovae and the interstellar medium*, eds M.M. Shapiro, R. Silberberg and J.P. Wefel, p. 313. Dordrecht: Kluwer Academic Publishers.)

been presented by Wdowczyk and Wolfendale (1989) in which they decompose the dependence of the cosmic ray flux upon galactic latitude into an isotropic and Galactic component. They write the intensity distribution as a function of galactic latitude as $I(b) = I_0[(1 - f_E) + f_E \exp(-b^2)]$. If f_E is zero, the distribution is isotropic; if f_E is unity, the distribution is strongly associated with the Galaxy. Fig. 20.14 shows the variation of f_E with increasing energy. It can be seen that, even up to energies $E \sim 10^{19}$ eV, there is a good case that these particles are associated with the Galaxy. The sum of these arguments leads to the conclusion that the particles in the energy range $10^{14} < E < 10^{19}$ eV are likely to be of Galactic origin.

Highest energies, $E > 10^{19}$ eV. At the very highest energies, the picture seems to change. From the figures in Table 20.4, it can be seen that protons with rigidity 10^{19} V have gyroradii ten times greater than the half-thickness of the disc. It is, therefore, impossible to confine them to the plane of the Galaxy. It is also unlikely that the particles can even be confined to the halo because the magnetic field is about an order of magnitude smaller than that in the plane. In the Galactic magnetic field, the highest energy cosmic rays are barely deflected, and consequently they should travel more or less directly from their sources to the Earth.

Fig. 20.14 shows that, at the very highest energies, the particles seem to be anticorrelated with the direction of the Galactic plane. Wolfendale (1991) interprets this observation in terms of an extragalactic component of cosmic rays, probably originating within the local supercluster of galaxies, the centre of which is located at high galactic latitudes. The local supercluster contains a number of powerful extragalactic radio sources, among which M87, the massive elliptical galaxy with the famous optical jet, is known to be a source of very high energy electrons. In his interpretation of the overall spectrum of cosmic rays, Wolfendale (1991) proposes that this extragalactic component only begins to make its presence known at the very highest energies, while, at lower energies, the extragalactic component is swamped by the Galactic component.

The anisotropy of the highest energy cosmic rays is, in my view, the strongest argument for an extragalactic component of cosmic rays.

Photo-pion and photo-pair production by the highest energy cosmic rays There are two further limits which can be set to the origin of the very highest energy cosmic rays. These have such large Lorentz factors that photons of the cosmic microwave background radiation have very high energies in the rest frame of the cosmic ray, so high in fact that photo-pion and photo-pair production can take place, which degrade the energy of the cosmic ray.

If a proton is bombarded with high energy γ-rays, pions are created, the threshold for this process being $\epsilon_t = 200$ MeV. The reactions are

$$\gamma + p \rightarrow n + \pi^+$$

$$\gamma + p \rightarrow p + \pi^0 \rightarrow p + \gamma + \gamma$$

$$\gamma + p \rightarrow p + n\pi$$

The cross-section for this process is about 250 microbarns. The cosmic microwave background permeates all space, and therefore the cosmic rays cannot escape from it. The energy of the photons is, on average, $\epsilon_0 = 6 \times 10^{-4}$ eV ($\nu = 1.5 \times 10^{11}$ Hz), and therefore, in the rest frame of the cosmic ray, they have energy

$$\epsilon = \epsilon_0 \gamma \left(1 + \frac{v}{c} \cos \theta \right) \approx \gamma \epsilon_0$$

Therefore, the threshold for pion production corresponds to an energy $E = \gamma m_{\rm p} c^2$ for protons, where $\gamma = \epsilon_t/\epsilon_0$, that is, $\gamma = 3 \times 10^{11}$, or $E = 3 \times 10^{20}$ eV. We should perform the integration properly over the Planck spectrum of the cosmic microwave background radiation and over all angles. We would then find that the threshold for the process decreases to 5×10^{19} eV, just within the energy range of cosmic rays which have been observed in extensive air-showers. The mean free path for a single scattering is then

$$\lambda = (\sigma_{\pi p} N_{\rm photon})^{-1}$$

Taking $N_{\rm photon} = 5 \times 10^8$ m^{-3} for the cosmic microwave background radiation and $\sigma_{\pi p} = 2.5 \times 10^{-32}$ m^{-2}, we find $\lambda \approx 10^{23}$ m, corresponding to a propagation length of 3 Mpc or a propagation time of 10^7 years. The energy of the pion created in this process is $\gamma m_\pi c^2$, and therefore the fractional loss of energy of a cosmic ray proton is $\Delta E/E \approx m_\pi/m_p \approx \frac{1}{10}$. Therefore, it is clear that the total mean free path for the cosmic ray proton to lose all its energy corresponds to a propagation time of 10^8 years. If cosmic rays of this energy permeated all space and had been present for

10^{10} years, there should be a cut-off in the cosmic ray energy spectrum at about 5×10^{19} eV, and this does not seem to have been observed. This observation suggests that the highest energy cosmic rays cannot have come from further than 30 Mpc. An origin in the Virgo supercluster would still be entirely acceptable.

One should add that this result depends upon the value of γ of the cosmic rays, and so, if it turned out that the highest energy cosmic rays were iron nuclei rather than protons, we would not expect there to be a cut-off at the highest energies. This possibility is discussed by Wdowczyk and Wolfendale (1989), who conclude, on the basis of studies of the fluctuations in the numbers of muons and electrons produced in extensive air-showers, that the cosmic rays with energies up to about 10^{17} eV are probably protons. Thus, by continuity, it might be argued that those with energies 10^{20} eV are also protons, but there is no direct evidence for this.

Exactly the same calculation can be performed for the photo-pair production process. We noted in Section 4.4 that the cross-section for this process in the ultrarelativistic limit was $\sigma_{pair} = 10^{-30}\,\mathrm{m}^2$ (expression (4.56)). Although this cross-section is larger than that for the production of pions, the pair produced takes away much less energy than in the photo-pion process. In fact, the ratio of the rest masses of the electron and the pion is $m_e/m_\pi = 1/280$. Consequently, this process is less important for cosmic rays of 3×10^{19} eV by a factor $(m_\pi/m_e)\times(\sigma_{\pi p}/\sigma_{pair}) \approx 6$.

Observations of the very highest energy cosmic rays are thus of great interest. The problem is that one of these primary cosmic rays is only observed once every two or three years and very large arrays are needed to observe them.

The energetics of sources of cosmic rays We showed in Section 19.5.1 that supernova in our Galaxy could account for the local energy density of cosmic rays, without making any extreme assumptions about the energy release or about the supernova rate.

Let us treat, first of all, the extreme extragalactic model, in which it is assumed that protons and nuclei are present throughout the Universe. We assume that each source of cosmic rays produces a total cosmic ray energy E_{CR} over a cosmological timescale ~ 1–2×10^{10} years and that N_0 is the space density of these sources. Then, assuming there are no losses, the local universal energy density of cosmic rays is

$$\epsilon_{CR} = E_{CR} N_0$$

where we require $\epsilon_{CR} = 10^6$ eV m^{-3}. Let us give two examples to show what this would mean. For strong radio galaxies, $N_0 \approx 10^{-5}$ Mpc^{-3} and hence $E_{CR} \approx 5\times10^{59}$ J; for normal galaxies, $N_0 \approx 10^{-2}$ Mpc^{-3} and hence $E_{CR} \approx 5\times10^{56}$ J. These are very large energies indeed. A galaxy of mass $M = 10^{11}M_\odot$ has available $Mc^2 = 2\times10^{58}$ J of rest mass energy. These figures look too extreme – for example, the radio galaxies would need to be about 100 times more massive than normal galaxies. This may not be too unreasonable because evidence has been accumulating to suggest that radio galaxies, which are among the most massive galaxies known, may well have masses up to $10^{13}M_\odot$. Even being generous, however, it is evident that one would have to discover an acceleration mechanism capable of 10% efficiency in converting rest mass energy into cosmic ray energy.

One way in which Ginzburg has expressed the energy requirements of the extragalactic hypothesis is to compare the average energy density of cosmic rays with the average rest mass energy density of matter in the Universe. For galaxies, the amount of visible matter corresponds to only about $\Omega_{vis} \sim 0.02$, where Ω_{vis} is a density parameter which is the ratio of the density of visible matter in the Universe to the critical cosmological density (Section 13.1.3). We find

$$\frac{\epsilon_{CR}}{\Omega_{vis}\rho_{crit}c^2} \approx 0.02$$

Ginzburg states that it is quite inconceivable that the acceleration mechanism could be so efficient as to convert 2% of the rest mass energy of the visible matter in the Universe into high energy particles. This is possibly a dangerous approach. The history of astronomy and physics is full of examples in which this type of argument leads to misleading results. In particular, the pulsar in the Crab Nebula is currently converting the gravitational energy of collapse of the neutron star into relativistic particle energy, with an efficiency exceeding 1%. Therefore, there *do* exist means for converting rest mass energy into cosmic ray energy with high efficiency. Black-hole models of galactic nuclei have become part of the conventional theoretical apparatus for studying galactic nuclei, and energy production corresponding to up to 42% seems feasible (Section 15.6.3). Having said this, the efficiencies still look very large to me, but perhaps I am just being conservative.

A less extreme hypothesis would be to adopt a more modest extragalactic confinement volume, say, on the scale of clusters or superclusters of galaxies. If superclusters occupy only one part in 10^2 or 10^3 of the total volume of space, the average energy released by each radio galaxy would be reduced by this factor to $\sim 10^{57}$ J over the lifetime of the Universe. How far this figure can be reduced depends upon what we think the smallest reasonable volume is which could contain the cosmic rays and a sufficient number of active galaxies. As an extreme example, suppose that cosmic rays were confined to the Local Group of galaxies. How much energy would our Galaxy and M31, the brightest members, have had to produce in the last 10^{10} years? Let us take the Local Group to be a sphere of radius 0.4 Mpc. Then we have to fill that volume (4×10^{66} m^3) with 10^6 eV m^{-3}, and therefore 10^{54} J of high energy particles have to be produced in the last 10^{10} years. Although these energies are still large, they are within the bounds of credibility.

Another question we should ask is how much energy we would need if we were to supply only the highest energy cosmic rays from extragalactic sources, say, only those of energy 10^{15} eV and greater. Because of the shape of the spectrum, most of the energy is in the lowest energy particles. For those with energies greater than 10^{15} eV, we require roughly 10^{-3}–10^{-4} of the energy in the low energy cosmic rays. It seems quite feasible that this energy could originate in extragalactic sources because the energy problem is relieved by a large factor. For example, it would be very difficult to argue, on energy grounds, that the highest energy cosmic rays could not originate within the local supercluster. The energy density of the cosmic rays with energies $E \geq 10^{19}$ eV is 1 eV m^{-3}, only about 10^{-6} of the total energy density of cosmic rays. It is, therefore, quite feasible that the highest energy cosmic rays are indeed extragalactic, and they could come from radio galaxies, alive or extinct within the local supercluster.

20.6   Summary

The discussion of Section 20.5 has both astrophysical and pedagogical significance. On the pedagogical side, it always comes as something of a surprise to find how difficult it sometimes is to exclude certain astrophysical hypotheses. In the end, much depends upon how extreme one is prepared to be in what one considers plausible astrophysics.

In the present instance, the most straightforward interpretation of the data is that the bulk of the cosmic rays are of Galactic origin but that the very highest energy particles, with $E \geq 10^{19}$ eV, may have an origin outside our Galaxy and probably within the local supercluster. The number of unambiguous arguments is, however, much smaller than one might imagine.

21

The acceleration of high energy particles

21.1   The problem

We have left to the last chapter of this volume one of the most intriguing problems in high energy astrophysics – the mechanisms by which high energy particles are accelerated to ultrarelativistic energies. In these first two volumes, sites where particles are accelerated include solar flares, the boundary of the Earth's magnetosphere, pulsar magnetospheres, supernovae and supernova remnants. In volume 3, we will find evidence for particle acceleration in active galactic nuclei and in extended radio sources. It is appropriate to consider the problem of the acceleration of charged particles at this point because a number of important features of the cosmic rays are common to the energy spectra of particles in other astrophysical environments.

The specific features of particle acceleration which we have to account for are as follows.

1 A power-law energy spectrum for particles of all types. The energy spectrum of cosmic rays and the electron energy spectrum of many non-thermal sources have the form

$$\mathrm{d}N(E) \propto E^{-x}\mathrm{d}E$$

where the exponent x lies in the range roughly 2.2–3. For the cosmic rays, $x = 2.5$–2.7 at energies ~ 1–10^3 GeV (Section 9.1), with slightly flatter spectra for primary nuclei such as iron. The typical spectra of radio sources correspond to electron spectra with $x \approx 2.6$ with a scatter of about 0.4 about this mean value. The continuum spectra of quasars in the optical and X-ray wavebands correspond to $x \sim 3$.

2 The acceleration of cosmic rays to energies $E \sim 10^{20}$ eV.

3 The acceleration mechanism should result in chemical abundances for the cosmic rays which are similar to the cosmic abundances of the elements.

It would be helpful if we could appeal to the physics of laboratory plasmas for some clues, but the evidence is somewhat contradictory. On the one hand, if we want to accelerate particles to very high energies, we have to go to a great deal of trouble to ensure that the particles remain within the region of the accelerating field, for example, in machines such as betatrons, synchrotrons, cyclotrons and so on. By very careful design, we obtain controllable beams of very high energy particles. It is not plausible that Nature goes to all this trouble to accelerate high energy particles – there must be a simpler mechanism. On the other hand, as soon as we try to build a machine to store high-temperature plasmas, the configurations are usually grossly unstable and, in the instability, particles are accelerated to suprathermal energies.

We will survey some of the mechanisms which have been discussed in the literature, with particular emphasis upon first order Fermi acceleration in strong shocks, which possesses many attractive features for accounting for some of the above features.

21.2 General principles of acceleration

The acceleration mechanisms may be classified as *dynamic, hydrodynamic* and *electromagnetic*. In many cases, there is no firm distinction between them because, being charged, the particles are closely coupled to magnetic field lines. In some models, the acceleration is purely dynamical; for example, in those cases in which acceleration takes place through the collision of particles with clouds. Hydrodynamic models can involve the acceleration of whole layers of plasma to high velocities. The electromagnetic processes include those in which particles are accelerated by electric fields, for example, in neutral sheets, in electromagnetic or plasma waves and in the magnetospheres of neutron stars.

The general expression for the acceleration of a charged particle in electric and magnetic fields is

$$\frac{\mathrm{d}}{\mathrm{d}t}(\gamma m \mathbf{v}) = e(\mathbf{E} + \mathbf{v} \times \mathbf{B}) \tag{21.1}$$

In most astrophysical environments, static electric fields cannot be maintained because of the very high electrical conductivity of ionised gases – any electric field is instantly short-circuited by the motion of free charges. Therefore, the acceleration mechanism can only be associated either with non-stationary electric fields, for example electromagnetic waves of very high energy density, or with time-varying magnetic fields. In a static magnetic field, no work is done on the particle, but, if the magnetic field is time-varying, work can be done by the induced electric field, that is, the electric field given by Maxwell's equation, curl $\mathbf{E} = -\partial \mathbf{B}/\partial t$. It has been suggested that phenomena such as the betatron effect might be applicable in some astrophysical environments. For example, the collapse of a cloud of ionised gas with a frozen-in magnetic field could lead to the acceleration of charged particles since they conserve their adiabatic invariants in a time-varying magnetic field (expressions (11.12), Section 11.2). There is no particularly obvious way in which this mechanism could play a role in the regions where we know that particles are accelerated to high energies, for example, in the shells of supernova remnants.

One interesting variant of acceleration by induced electric fields occurs in neutral sheets. The physics of neutral sheets in the context of solar flares was discussed in some detail in Section 12.4, and their possible relevance to particle acceleration in these regions was discussed in Section 12.6. It seems beyond question that there are indeed strong induced electric fields in these regions, but their efficacy in accelerating particles to high energies is not clear. The discussion of section 12.6 outlines the uncertainties involved.

In recent years, most effort has been devoted to the study of particle acceleration in strong shock waves, and we look at these mechanisms in some detail in the next two sections.

21.3   Fermi acceleration – first version

The Fermi mechanism was first proposed by Fermi in 1949 as a stochastic means by which particles colliding with clouds in the interstellar medium could be accelerated to high energies. We will consider two versions of the mechanism. In this section, we consider Fermi's original version of the theory, the problems it encounters and how it can be reincarnated in a modern guise. The analysis contains some features which are important for particle acceleration in general.

In Fermi's original picture, charged particles are reflected from 'magnetic mirrors' associated with irregularities in the Galactic magnetic field. The mirrors are assumed to move randomly with typical velocity V, and Fermi showed that the particles gain energy statistically in these reflections. If the particles only remain within the acceleration region for some characteristic time τ_{esc}, a power-law distribution of particle energies is found.

Let us repeat Fermi's original calculation, in which the collision between the particle and a mirror, or massive cloud, takes place such that the angle between the initial direction of the particle and the normal to the surface of the mirror is θ, as illustrated in Fig. 21.1(a). Let us work out the change of energy of the particle in a single collision. It is important to carry out a proper relativistic analysis.

We suppose the cloud is infinitely massive so that its velocity is unchanged in the collision. The centre of momentum frame is therefore that of the cloud moving at velocity V. The energy of the particle in this frame is

$$E' = \gamma_V(E + V p \cos\theta) \tag{21.2}$$

where

$$\gamma_V = \left(1 - \frac{V^2}{c^2}\right)^{-1/2}$$

The x component of the relativistic three-momentum in the centre of momentum frame is

$$p'_x = p' \cos\theta' = \gamma_V \left(p \cos\theta + \frac{VE}{c^2}\right) \tag{21.3}$$

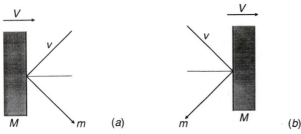

Figure 21.1. Illustrating the collision between a particle of mass m and a cloud of mass M. (a) A head-on collision; (b) a following collision. The probabilities of head-on and following collisions are proportional to the relative velocities of approach of the particle and the cloud, namely, $v + V\cos\theta$ for (a) and $v - V\cos\theta$ for (b). Since $v \approx c$, the probabilities are proportional to $1 + (V/c)\cos\theta$, where $0 < \theta < \pi$.

In the collision, the particle's energy is conserved, $E'_{\text{before}} = E'_{\text{after}}$, and its momentum in the x direction is reversed, $p'_x \rightarrow -p'_x$. Therefore, transforming back to the observer's frame, we find

$$E'' = \gamma_V(E' + Vp'_x) \tag{21.4}$$

Substituting equations (21.2) and (21.3) into equation (21.4) and recalling that $p_x/E = v\cos\theta/c^2$, we can find the change in energy of the particle

$$E'' = \gamma_V^2 E \left[1 + \frac{2Vv\cos\theta}{c^2} + \left(\frac{V}{c}\right)^2 \right] \tag{21.5}$$

Expanding to second order in V/c, we find

$$E'' - E = \Delta E = \frac{2Vv\cos\theta}{c^2} + 2\left(\frac{V}{c}\right)^2 \tag{21.6}$$

We now have to average over the angle θ. Because of scattering by hydromagnetic waves or irregularities in the magnetic field, it is likely that the particle is randomly scattered in pitch angle between encounters with the clouds, and we can therefore work out the mean increase in energy by averaging over the angle θ in the expression (21.6). A crucial point is that there is a slightly greater probability of head-on encounters as opposed to the following collisions (Fig. 21.1). It will be observed that the probability of encounters taking place at an angle of incidence θ is given by exactly the same reasoning which led to rate of arrival of photons at an angle θ in our analysis of inverse Compton scattering (see expression (4.28)). The only difference is that the particles move at a velocity v rather than c. For simplicity, let us consider the case of a relativistic particle with $v \approx c$, in which case the probability of a collision at angle θ is proportional to $\gamma_V[1 + (V/c)\cos\theta]$. Recalling that the probability of the pitch angle lying in the angular range θ to $\theta + d\theta$ is proportional to $\sin\theta\,d\theta$, we find on averaging over all angles in the range 0 to π that the first term in expression (21.6) in the limit $v \to c$ becomes

$$\left\langle \frac{2V \cos\theta}{c} \right\rangle = \left(\frac{2V}{c}\right) \frac{\displaystyle\int_{-1}^{1} x[1 + (V/c)x]\,\mathrm{d}x}{\displaystyle\int_{-1}^{1} [1 + (V/c)x]\,\mathrm{d}x} = \frac{2}{3}\left(\frac{V}{c}\right)^2$$

where $x = \cos\theta$. Thus, in the relativistic limit, the average energy gain per collision is

$$\left\langle \frac{\Delta E}{E} \right\rangle = \frac{8}{3}\left(\frac{V}{c}\right)^2 \tag{21.7}$$

This illustrates the famous result derived by Fermi that the average increase in energy is only *second order* in V/c. It is also immediately apparent that this result leads to an exponential increase in the energy of the particle since the same fractional increase occurs per collision. Before looking at this part of the calculation a little more deeply, let us complete the essence of Fermi's original argument. If the mean free path between clouds along a field line is L, the time between collisions is $L/(c\cos\phi)$, where ϕ is the pitch angle of the particle with respect to the magnetic field direction. We need to average $\cos\phi$ over the pitch angle ϕ to find the average time between collisions, which is just $2L/c$. Therefore, we find a typical rate of energy increase

$$\frac{\mathrm{d}E}{\mathrm{d}t} = \frac{4}{3}\left(\frac{V^2}{cL}\right)E = \alpha E \tag{21.8}$$

It is assumed that the particle remains in the accelerating region for a characteristic time τ_{esc}. We now write down the diffusion-loss equation (19.13) and find the solution for $N(E)$ in equilibrium, that is,

$$\frac{\mathrm{d}N}{\mathrm{d}t} = D\nabla^2 N + \frac{\partial}{\partial E}[b(E)N(E)] - \frac{N}{\tau_{\mathrm{esc}}} + Q(E) \tag{21.9}$$

We are interested in the steady-state solution and, hence, $\mathrm{d}N/\mathrm{d}t = 0$. We are not interested in diffusion and, hence, $D\nabla^2 N = 0$, and we assume there are no sources, $Q(E) = 0$. The energy loss term is $b(E) = -\mathrm{d}E/\mathrm{d}t$, which in our case is $-\alpha E$. Therefore, equation (21.9) reduces to

$$-\frac{\mathrm{d}}{\mathrm{d}E}[\alpha E N(E)] - \frac{N(E)}{\tau_{\mathrm{esc}}} = 0 \tag{21.10}$$

Differentiating and rearranging this equation, we find

$$\frac{\mathrm{d}N(E)}{\mathrm{d}E} = -\left(1 + \frac{1}{\alpha\tau_{\mathrm{esc}}}\right)\frac{N(E)}{E}$$

Therefore

$$N(E) = \text{constant} \times E^{-x} \tag{21.11}$$

where $x = 1 + (\alpha\tau_{\mathrm{esc}})^{-1}$. It can be seen that we have succeeded in deriving a power-law energy spectrum.

In Fermi's original paper, it was assumed that collisions with interstellar clouds would be the main source of energy for the particles. This, however, leads to problems.

(i) The random velocities of interstellar clouds in the Galaxy are very small in comparison with the velocity of light, $V/c \leq 10^{-4}$. Furthermore, we have estimated the mean-free path of cosmic rays in the interstellar medium to be of the order of 1 pc (Section 20.3), and so the number of collisions would be roughly one per year, resulting in a very slow gain of energy by the particles. This means that there is very little hope of gaining significant acceleration from the original Fermi mechanism. We might do rather better if we restrict our attention to regions where there is small-scale turbulence. A good example of this might be the shells of young supernova remnants where there is certainly a great deal of small-scale turbulent motion present.

(ii) We have not yet considered the effect of energy losses upon the acceleration process. In particular, ionisation losses hamper the acceleration of particles from low energies. The form of the ionisation loss rate as a function of kinetic energy is compared qualitatively with a uniform acceleration rate in Fig. 21.2. Thus, if the acceleration mechanism is to be effective, the particles must either be injected into the acceleration region with energies greater than that corresponding to the maximum enegy loss rate, or else the initial acceleration process must be sufficiently rapid to overcome the energy losses. This is a quite general problem for all acceleration mechanisms.

(iii) There is nothing in the theory which tells us why the exponent x of the energy spectrum should be roughly 2.5 everywhere we look. It would be remarkable if the mechanism of acceleration in very diverse types of source were such that the product of the characteristic escape time, τ_{esc}, and the rate of energy gain, as represented by the constant α, conspired to be a universal constant.

This simplified version of second-order Fermi acceleration disguises a key aspect of the acceleration process. Inspection of the expression (21.6) shows that, to first order in V/c, the particle does not gain energy. The particle's energy is, however, changing all the time stochastically, and, if we were to inject particles with a single energy, the energy distribution would be broadened by random encounters with interstellar clouds. In fact, we see that, on average, the typical root mean square change of energy of the particle is $O(V/c)$, whereas the systematic energy increase is only $O(V/c)^2$. What this means is that, in a proper calculation, we have to take account of the statistical nature of the acceleration process as well as the average systematic increase in energy.

The problem is exactly the same as that discussed in connection with Compton scattering in Section 4.3.4 and illustrated in Fig. 4.14. That figure shows the broadening of the photon spectrum which necessarily accompanies the systematic increase in energy. As a result, we can only find approximate answers using the arguments which led to the expression (21.11). The full treatment has to take account explicitly of the stochastic nature of the acceleration process and the spreading of the energy spectrum by scattering. This is accomplished by developing a Fokker–Planck equation for the diffusion of the particles in momentum space. This approach is described in the review by Blandford and Eichler (1987), who show that,

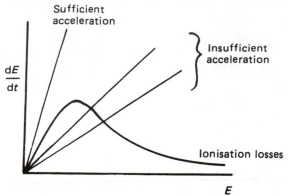

Figure 21.2. Comparison of the acceleration rate and energy loss rate due to ionisation losses for a high energy particle.

for an isotropic distribution of scatterers, the differential energy spectral index is

$$x = \frac{3}{2}\left(1 + \frac{4cL}{3\langle V^2\rangle \tau_{\text{esc}}}\right)^{1/2} - \frac{1}{2} \tag{21.12}$$

This result reinforces the point that the value of x is sensitive to assumptions about the adopted values of L and τ_{esc}. Note, however, that the result is of quite different form from that of the simple Fermi argument, except in the limit of small accelerations. The cause of this is that the argument does not take proper account of the stochastic nature of the diffusion of the particles in momentum space.

The origin of this difference can be illustrated by the following argument. The Fokker–Planck equation can be reduced to a diffusion-loss equation of form similar to that derived in Section 19.13 but with an additional term to describe the diffusion of the particles in momentum space (or energy space in the case of isotropic scatterers):

$$\frac{dN}{dt} = D\nabla^2 N + \frac{\partial}{\partial E}[b(E)N(E)] - \frac{N}{\tau_{\text{esc}}} + Q(E) + \frac{1}{2}\frac{\partial^2}{\partial E^2}[d(E)N(E)] \tag{21.13}$$

$d(E)$ is the mean square change in energy per unit time:

$$d(E) = \frac{d}{dt}\langle(\Delta E)^2\rangle$$

As before, we can find an expression for the average value of $(\Delta E)^2$ from expression (21.6). To second order in V/c, the only term which survives is $(\Delta E)^2 = 4E^2(V/c)^2 \cos^2\theta$, and, averaging over angle as before, we find

$$\langle(\Delta E)^2\rangle = 4E^2\left(\frac{V}{c}\right)^2 \frac{1}{2}\int_0^{\pi} \cos^2\theta \sin\theta \, d\theta = \frac{4}{3}E^2\left(\frac{V}{c}\right)^2 \tag{21.14}$$

Thus, there is a very close relation between the mean square energy change and the average increase in energy per collision. Comparing the expressions (21.7) and (21.14), we see that $d(E) = -Eb(E)/2 = \alpha E^2/2$, recalling that $b(E)$ is defined to be the rate of loss of energy. We seek steady-state solutions of equation (21.13), but including the stochastic acceleration term, that is,

$$\frac{\partial}{\partial E}[b(E)N(E)] - \frac{N}{\tau_{esc}} + \frac{1}{2}\frac{\partial^2}{\partial E^2}[d(E)N(E)] = 0 \tag{21.15}$$

Substituting for $b(E)$ and $d(E)$ and seeking solutions of power-law form, $N \propto E^{-x}$, we find that the partial differential equation reduces to the quadratic equation

$$x^2 + x - \left(2 + \frac{4}{\alpha\tau_{esc}}\right) = 0$$

which has solution

$$x = \frac{3}{2}\left(1 + \frac{16}{9\alpha\tau_{esc}}\right)^{1/2} - \frac{1}{2} \tag{21.16}$$

which is exactly the result derived by Blandford and Eichler (1987) if we adopt the expression $\alpha = \frac{4}{3}(V^2/cL)$ derived above. The importance of including the diffusion term in the analysis can be appreciated by comparing expressions (21.11) and (21.16). It is apparent that, in these second order acceleration mechanisms, the value of α to be used in the formulae for $b(E)$ and $d(E)$ is model dependent.

In the modern version of the Fermi second order acceleration, the particles interact with various types of plasma wave and gain energy by being scattered stochastically by these waves. This process is similar to that described in Section 20.4 on the interaction between high energy particles and waves or irregularities in the interstellar magnetic field.

21.4 Particle acceleration in strong shocks

We can rewrite the essence of the Fermi mechanism in a rather simpler fashion if we let $E = \beta E_0$ be the average energy of the particle after one collision and P be the probability that the particle remains within the accelerating region after one collision. Then, after k collisions, there are $N = N_0 P^k$ particles with energies $E = E_0\beta^k$. If we eliminate k between these quantities,

$$\frac{\ln(N/N_0)}{\ln(E/E_0)} = \frac{\ln P}{\ln \beta}$$

and hence

$$\frac{N}{N_0} = \left(\frac{E}{E_0}\right)^{\ln P/\ln \beta}$$

In fact, this value of N is $N(\geq E)$ since this number reach energy E and some fraction of them go on to higher energies. Therefore

$$N(E)dE = \text{constant} \times E^{-1+(\ln P/\ln \beta)}dE \tag{21.17}$$

It is clear in this formulation that we have again recovered a power law. To make the equivalence between the first and second versions of Fermi acceleration complete, we see that, from equation (21.11) and the definition of β, $\beta = 1+(\alpha/M)$, where α/M is the increment in energy per collision and P is related to τ.

In the version of the Fermi mechanism described in Section 21.3, α is proportional to $(V/c)^2$, because of the decelerating effect of the following collisions.

The original version of Fermi's theory is therefore known as *second order Fermi acceleration* and clearly is a very slow process. We would do much better if there were only head-on collisions. Inspection of expression (21.6) shows that, in this case, the energy increase is $\Delta E/E \propto 2V/c$, that is, first order in V/c, and, appropriately, this is called *first order Fermi acceleration.*

A very attractive version of first order Fermi acceleration in the presence of strong shock waves was discovered independently by a number of workers in the late 1970s. The papers by Axford, Leer and Skadron (1977), Krymsky (1977), Bell (1978) and Blandford and Ostriker (1978) stimulated an enormous amount of interest in this process for the many environments in which high energy particles are found in astrophysics. There are two different ways of tackling the problem, one starting from the diffusion equation for the evolution of the momentum distribution of high energy particles in the vicinity of a strong shock (for example, Blandford and Ostriker (1978)) and the other, a more physical approach, in which the behaviour of individual particles is followed (for example, Bell (1978)). I will adopt Bell's version of the theory, which makes the essential physics clear and indicates why this version of first order Fermi acceleration results remarkably naturally in a power-law energy spectrum of high energy particles.

To illustrate the basic physics of the acceleration process, let us consider the case of a strong shock, for example, that caused by a supernova explosion, propagating through the interstellar medium. A flux of high energy particles is assumed to be present both in front of and behind the shock front. The particles are considered to be of very high energy, and so the velocity of the shock is very much less than the velocities of the high energy particles. The key point about the acceleration mechanism is that the high energy particles hardly notice the shock at all, since its thickness will normally be very much smaller than the gyroradius of a high energy particle. Because of turbulence behind the shock front and irregularities ahead of it, when the particles pass though the shock in either direction, they are scattered so that their velocity distribution rapidly becomes isotropic on either side of the shock front. The key point is that the distributions are isotropic with respect to the frames of reference in which the fluid is at rest on either side of the shock.

Let us consider the case of a strong shock which was analysed in Section 10.6. This is the case, for example, for the material ejected in supernova explosions, where the velocities can be up to about 10^4 km s^{-1}, compared with the sound and Alfvén speeds of the interstellar medium, which are at most about 10 km s^{-1}. In the case of a strong shock, the shock wave travels at a highly supersonic velocity $U \gg c_s$, where c_s is the sound speed in the ambient medium (Fig. 21.3(a)). It is often convenient to transform into the frame of reference in which the shock front is at rest, and then the upstream gas flows into the shock front at velocity $v_1 = U$ and leaves the shock with a downstream velocity v_2 (Fig. 21.3(b)). The equation of continuity requires mass to be conserved through the shock, and so

$$\rho_1 v_1 = \rho_2 v_2$$

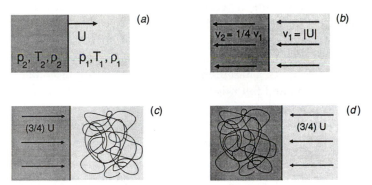

Figure 21.3. The dynamics of high energy particles in the vicinity of a strong shock wave. (*a*) A strong shock wave propagating at a supersonic velocity, U, through stationary interstellar gas with density ρ_1, pressure p_1 and temperature T_1. The density, pressure and temperature behind the shock are ρ_2, p_2 and T_2, respectively. The relations between the variables on either side of the shock front are given by the relations (10.30), (10.31) and (10.32). (*b*) The flow of interstellar gas in the vicinity of the shock front in the reference frame in which the shock front is at rest. In this frame of reference, the ratio of the upstream to the downstream velocity is $v_1/v_2 = (\gamma + 1)/(\gamma - 1)$. For a fully ionised plasma, $\gamma = \frac{5}{3}$ and the ratio of these velocities is $v_1/v_2 = 4$ as shown. (*c*) The flow of gas as observed in the frame of reference in which the upstream gas is stationary and the velocity distribution of the high energy particles is isotropic. (*d*) The flow of gas as observed in the frame of reference in which the downstream gas is stationary and the velocity distribution of high energy particles is isotropic.

In the case of a strong shock, $\rho_2/\rho_1 = (\gamma + 1)/(\gamma - 1)$, where γ is the ratio of specific heats of the gas (Section 10.6.1). Taking $\gamma = \frac{5}{3}$ for a monatomic or fully ionised gas, we find $\rho_2/\rho_1 = 4$, and so $v_2 = \frac{1}{4}v_1$.

Now let us consider the high energy particles ahead of the shock. Scattering ensures that the particle distribution is isotropic in the frame of reference in which the gas is at rest. It is instructive to draw diagrams illustrating the dynamical situation so far as typical high energy particles upstream and downstream of the shock are concerned. Let us consider the upstream particles first. The shock advances through the medium at velocity U, but the gas behind the shock travels at a velocity $\frac{3}{4}U$ relative to the upstream gas (Fig. 21.3(*c*)). When a high energy particle crosses the shock front, it obtains a small increase in energy of the order $\Delta E/E \sim U/c$, as we will show below. The particles are then scattered by the turbulence behind the shock front so that their velocity distributions become isotropic with respect to that flow.

Now let us consider the opposite process of the particle diffusing from behind the shock to the upstream region in front of the shock (Fig. 21.3(*d*)). Now the velocity distribution of the particles is isotropic behind the shock, and, when they cross the shock front, they encounter gas moving towards the shock front, again with the same velocity, $\frac{3}{4}U$. In other words, the particle undergoes exactly the same process of receiving a small increase in energy ΔE on crossing the shock from downstream to upstream as it did in travelling from upstream to downstream. This is the clever aspect of this acceleration mechanism. Every time the particle crosses the shock front it receives an increase of energy – there are never crossings in which the particles lose energy – and the increment in energy is the same going in both directions. Thus, unlike the standard Fermi mechanism in which there are both head-on and following collisions, in the case of the shock front, the collisions are always head on and energy is transferred to the particles. The beauty of the mechanism is the complete symmetry between the passage of the particles from upstream to downstream and from downstream to upstream through the shock wave.

Let us now be somewhat more quantitative about the actual process of acceleration. By simple arguments, due originally to Bell (1978), we can work out both β and P for this cycle. First, we evaluate the average increase in energy of the particle on crossing from the upstream to the downstream sides of the shock. The gas on the downstream side approaches the particle at a velocity $V = \frac{3}{4}U$ and so, performing a Lorentz transformation, the particle's energy when it passes into the downstream region is

$$E' = \gamma_V(E + p_x V) \tag{21.18}$$

where we take the x coordinate to be perpendicular to the shock. We assume that the shock is non-relativistic, $V \ll c$, $\gamma_V = 1$, but that the particles are relativistic, so that we can write $E = pc$, $p_x = (E/c)\cos\theta$. Therefore,

$$\Delta E = pV\cos\theta \quad ; \quad \frac{\Delta E}{E} = \frac{V}{c}\cos\theta$$

We now seek the probability that the particles which cross the shock arrive at an angle θ per unit time. This is a standard piece of kinetic theory. The number of particles within the angles θ to $\theta + d\theta$ is proportional to $\sin\theta\, d\theta$, but the rate at which they approach the shock front is proportional to the x component of their velocities, $c\cos\theta$. Therefore the probability of the particle crossing the shock is proportional to $\sin\theta\cos\theta\, d\theta$. Normalising so that the integral of the probability distribution over all the particles approaching the shock is equal to unity, that is, those with θ in the range 0 to $\pi/2$, we find

$$p(\theta) = 2\sin\theta\cos\theta\, d\theta \tag{21.19}$$

Therefore, the average gain in energy on crossing the shock is

$$\left\langle \frac{\Delta E}{E} \right\rangle = \frac{V}{c} \int_0^{\pi/2} 2\cos^2\theta\sin\theta d\theta = \frac{2}{3}\frac{V}{c}$$

The particle's velocity vector is randomised without any energy loss by scattering in the downstream region and it then recrosses the shock, as illustrated in Fig. 21.2(*d*), when it gains another fractional increase in energy $\frac{2}{3}(V/c)$ so that, in making one round trip across the shock and back again, the fractional energy increase is, on average,

$$\left\langle \frac{\Delta E}{E} \right\rangle = \frac{4}{3}\frac{V}{c} \tag{21.20}$$

Consequently,

$$\beta = \frac{E}{E_0} = 1 + \frac{4V}{3c}$$

in one round trip.

To work out the escape probability P, we use a clever argument due to Bell (1978). According to classical kinetic theory, the number of particles crossing the shock is $\frac{1}{4}Nc$, where N is the number density of particles. This is the average number of particles crossing the shock in either direction, since, as noted above, the particles scarcely notice the shock. Downstream, however, the particles are swept away, or 'advected', from the shock because the particles are isotropic in that frame. Referring to Fig. 21.3(b), it can be seen that the particles are removed from the region of the shock at a rate $NV = \frac{1}{4}NU$. Thus, the fraction of the particles lost per unit time is $\frac{1}{4}NU / \frac{1}{4}Nc = U/c$. Since we assume that the shock is non-relativitistic, it can be seen that only a very small fraction of the particles is lost per cycle. Thus, $P = 1 - (U/c)$. This solves the problem since we need $\ln \beta$ and $\ln P$ to insert into expression (21.17). Therefore, since

$$\ln P = \ln\left(1 - \frac{U}{c}\right) = -\frac{U}{c} \quad \text{and} \quad \ln \beta = \ln\left(1 + \frac{4V}{3c}\right) = \frac{4V}{3c} = \frac{U}{c}$$

we find

$$\frac{\ln P}{\ln \beta} = -1$$

and, hence, the differential energy spectrum of the high energy particles is

$$N(E)\mathrm{d}E \propto E^{-2}\mathrm{d}E \tag{21.21}$$

This is the result we have been seeking. It may be objected that we have obtained a value of 2 rather than 2.5 for the exponent of the differential energy spectrum, and that problem cannot be neglected. However, the reason that this mechanism has excited so much interest is that, for the first time, there are excellent physical reasons why power-law energy spectra with a unique spectral index should occur in diverse astrophysical environments. In this simplest version of the theory, the only requirements are the presence of strong shock waves and that the velocity vectors of the high energy particles be randomised on either side of the shock. It is entirely plausible that there are strong shocks in most sources of high energy particles, supernova remnants, active galactic nuclei and the diffuse components of extended radio sources.

One important feature of the model is that the particles must be scattered in both the upstream and downstream regions. Behind the shock, it is entirely plausible that there are turbulent motions which can scatter the particles as described in Section 11.4.2. In Bell's original proposal, the particles which recross the shock from downstream to upstream result in bulk streaming of the relativistic particles through the unperturbed interstellar medium, and, consequently, he argued that the particles are scattered by the generation of Alfvén and hydromagnetic waves, just as described in Section 20.5. It is therefore expected that the high energy particles will be confined within some characteristic distance in front of the shock.

The number of high energy particles is expected to decrease exponentially ahead of the shock wave.

It is also apparent that there is an upper limit to the energy to which particles can be accelerated by this mechanism. The precise value of the upper limit has been the subject of some disagreement. Lagage and Cesarsky (1983) describe in detail the processes which can limit the acceleration process. The basic problem is that, although the first order acceleration mechanism is a distinct improvement upon the original Fermi mechanism, it is still not a rapid process. The particles have to diffuse back and forth across the shock wave and, in the case of supernova remnants, increase their energy by typically about one part in 100 in each crossing. Supernova remnants decelerate once they have swept up roughly their own mass of interstellar gas, and enter the blast wave phase of evolution. Typically, the acceleration phase can only last about 10^5 years. In their detailed analysis of the problem, Lagage and Cesarsky find that the upper limit to the energy of particles which can be accelerated in typical supernova explosions is about 10^5 GeV nucleon^{-1}, and this is probably a generous upper limit. It is significant that the spectrum of Galactic cosmic rays extends well beyond this upper limit, and so shock acceleration in supernovae on its own cannot account for the complete range of energies observed in the cosmic rays.

There is one further beautiful aspect of this process. It can be seen that the particles are accelerated where they are needed, and this enables us to overcome the problem of adiabatic losses, which haunts many theories of non-thermal sources (see Section 19.5.2). For example, in supernova remnants such as Cassiopeia A and Tycho's supernova, the particles are accelerated *in situ*, and the energy for accelerating them is extracted from the kinetic energy of the expanding supernova shell. Thus, as suggested in Section 19.5.2, there are good reasons why both the magnetic field energy *and* the particle energy in shell-like supernova remnants are derived from the kinetic energy of expansion of the supernova. It is also suggestive that the radio spectral indices of many young supernova remnants, such as Tycho's supernova remnant, are $\alpha \approx 0.6$, corresponding to $x = 2.2$. On the other hand, the radio spectral index of Cassiopeia A is steeper, $\alpha = 0.77$, corresponding to $x = 2.54$. The recent discovery of intense γ-rays from radio-band compact quasars by the Compton Gamma-ray Observatory has revealed that the γ-ray spectra of these compact sources have spectral indices corresponding to $x = 2$, which may be associated with the same type of acceleration process.

The model can be directly tested by measuring the fluxes of particles in the vicinity of shocks propagating through the interplanetary medium. Fig. 21.4 is an example of such a shock, which was observed by the ISEE-3 satellite on November 12, 1978. The passage of the shock wave past the spacecraft was determined by the abrupt increase in the Solar Wind velocity from just less than 400 km s^{-1} to about 700 km s^{-1}. The fluxes of energetic protons increased roughly exponentially as the spacecraft approached the shock and then remained at roughly a constant level in the downstream region. The more energetic protons have a longer length scale in front of the shock than the lower energy particles.

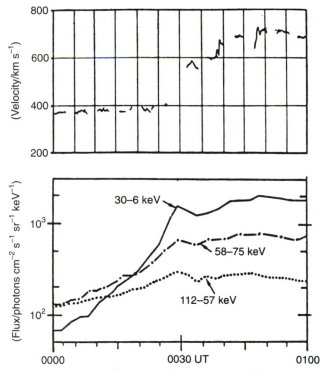

Figure 21.4. The distribution of energetic particles in the vicinity of an interplanetary shock wave observed by the ISEE-3 satellite on November 12, 1978. The upper diagram shows the Solar Wind velocity, which suddenly increases at 0028:16 UT as the shock passes the spacecraft. The energetic proton flux increases roughly exponentially ahead of the shock, and the length scale increases with increasing particle energy. After the passage of the shock, the fluxes of energetic particles remain roughly constant on the downstream side of the shock. (From C.F. Kennel, F.V. Coroniti, F.L. Scarf, W.A. Livesey, C.T. Russell, E.J. Wenzel and M. Scholer (1986). *J. Geophys. Res.*, **91**, 11, 917.)

These observations are exactly what would be expected according to the standard theory of shock acceleration. According to Völk (1987), the shock mechanism of acceleration can give a good account of the distributions of fast particles observed in shocks which propagate along the magnetic field direction but is less successful for oblique shocks.

21.5   Beyond the standard model

The subject of particle accelertion in strong shocks has developed dramatically since the results derived in the preceding section were established. Detailed reviews of these developments have been presented by Drury (1983), Blandford and Eichler (1987) and Völk (1987). The most interesting question concerns how the spectral index of the particle energy distribution, x, changes as the properties

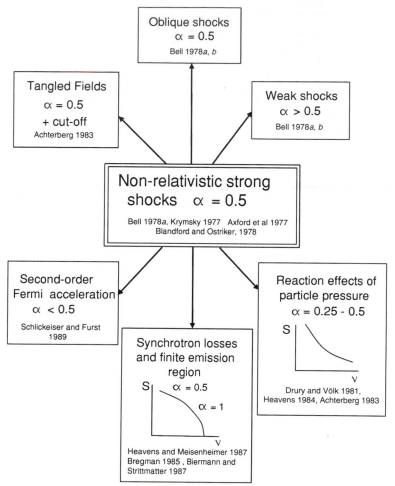

Figure 21.5. Modifications to the standard model of shock acceleration for non-relativistic shock waves (courtesy of Dr Alan Heavens). (References: A.A. Achterberg (1983). *Astron. Astrophys.*, **119**, 274. A.R. Bell (1978*a*). *Mon. Not. R. Astron. Soc.*, **182**, 147; A.R. Bell (1978*b*). *Mon. Not. R. Astron. Soc.*, **182**, 443; G.F. Krymsky (1977). *Dok. Acad. Nauk. USSR*, **234**, 1306; W.I. Axford, E. Leer and G. Skadron (1977). *Proc. 15th Intl Cosmic ray Conf.*, **11**, 132; R.D. Blandford and J.P. Ostriker (1978). *Astrophys. J.*, **221**, L29; R. Schlickeiser and E. Furst (1989). *Astron. Astrophys.*, **219**, 192; L.O'C. Drury and H.J. Völk (1981). *Astrophys. J.*, **248**, 344; A.F. Heavens (1984). *Mon. Not. R. Astron. Soc.*, **211**, 195; A.F. Heavens and K. Meisenheimer (1987). *Mon. Not. R. Astron. Soc.*, **225**, 335; J.N. Bregman (1985). *Astrophys. J.*, **288**, 32; P.L. Biermann and P.A. Strittmatter (1987). *Astrophys. J.*, **322**, 643.)

of the shock wave change. The following summary is based upon a survey kindly provided by Dr Alan Heavens, University of Edinburgh, UK.

Fig. 21.5 summarises how the spectral index of the radio emission of optically thin synchrotron radiation changes for different assumptions about the physical conditions in non-relativistic shock waves. The results derived in Section 21.4

apply only for strong non-relativistic shocks in which the pressure of the accelerated particles can be neglected. If there is a magnetic field present, the standard
results are applicable when the field is uniform and the field direction is perpendicular to the shock. Bell (1978) showed that the predicted electron energy spectral
index remains the same, $x = 2$, $\alpha = 0.5$, in the case in which the strong shock propagates at an angle to the magnetic field direction. If the shock is weak, the compression ratio $\rho_2/\rho_1 < 4$ and the velocity discontinuity is smaller. The same calculation as in Section 21.4 can be repeated for the case $r = v_1/v_2 < 4$, and then the
spectral index of the synchrotron radiation becomes $\alpha = 3/[2(r - 1)]$. Therefore,
steeper spectra are expected for weak shock waves. The effect of the relativistic
particle pressure is to flatten the spectra, as illustrated schematically in Fig. 21.5.

One way of steepening the electron energy spectra is to invoke synchrotron
losses of the accelerated particles, as described in Section 19.3.3. For example,
in the model of Heavens and Meisenheimer (1987), the accelerated electrons are
swept downstream, where they suffer synchrotron losses. If the synchrotron loss-
time of the electrons is less than the age of the source, the steady-state spectrum
steepens by $\Delta\alpha = 0.5$, as shown in Section 19.2. Notice that, in this case, the energy
spectrum of the particles at a given distance behind the shock front has an abrupt
cut-off at that energy at which the lifetime of the particles to synchrotron losses is
equal to the time since they were accelerated in the shock front. However, when
the spectra of all the particles at different distances behind the shock are summed,
the standard result is obtained. This steepening may account for the forms of
spectra observed in some of the 'hot-spots' in extragalactic radio sources, which
have also been observed in the near infrared and optical wavebands (Fig. 21.6).
At the very highest energies, there is also expected to be an abrupt cut-off for
those particles for which the synchrotron radiation loss-time is equal to the
characteristic time for acceleration of the particles in the vicinity of the shock
front. At low frequencies, the standard spectral index $\alpha = 0.5$ is expected.

The case of relativistic shocks is more complicated, and the results of various
analyses are displayed in Fig. 21.7. The relativistic equivalent of the calculations
carried out in Section 21.4 result in radio spectral indices in the range $\alpha = 0.35$–0.6,
still somewhat flatter than the typical spectra of cosmic rays and extragalactic
radio sources. As in the case of non-relativistic shocks, steeper spectra are found
in weak relativistic shocks. The case of oblique shocks becomes more complicated
because, at small enough angles between the shock normal and the magnetic field
direction, the shock propagates along the magnetic field lines superluminally, and
so the particles gyrating about the field lines cannot recross the shock. Kirk and
Heavens (1989; see caption to Fig. 21.7 for reference) find values of the radio
spectral index in the range $\alpha = 0$–0.5.

It is now possible to test some of these models directly by following the
trajectories of individual particles as they propagate back and forth across the
shock front. For example, Ballard and Heavens (1992; see caption to Fig. 21.8
for reference) have studied the acceleration of particles in relativistic shocks in
which it is assumed that the magnetic field is tangled on either side of the shock.

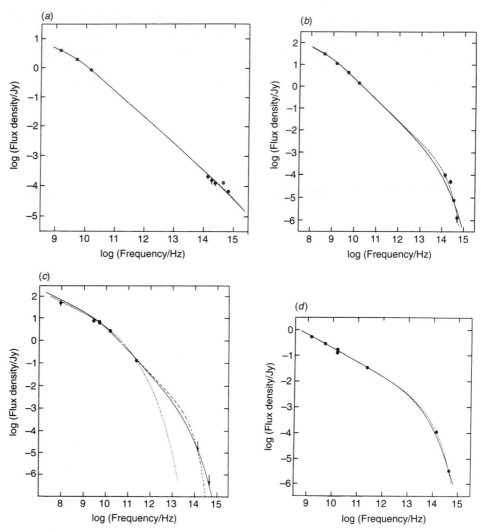

Figure 21.6. The spectra of the hot-spots in the radio sources (*a*) Pictor A (west), (*b*) 3C 273, (*c*) 3C 123 (east) and (*d*) 3C 33 (south). These forms of spectra can be accounted for by the standard theory of acceleration, including the effects of synchrotron losses. For the definition of jansky (Jy), see Appendix A2. (From K. Meisenheimer, H.-J. Röser, P.R. Hiltner, M.G. Yates, M.S. Longair, R. Chini and R.A. Perley (1989). *Astron. Astrophys.*, **219**, 63.)

Fig. 21.8 shows an example of the trajectory of a particle which crosses and then recrosses the shock. By averaging over large numbers of particles, the energy spectra of the particles can be found. In their computations, Ballard and Heavens found the standard value $\alpha \sim 0.5$ for shock velocities $v \le 0.5c$, but the spectra steepen to $\alpha \sim 0.6$–1.1 for greater shock velocities. These results may be relevant to the spectra of the jets in extragalactic radio sources in which the most luminous sources have the steepest radio spectra. Notice that, in the case of the relativistic

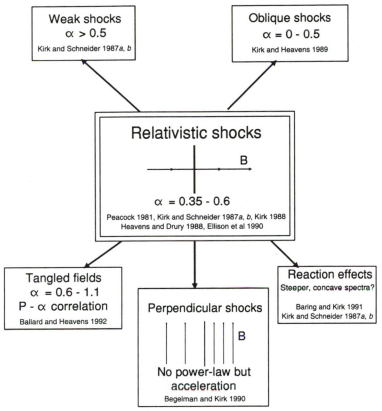

Figure 21.7. Modifications to the standard model of shock acceleration for relativistic shock waves (courtesy of Dr Alan Heavens). (References: J.G. Kirk and P. Schneider (1987*a*). *Astrophys. J.*, **315**, 425; J.G. Kirk and P. Schneider (1987*b*). *Astrophys. J.*, **322**, 256; J.G. Kirk and A.F. Heavens (1989). *Mon. Not. R. Astron. Soc.*, **239**, 995; J.A. Peacock (1981). *Mon. Not. R. Astron. Soc.*, **196**, 135; J.G. Kirk (1988). Habilitationschrift, University of Munich, Germany; A.F. Heavens and L.O'C. Drury (1988). *Mon. Not. R. Astron. Soc.*, **235**, 997; D.C. Ellison, F.C. Jones and S.P. Reynolds (1990). *Astrophys. J.*, **360**, 702; K.R. Ballard and A.F. Heavens (1992). *Mon. Not. R. Astron. Soc.*, **259**, 89; M.C. Begelman and J.G. Kirk (1990). *Astrophys. J.*, **353**, 66; M.G. Baring and J.G. Kirk (1991). *Astron. Astrophys.*, **241**, 329.)

shocks, the particles make large energy gains each time they cross the shock front,

$$E' = \gamma_V E [1 + (V/c) \cos \theta].$$

21.6   Pulsar magnetospheres

We have already discussed in Section 15.4 the fact that pulsars possess very strong magnetic fields, up to 3×10^8 T, and are rotating rapidly. The fastest rotators, the millisecond pulsars, have only weak fields, and we have argued that this is at least partly because a weak field enables the spin-up of a neutron star to be more effective since the Alfvén surface is closer to the surface of the neutron

Particle orbits: xz plane

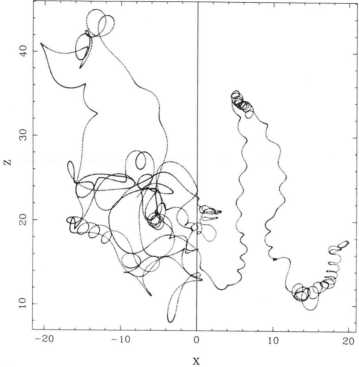

Figure 21.8. A typical orbit of a relativistic particle in a disordered magnetic field. The particle begins its motion to the right of the diagram and crosses the relativistic shock at $x = 0$. The random path of the particle in the compressed disordered field eventually results in it recrossing the shock. (From K.R. Ballard and A.F. Heavens (1992). *Mon. Not. R. Astron. Soc.*, **259**, 89.)

star. However, there are some pulsars with strong magnetic fields and short periods, for example the Crab pulsar. As was argued in Section 15.4, the result is that very strong electric potentials are developed within the magnetospheres of pulsars, and acceleration of particles to high energies is feasible.

We have already discussed the possibility of particles being accelerated at the poles of neutron stars along the open field lines, but there is also the possibility that the rotating magnet emits magnetic dipole radiation. The problem is that the frequency of the emitted waves is only 30 Hz, which is well below the plasma frequency of the interstellar plasma. How then can the waves propagate? The answer is that the waves are very strong indeed, and our conventional ideas of plasma physics are no longer applicable.

It is a simple sum to work out the strength of the electromagnetic waves emitted by the pulsar. The magnetic field strength in the wave at the bottom of the wave zone is 100 T. If we extrapolate back to the strength of the field when the pulsar first formed, it must have been 10^6 T. It is, therefore, clear that we are dealing with the opposite of the situation normally encountered in physics, where the field of the wave is only a small perturbation on the motion of the particle. Normally, one considers only the effect of the electric field of the wave on the motion of a test particle. Now, however, the magnetic field cannot be neglected under any circumstance.

The strength of the electric field at the bottom of the wave zone is

$$E = cB = 3 \times 10^{14} \text{ V m}^{-1}$$

Let us now work out how long it takes a particle to be accelerated to a velocity close to that of light:

$$m\frac{\mathrm{d}v}{\mathrm{d}t} = eE$$

Therefore,

$$\frac{mc}{T} = eE$$

that is,

$$T \approx \frac{mc}{eE} \approx 10^{-14} \text{ s}$$

This is negligible in comparison with the minimum possible period of the pulsar, $t \approx 10^{-3}$ s. Therefore, there can be no such thing as a non-relativistic particle in the electromagnetic field emitted by a rotating, magnetised neutron star. As soon as the particle's velocity becomes comparable to the velocity of light, we must take into account the magnetic field, which then twists the direction of motion round into the direction of the wave, and the particle is continually accelerated. In fact, it travels so close to the velocity of light that it is more or less riding at constant phase with a wave. Detailed trajectories of these particles have been worked out by a number of authors, the first important work following the discovery of pulsars being that of Gunn and Ostriker (1969). The actual behaviour of the particle is strongly dependent upon the phase at which the particle is injected into the strong wave. We cannot go into the details of the dynamics of particles, but what is important is that there seems to be no reason why particles cannot be accelerated to high energies in these systems. The limitation on the energy of the particles is actually set by the radiation reaction term becoming important.

Syrovatskii has given a general argument which suggests that it might be possible to account for even the highest energy cosmic rays by processes taking place in the vicinity of rotating neutron stars. He derives the maximum amount of energy which a particle can attain in a magnetic field of strength B and scale L.

Let us use Maxwell's first equation for the electromagnetic field to find the maximum electric field obtainable by changing the magnetic flux density, B, as fast as possible in the region of dimension L:

$$\nabla \times \mathbf{E} = -\frac{\partial \mathbf{B}}{\partial t}$$

Therefore,

$$\frac{E}{L} = \frac{B}{L/c} \quad ; \quad E = Bc$$

Then, the total energy given to the particle is

$$\int zE\mathrm{d}x = zeBcL$$

that is,

$$\gamma mc^2 \approx zeBcL$$

Let us put in appropriate values for protons close to the surface of a neutron star, $B = 10^6$ T, $L \approx 100$ km,

$$E_{max} = 5 \text{ J} = 3 \times 10^{19} \text{ eV}$$

Thus, in principle, particles could be accelerated to the highest energies in the vicinity of pulsars. Whether or not this is possible requires a much more refined calculation.

21.7   Summary

We have by no means solved the problem of the acceleration of particles to cosmic ray energies, but the situation is not as unpromising as it seemed a few years ago. There now exist mechanisms by which a power-law spectrum of particle energies can be produced, acceleration in strong shock waves dominating much of current thinking in high energy astrophysics. This is unlikely to be the whole story, however, and it is is most unlikely that this process can account for the highest energy cosmic rays, $E > 10^{15}$ eV, for which the gyroradii are of the same size as the supernova remnant itself. Furthermore, it has yet to be demonstrated that the observed power-law spectral indices can be obtained by this process. Probably another mechanism has to be sought for the highest energy cosmic rays, and pulsar magnetospheres are one possible candidate. We have suggested that the very highest energy cosmic rays may be of cosmological origin, and we will return to this question when we consider the acceleration of charged particles in active galactic nuclei.

Appendices – astronomical nomenclature

A1 Distances in astronomy

The unit of distance used in astronomy is the *parallax-second*, or *parsec*. It is defined to be the distance at which the mean radius of the Earth's orbit about the Sun subtends an angle of one second of arc. In metres, the parsec, abbreviated to pc, is 3.0856×10^{16} m. For many purposes, it is sufficiently accurate to adopt 1 pc $= 3 \times 10^{16}$ m. The parsec is a recognised SI unit and it is often convenient to work in kiloparsecs (1 kpc $= 1000$ pc $= 3 \times 10^{19}$ m), megaparsecs (1 Mpc $= 10^6$ pc $= 3 \times 10^{22}$ m) or even gigaparsecs (where 1 Gpc $= 10^9$ pc $= 3 \times 10^{25}$ m).

Sometimes, it is convenient to measure distances in *light-years*, which is the distance light travels in one year: 1 light-year $= 9.4605 \times 10^{15}$ m. Thus, 1 pc $= 3.26$ light-years.

Another commonly used distance unit in astronomy is the *astronomical unit*, abbreviated to AU, which is the mean radius of the Earth's orbit about the Sun: 1 AU $= 1.49578 \times 10^{11}$ m. The very nearest stars to the Earth are at a distance of about 1 pc, and so they are about 2×10^5 times as far away as the Earth is from the Sun.

Accurate distances are among the most difficult measurements to make in astronomy, and there must, therefore, exist corresponding uncertainties in all the derived physical properties of astronomical objects. The most accurate direct distance measurements are derived from measurements of the *parallaxes* of nearby stars, that is, the apparent motion of nearby stars against the background of very distant stars due to the motion of the Earth in its orbit about the Sun. From the definition of the parsec given above, it can be appreciated that a star at a distance of 1 pc is observed to move an angular distance of one arcsecond (1 arcsec) relative to the background stars when the Earth moves a distance equal to the Earth's mean radius about the Sun, perpendicular to the line of sight to the star. Even for the stars closest to the Sun, few distances are known with precision much better than about 5%. The accuracy with which this local distance scale is known will improve by a factor of about ten when the survey being conducted by the *Hipparcos* astrometry satellite of the European Space Agency is completed. This satellite was launched in 1989, and it is expected that the positions of about 120 000 preselected stars will be measured with an accuracy of 2 milliarcsec and parallaxes will be obtained with the same accuracy. The catalogue should become available in 1995.

The standard procedure for extending the distance scale from nearby stars to Galactic and extragalactic distances depends upon being able to find *distance indicators* which provide *relative distances* for distant objects as compared to similar objects nearby. The complexities of undertaking these observations to extend the distance scale to cosmological distances has been reviewed by Rowan-Robinson (1985, 1988). The range distances over which the different distance indicators can be used is shown in Fig. A1. To give just one example of the application of these distance indicators, *Cepheid variables* are used to extend the local distance scale from our own Galaxy to nearby galaxies. The Cepheid variable stars are pulsating stars which have characteristic light curves, by which is meant the variation of their luminosities with the phase of the pulsation. In 1912, Henrietta Leavitt showed that there exists a tight correlation between the intrinsic luminosity of the variable star and its pulsation period. Therefore, if the pulsation periods of Cepheid variables in nearby galaxies can be measured, their intrinsic luminosities, L, can be measured, and then, by measuring their flux densities, S, their distances, r, can be found from the inverse square law $r = (L/4\pi S)^{1/2}$.

The use of distance indicators can be extended to extragalactic distances using other 'standard' properties of galaxies or the brightest star systems in them. On the very largest scale, relative distances can be measured using the *velocity–distance relation* for galaxies. In 1929, Hubble showed that the whole system of galaxies is not stationary but is expanding uniformly such that, nearby, the distance of a galaxy, r, from our own Galaxy is proportional to its velocity of recession, v, away from it, $v = H_0 r$, where H_0 is Hubble's constant (see Section 13.1.2). The value of Hubble's constant is a hotly disputed topic, one school of thought favouring a value close to 50 km s^{-1} Mpc^{-1} while others favour somewhat larger values, about 80 km s^{-1} Mpc^{-1}. The reason for this discrepancy partly reflects the difficulty of measuring accurate extragalactic distances and partly the major problems in ensuring that systematic selection effects are taken into account. For this reason, Hubble's constant is often written $H_0 = 100h$ km s^{-1} Mpc^{-1}, and the parameter h can take the value preferred by the reader. Notice that *relative distances*, can be measured with much greater accuracy than absolute distances in extragalactic astronomy.

The problem of the accumulating errors in using distance indicators could be avoided if some physical method were discovered of measuring physical sizes and distances directly at large distances. If the angular size, $\Delta\theta$, of some object is measured for which the physical size, d, can be estimated by physical arguments, the distance r is found immediately from $r = d/\Delta\theta$. An interesting application of this technique is the use of what is known as the *Baade–Wesselink method*

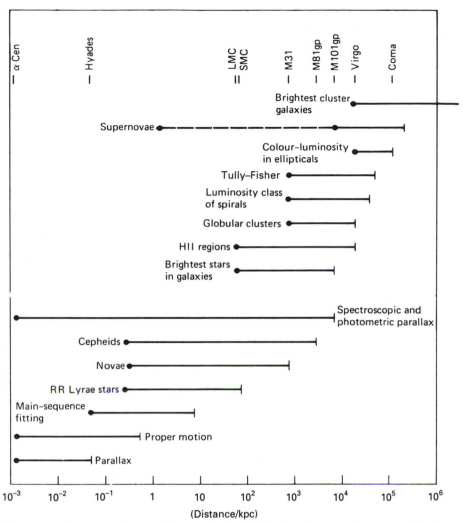

Figure A1. The cosmological 'distance ladder' indicating the approximate range of distances over which different distance indicators can be used. (After M. Rowan-Robinson (1985). *The cosmological distance ladder*, p. 296. New York: W.H. Freeman and Co.)

to supernova explosions in distant galaxies. In a simple approximation, the expanding sphere may be considered a black body at temperature T_1. As the sphere expands from diameter d_1 to d_2, its luminosity changes because the sphere has a larger surface area at some new temperature T_2. By measuring the change in temperature and flux density of the supernova, the change in *angular size*, $\Delta\theta$, of the supernova can be measured using Stefan–Boltzmann's law since at t_1 the flux density is $S_1 = (\sigma T_1^4/4)(d_1/r)^2 = (\sigma T_1^4/4)\theta_1^2$, and at t_2 the flux density is $S_1 = (\sigma T_2^4/4)\theta_2^2$. The *physical size* of the expansion $(d_2 - d_1)$ can also be measured from the velocity of expansion of the shell, v, and the times, t_1 and t_2, between which the temperatures T_1 and T_2 were measured, $(d_2 - d_1) = 2v(t_2 - t_1)$. Hence, the distance r can be found. This method can potentially be used to measure very large cosmological distances directly.

Another possible method for measuring directly physical sizes for distant objects is to use the Sunyaev–Zeldovich effect as applied to the very hot gas observed in clusters of galaxies (see Section 4.3.4). In this case, the temperature, pressure and bremsstrahlung emission of the hot gas can all be measured, enabling the physical size of the hot gas cloud to be determined. Then, by measuring its angular size, the distance of the cluster can be found. Another direct physical method of measuring extragalactic distances is to use the properties of *gravitational lenses*. If the background source is variable, then the different images of the source will display the same variability but with time delays depending upon the geometry of the lensing galaxy. From these observations, the physical dimensions of the lensed image can be found and hence the distance to the lens.

The great advantage of the last three methods is that they eliminate the necessity of measuring distances through the hierarchy of the cosmological distance ladder but measure them directly without any intermediate steps.

A2   Masses and luminosities in astronomy

It is convenient to describe the masses of many celestial objects in terms of the mass of the Sun. One solar mass, written $1M_\odot$, is $1.989 \times 10^{30} \approx 2 \times 10^{30}$ kg. Thus, Jupiter has mass about $0.001M_\odot$, our own Galaxy has mass about $10^{11}M_\odot$ and the Coma cluster of galaxies has mass about $3 \times 10^{15}M_\odot$.

All direct methods of measuring masses in astronomy involve the combination of Newton's laws of motion with Newton's law of gravity. In the simple case of the Sun, its mass is found by equating the centripetal accelerations of the planets to their gravitational accelerations due to the Sun: $v^2/r = GM_\odot/r^2$, $M_\odot = v^2r/G$. In the case of binary star systems, similar methods can be used. For a cluster of galaxies, the total mass is found using the Viral Theorem, which states that, for a system in dynamical equilibrium, the internal kinetic energy is equal to half its gravitational potential energy, $\frac{1}{2}M\langle v^2 \rangle = \frac{1}{2}GM^2/r_s$, $M \approx r_s \langle v^2 \rangle /G$, where $\langle v^2 \rangle$ is the mean square velocity and r_s is some suitably chosen radius (see Volume 3). Many cases are described in the text in which masses have to be estimated, and, in each of them, it is necessary to measure some characteristic velocity v and a physical radius r_s.

It is also convenient to measure the luminosities of celestial objects in terms of the *bolometric luminosity* of the Sun. The bolometric luminosity of the Sun, written as $L_\odot$, is the total luminosity of the Sun integrated over all wavelengths. Its value is $L_\odot = 3.90 \times 10^{26}$ W. Notice that, for the Sun, the luminosity, $L_\odot$, is almost entirely emitted in the ultraviolet, optical and near infrared wavebands. Astronomers often use units of $L_\odot$ to describe luminosities far outside those wavebands in which the Sun's luminosity is emitted. For example, the far infrared luminosities of regions of star formation and the X-ray luminosities of binary X-ray sources are often quoted in units of $L_\odot$, and this provides a useful measure of the energetics of these sources relative to the optical/infrared luminosity of the Sun.

A3   Magnitudes and colours in astronomy

The standard measure of the observed intensity of a celestial object is its *flux density*, S_ν, which is the energy incident per second per unit area per unit frequency band at the Earth. Thus, if an energy ΔE is detected in a bandwidth ν to $\nu + \Delta \nu$ in time Δt and A is the collecting area of the telescope, which is assumed to be 100% efficient at registering the radiation, the flux density of the source at frequency ν is defined to be $S_\nu = \Delta E / (\Delta t \Delta \nu A)$. The SI units of flux density are watts per square metre per hertz. If this flux density originates within solid angle $d\Omega$ on the sky, the *intensity* of the source is $I_\nu = S_\nu / \Delta \Omega$; the SI units of intensity are watts per square metre per hertz per steradian. The typical flux densities of astronomical objects are very small, and so the radio astronomers introduced the unit 10^{-26} W m^{-2} Hz^{-1}, which is known as 1 jansky, abbreviated to Jy. Smaller flux densities are referred to in millijanskys (10^{-3} Jy $= 10^{-29}$ W m^{-2} Hz^{-1}) and microjanskys (10^{-6} Jy $= 10^{-32}$ W m^{-2} Hz^{-1}). In X-ray and γ-ray astronomy, flux densities are often quoted in terms of the flux of photons rather than of energy, so that the units photons per second per unit area per unit energy range (for example, photons s^{-1} m^{-2} keV^{-1}) are often used. Notice that flux densities and intensities are often quoted per unit wavelength I_λ rather than per unit frequency band. Since $I_\nu d\nu = I_\lambda d\lambda$, it follows that $I_\nu = cI_\lambda / \nu^2$.

In optical astronomy, although flux densities are becoming commoner, the traditional system of *optical magnitudes* is still commonly used. The concept of magnitudes was introduced by the Greek astronomers, who placed stars into five 'magnitude' classes, the magnitudes being assigned on the basis of the brightnesses of the stars to the naked eye. Fifth-magnitude stars were the faintest objects visible. The system was put on a systematic basis by Pogson in 1854, who showed that the magnitude scale was logarithmic in flux density, like many physiological sensations, and that the magnitude scale of the ancients could be well approximated by a rule in which five logarithmic magnitudes correspond to a factor of 100 in flux density.

Observed flux densities are converted to *apparent magnitudes*, *m*, which are negative logarithmic measures of flux density defined by

$$m = \text{constant} - 2.5 \log_{10} S$$

where S is the flux density of the object. Thus, a difference of 5 magnitudes corresponds exactly to a factor of 100 in flux density. The magnitude system is normalised so that a standard star, chosen to be the bright star Vega (or α-Lyrae) in the constellation of Lyra, has zero magnitude at all wavelengths. In this way, magnitudes can be defined at all wavelengths (see Table 8.2). The very brightest stars in the sky have magnitude about 0. The faintest stars which can be seen with the naked eye have $m \approx 5$. The faintest stars which can now be observed

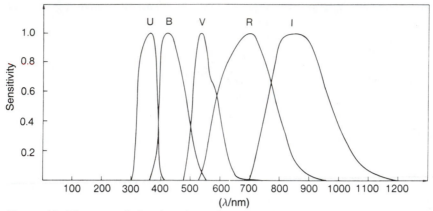

Figure A2. The transmission functions of some of the standard filters for the optical and near infrared wavebands (see also Table 8.2). (From H. Karttunen, P. Kroger, H. Oja, M. Pountanen and K.J. Donner (1987). *Fundamental astronomy*, p. 93. Heidelberg: Springer-Verlag.)

with a 4-metre telescope in a 5-minute observation using a CCD camera have magnitudes about 25.

In practice, the flux density S is measured within some range of frequencies, v_1 to v_2, determined by the transmission of the atmosphere, the properties of the telescope mirror and the instrumentation of the telescope as a function of frequency. In general, observations are made through a filter with a certain transmission function T_v which describes the fraction of the energy incident upon the filter which is transmitted to the detector as a function of frequency, v. Thus, for a particular filter i, $S_i = \int_0^\infty T_i(v)S_v dv$. If $T_i(v) = 1$ for all frequencies, the flux density S_i is called a *bolometric flux density* or, converting to magnitudes, a *bolometric apparent magnitude*. This magnitude corresponds to the total radiation emitted by the source at all wavelengths. In practice, bolometric magnitudes are taken to mean the total flux density within a given waveband or associated with a particular physical process.

At optical and infrared wavelengths, a number of standard magnitudes systems have been established by adopting particular filters as standards. In the days of photographic astronomy, the filters were determined largely by the sensitivity response of the photographic emulsions and the available filter materials. In the early 1950s, a standard system of U, B, V photometry was introduced by Johnson and Morgan in which the transmission functions of the filters were precisely defined. The letters U, B, V refer to thc ultraviolet, blue and visual regions of the optical spectrum, respectively, and the filter transmission functions are shown in Fig. A2. For each filter, it is often convenient to define an effective central wavelength (or frequency) and its effective bandpass $(\lambda_{eff}, \Delta\lambda_{eff})$. For a zero-magnitude star at all wavelengths, that is, for α-Lyrae, the central wavelengths and bandpasses for the U, B, V filters in nanometres are $U(365, 68)$,

$B(440, 98)$ and $V(550, 89)$ (see Table 8.2). Notice that the mean energy of the photons detected within a given waveband depends upon the spectrum of the star, and, in general, this will be quite different from the spectrum of α-Lyrae. In precise work, it is essential to use the full transmission function, $T_i(v)$, and the spectrum of the star, $S(v)$, to derive precise magnitudes for comparison with theory.

The type of photocathode used to define the *UBV* system (an S4 photocathode) is now obsolete, but many astronomers still find it helpful to relate their measurements to the *UBV* system. Many new magnitude systems have been devised which are more appropriate for use with the new generations of fine grain photographic emulsions, such as the IIIaJ and IIIaF emulsions produced by Kodak, and of electro-optical detectors for the optical and infrared wavebands, such as CCD detectors and similar infrared devices. A list of some of the commoner filters used in astronomy is given in Table 8.2, which includes a conversion to flux density, measured in janskys, for a zero-magnitude star. This table is useful for rough calculations, but it must be stressed that these figures assume that the spectrum of the object within the filter passband is the same as that of α-Lyrae. For precise work, a correction must be made to the calibration to take account of the actual spectrum of the object within the filter passband.

To measure the intrinsic luminosities of celestial objects, astronomers often use *absolute magnitudes*, $\mathcal{M}$. These are defined to be the magnitudes which the objects would have if they were placed at a distance of 10 pc. An object of intrinsic luminosity L has flux density $S = L/4\pi r^2$, and therefore, for any object,

$$\mathcal{M} = m - 5\log_{10}(r/10)$$

where the distance r is measured in parsecs. For stars of different luminosities,

$$\mathcal{M} = \mathcal{M}_\odot - 2.5\log_{10}(L/L_\odot)$$

Just as in the case of apparent magnitudes, the absolute magnitudes can either refer to a specific wavelength or waveband or else to a bolometric absolute magnitude, which corresponds to the total luminosity of the object integrated over all wavelengths.

The absolute bolometric magnitude of the Sun is $\mathcal{M}_\odot = 4.75$, and hence in general we can write

$$\mathcal{M} = 4.75 - 2.5\log_{10}(L/L_\odot)$$

where L is the bolometric luminosity of the object.

Astronomers define the *colours* of stars in terms of the differences in their magnitudes at different wavelengths. For example, a commonly used colour is the difference between the magnitudes of a star in the blue (B) waveband (centred at 440 nm) and the visual (V) waveband (centred at 550 nm). This *colour index* or *colour* $(B - V)$ is a measure of how blue or red the star is. Since stars may be thought of as black-body emitters in the simplest approximation, blue stars are hot and red stars are cool. This statement is quantified by working in terms of colours such as $B - V$. The use of colour is a means of discriminating the general spectral shapes of different classes of object. A particularly common way of displaying this information is in terms of a two-colour diagram in which, for example, $(U-B)$ is plotted against $(B-V)$. Fig. A3 is an example of a two-colour diagram for 46000 stars. Notice also that a colour such as $(U-B)$ or $(B-V)$ is a strong function of the surface temperature of the star, and hence, in plotting a luminosity–temperature diagram, it is much more convenient to plot one of these colours rather than the surface temperature of the star, which requires a detailed knowledge of radiative transport in the stellar atmosphere.

A4   Coordinate systems

We need say little about coordinate systems in this text, and for more details we refer the interested reader to the text-books by Smart (1977) and Murray (1983). The complexities of defining precisely the celestial system of coordinates and of time are enormous, and go far beyond what is needed in this text. The complications arise because of the fact that the Earth, in its orbit about the Sun, does not move in a perfectly elliptical orbit but is subject to the perturbing influences of the Moon and the planets. These introduce small but important wobbles and perturbations upon the motion of the Earth about the Sun.

The important points for our purposes are that it is possible to relate the positions of celestial objects to a fixed set of celestial coordinates on the sky, which are known as *right ascension* (RA or α) and *declination* (Dec or δ). The North Celestial Pole (NCP) is taken to be the mean direction of the rotation axis of the Earth, and declination is the polar angle measured from the equator ($\delta = 0°$) towards the NCP ($\delta = 90°$) (Fig. A4). The South Celestial Pole (SCP) has declination $\delta = -90°$. The zero point of right ascension is taken to be the point at which the celestial equator intersects the plane of the ecliptic, that is, the plane of the Earth's orbit about the Sun. This point is known as the *first point in Aries* or the *vernal equinox* and is taken to be zero hours of right ascension. Lines of constant longitude around the celestial equator are defined in terms of hours, minutes and seconds of time, such that 24 hours corresponds to completing a great circle of 360°. Thus, the units of position on the celestial sphere are different in right ascension and declination. The angles in declination can be directly related to arcseconds but, in right ascension, a difference of $\Delta\alpha$ seconds of time corresponds to an angle of $15\,\Delta\alpha \cos\delta$ arcseconds at declination δ. The international convention of defining the positions of objects through their nomenclature is illustrated by the designation of the binary pulsar PSR 1913 + 16, which means the pulsar with right ascension $19^h\,13^m$ and declination $+16°$.

The other system of coordinates which is frequently used in astronomy is that of *galactic coordinates*. Fig. 1.8(d) shows the Milky Way drawn in galactic coordinates. The images shown in Fig. 1.8 are all presented in galactic coordinates, and provide convincing evidence that we live in a highly flattened galaxy. The definition of the system of galactic coordinates is illustrated in Fig. A5. The

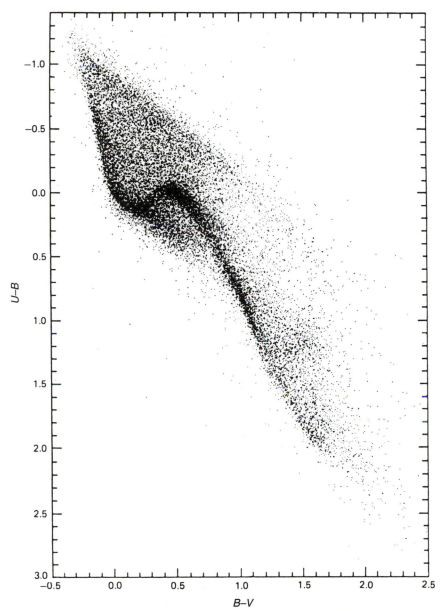

Figure A3. A two-colour plot of $(U - B)$ against $(B - V)$ for 46 000 stars. Most stars lie along the S-shaped locus, which is defined by stars lying on the main sequence. Hot blue stars are located at the top left of the diagram and cool red stars are to the bottom right. If stars radiated like black bodies, they would lie on a locus stretching smoothly from the bottom right to top left of the diagram. The 'kink' in the relation is due to the effect of the Balmer continuum absorption, which causes the spectra to deviate from those of black bodies. The effect of reddening by interstellar dust is to move the colours of each star towards the bottom right corner of the diagram. The rather sharp upper envelope to the distribution of hot stars above the main sequence follows precisely this reddening line. (From B. Nicolet (1980). *Astron. Astrophys. Suppl.*, **42**, p. 283.)

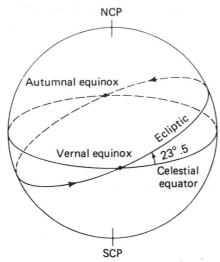

Figure A4. Illustrating the definition of the system of celestial coordinates used in astronomy. (From G. Walker (1987). *Astronomical observations – an optical perspective*, p. 55. Cambridge: Cambridge University Press.)

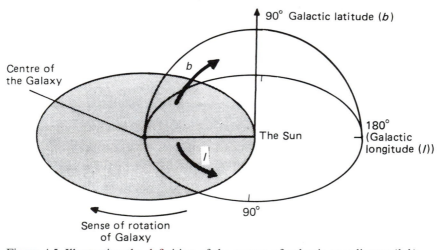

Figure A5. Illustrating the definition of the system of galactic coordinates (*l*, *b*).

direction of the centre of our Galaxy is defined by radio observations because of the lack of obscuration by dust at these wavelengths and because of the ability of VLBI techniques to measure extremely accurate positions. The Galactic centre is therefore taken as the zero of *galactic latitude* (*b*) and *longitude* (*l*). The Galactic equator is taken to be the great circle which passes through the centre of the Galactic plane as seen in Fig. 1.8(*d*). Galactic longitude is taken in the sense that it increases from right to left as we look at Fig. 1.8(*d*) starting at *l* = 0° at the Galactic centre and coming back to zero after 360°. The Galactic North Pole is taken to be the direction perpendicular to the plane of the Galaxy at the Sun's position in the Galaxy. The Galactic equator is at $b = 0°$ of galactic latitude and the Galactic North Pole at $b = +90°$. The Galactic South Pole lies at $b = -90°$. These are indicated schematically in Fig. A5. The North Galactic Pole lies in the Northern Celestial Hemisphere at coordinates $\alpha = 12^h 49^m$; $\delta = +27°24'$. The projections of the galactic coordinate system onto a plane shown in Fig. 1.8 are known as an *Aitoff projection*, in which, although equal areas are preserved at different declinations, there are strong distortions of the orthogonal coordinate grid towards the poles.

Further reading and references

Remarks about the astronomical literature

The literature on astronomy, astrophyiscs and cosmology is vast, and continues to expand at an amazing rate. The present volume should be considered a snapshot of high energy astrophysics in 1993. For each of the chapters of this volume, a list of books for further reading is provided as well as specific references to papers cited in the text.

To obtain an introduction to any specialised topic discussed in the text, a good starting point is to consult the volumes of *Annual Reviews of Astronomy and Astrophysics*. The series began in 1963, the 1993 volume being Volume 31. The reviews are authoritative, and, over a period of a few years, span most of the major fields of astronomy, astrophysics and cosmology. These volumes can be supplemented by the many volumes which record the proceedings of international symposia. To begin with, it is recommended that invited reviews in conference proceedings be consulted since they are normally of broader interest than contributed papers.

To track down specific papers, the series of *Astronomy and Astrophysics Abstracts*, published each year by Springer-Verlag, are quite invaluable. Papers can be traced by author, subject or astronomical object.

Chapter 13

General introductory texts

These books provide a broad overview of the whole of astronomy and astrophysics. They should be primarily used for general reading and as background references. There are many such introductory books – this list shows the books I refer to most regularly for general information.

Abell, G.O., Morrison, D. and Wolff, S.C. (1987). *Exploration of the universe*. Philadelphia: Saunders College Publishing.

Audouze, J. and Israël, G., eds (1988). *The Cambridge atlas of astronomy*, 2nd edition. Cambridge: Cambridge University Press.

Davies, P.C.W. ed. (1989). *The new physics*. Cambridge: Cambridge University Press. This volume contains several accessible reviews on subjects of astrophysical interest, specifically: C. Will, *The renaissance of general relativity*, p. 7; A. Guth and P. Steinhardt, *The inflationary universe*, p. 34; S. Hawking, *The edge of spacetime*, p. 61; C. Isham, *Quantum gravity*, p. 70; M.S. Longair *The new astrophysics*, p. 94.

Harrison, E.R. (1981). *Cosmology*. Cambridge: Cambridge University Press.

Harwit, M. (1988). *Astrophysical concepts*, 2nd edition. New York: Springer-Verlag.

Karttunen, H., Kröger, P., Oja, H., Poutanen, M. and Donner, K.J. (1987). *Fundamental astronomy*. Berlin: Springer-Verlag.

Kaufmann, W.J., III, (1991). *Universe*, 3rd edition. Oxford: W.H. Freeman and Co.

Lang, K.R. (1980). *Astrophysical formulae*, 2nd edition. Berlin: Springer-Verlag.

Maran, S.P., ed. (1992). *The astronomy and astrophysics encyclopedia*. New York: Van Nostrand Reinhold; Cambridge: Cambridge University Press.

Shu, F.H. (1982). *The physical universe: an introduction to astronomy*. Mill Valley, CA: University Science Books.

Tayler, R.J. (1974). *The stars: their structure and evolution*. London: Wykeham Publications (London) Ltd.

Unsöld, A. and Baschek, B. (1991). *The new cosmos*, 4th edition. Berlin: Springer-Verlag.

Wynn-Williams, C.G. (1992). *The fullness of space*. Cambridge: Cambridge University Press.

More specialist texts

These more specialist books provide more details on the contents of the main section of this chapter.

Böhm-Vitense, E. (1989–92). *Introduction to stellar astrophysics*, vols 1–3. Cambridge: Cambridge University Press.

Kippenhahn, R. and Weigert, A. (1990). *Stellar structure and evolution*. Berlin: Springer-Verlag.

Longair, M.S. (1989). *Galaxy formation*, in *Evolution of galaxies – astronomical observations*, eds I. Appenzeller, H.J. Habing and P. Lena, pp. 1–93. Heidelberg: Springer-Verlag.

Longair, M.S. (1992). *Theoretical concepts in physics*. Cambridge: Cambridge University Press.

Michalas, D. and Binney, J. (1981). *Galactic astronomy: structure and kinematics*. New York: W.H. Freeman and Co.

Peebles, P.J.E. (1993). *Principles of physical cosmology*. Princeton: Princeton University Press.

Rubin, V.R. and Coyne, G.V., eds. (1988). *Large-scale motions in the universe*. Vatican City: Pontificia Academia Scientiarum.

References

Gott, J.R., III, Melott, A.L. and Dickenson, M. (1986). *Astrophys. J.*, **306**, 341.

Hubble, E. (1929). *Proc. Natl Acad. Sci.*, **15**, 168.

Mather, J.C. (1994). In *Frontiers of space and ground-based astronomy*, eds W. Wamsteker, M.S. Longair and Y. Kondo. Dordrecht: Kluwer Academic Publishers (in press).

Melott, A.L., Weinberg, D.H. and Gott, J.R., III (1988). *Astrophys. J.*, **328**, 550.

Chapter 14

General texts

Bahcall, J.N. (1989). *Neutrino astrophysics*. Cambridge: Cambridge University Press.

Kippenhahn, R. and Weigert, A. (1990). *Stellar structure and evolution*. Berlin: Springer-Verlag.

Schwarzschild, M. (1958). *Structure and evolution of the stars*. Princeton: Princeton University Press.

Tayler, R.J. (1974). *The stars: their structure and evolution.* London: Wykeham Publications (London) Ltd.

References

Abasov, A.I., Anosov, O.L., Bowles, T.J., Cherry, M.L., Cleveland, B.T., Davis, Jr,R., Elliott, S.R., Faizov, E.L., Gavrin, V.N., Kalikhov, A.V., Knodel, T.V., Knyshenko, I.I., Kornoukhov, V.N., Kouzes, R.T., Lande, K., Mezentseva, S.A., Mirmov, I.N., O'Brien, H.A., Ostrinsky, A.V., Pshukov, A.M., Revzin, N.E., Shikhin, A.A., Timofeyev, P.V., Veretenkin, E.P., Vermul, V.M., Wark, D.L., Wilkerson, J.F. and Zatsepin, G.T. (1991). *Proc. 22nd International Cosmic Ray Conference, Dublin,* **3**, 724.

Bahcall, J.N. and Bethe, H.A. (1990). *Phys. Rev. Lett.* **65**, 2233.

Burbidge, E.M., Burbidge, G.R., Fowler, W.A. and Hoyle, F. (1957). *Rev. Mod. Phys.,* **54**, 547.

Deubner, F.L. and Gough, D.O. (1984). *Ann. Rev. Astron. Astrophys.,* **22**, 593.

Elsworth, Y., Howe, R., Isaak, G.R., McLeod, C.P. and New, R. (1990). *Nature,* **347**, 536.

Gough, D.O. and Toomre, J. (1991). *Ann. Rev. Astron. Astrophys.,* **29**, 627.

Griffin, R.F. (1985). In *Interacting binaries,* eds P.P. Eggleton and J.E. Pringle, p. 1. Dordrecht: D. Reidel Publishing Company.

Lamb, H. (1932). *Hydrodynamics,* 6th edition. Cambridge: Cambridge University Press.

Leighton, R.B. (1960). In *Aerodynamic phenomena in stellar atmospheres,* ed. R.N. Thomas, p. 321. Bologna: Nicola Zanichelli – Editore.

Turck-Chièze, S., Cahen, S., Cassé, M. and Doom, C. (1988). *Astrophys. J.,* **335**, 415.

Chapter 15

General texts

Bahcall, J.N. (1989). *Neutrino astrophysics.* Cambridge: Cambridge University Press.

Chandrasekhar, S. (1985). *The mathematical theory of black holes.* Cambridge: Cambridge University Press.

Hawking, S.W. and Ellis, G.F.R. (1973). *The large-scale structure of space-time.* Cambridge: Cambridge University Press.

Kippenhahn, R. and Weigert, A. (1990). *Stellar structure and evolution.* Berlin: Springer-Verlag.

Longair, M.S. (1992). *Theoretical concepts in physics.* Cambridge: Cambridge University Press.

Lyne, A.G. and Graham-Smith, F. (1990). *Pulsar astronomy.* Cambridge: Cambridge University Press.

Manchester, R.N. and Taylor, J.H. (1977). *Pulsars.* San Francisco: W.H. Freeman and Co.

Misner, C.W., Thorne, K. and Wheeler, J.A. (1973). *Gravitation.* San Francisco: W.H. Freeman and Co.

Murdin, P. and Murdin, L. (1985). *Supernovae.* Cambridge: Cambridge University Press.

Rindler, W. (1977). *Essential relativity.* New York: Springer-Verlag.

Shapiro, S.I. and Teukolsky, S.A. (1983). *Black holes, white dwarfs and neutron stars: the physics of compact objects.* New York: Wiley Interscience.

References

Baade, W. and Zwicky, F. (1934). *Proc. Natl Acad. Sci.,* **20**, 254.

Bell-Burnell, S.J. (1983). *Serendipitous discoveries in radio astronomy,* eds. K. Kellermann and B. Sheets, p. 160. Green Bank: NRAO Publications.

Chevalier, R. (1992). *Nature,* **355**, 691.

Cowley, A.P. (1992). *Ann. Rev. Astron. Astrophys.,* **30**, 287.

Feynman, R.F. (1972). *Statistical mechanics,* chap. 11. Reading, MA: W.A. Benjamin, Inc.

Gold, T. (1968). *Nature,* **205**, 787.

Goldreich, P. and Julian, W.H. (1969). *Astrophys. J.,* **157**, 839.

Hawking, S.W. (1975a). *Comm. Math. Phys.,* **43**, 199.

Hawking, S.W. (1975b). In *quantum gravity: an Oxford symposium*, eds C.J. Isham, R. Penrose and D.W. Sciama, p. 219. Oxford: Oxford University Press.

Hewish, A. (1986). *Q. J. R. Astron. Soc.*, **27**, 548.

Hewish, A., Bell, S.J., Piklington, J.D.H., Scott, P.F. and Collins, R.A. (1968). *Nature*, **217**, 709.

Kerr, R.P. (1963). *Phys. Rev. Letts.*, **11**, 237.

McClintock, J.E. (1992). *Proc. Texas/ESO-CERN symposium on relativistic astrophysics, cosmology and fundamental particles*, eds J.D. Barrow, L. Mestel and P.A. Thomas, pp. 495–502. New York: New York Academy of Sciences.

Michell, J. (1783). *Phil. Trans. Roy. Soc. Lond.*, **74**, 41.

Novikov, I.D. and Frolov, V.P. (1989). *The physics of black holes*. Dordrecht: Kluwer Academic Publishers

Pacini, F. (1967). *Nature*, **216**, 567.

Panagia, N., Gilmozzi, R., Macchetto, F., Adorf, H-M. and Kirshner, R.P. (1991). *Astrophys. J.*, **380**, L23.

Radhakrishnan, V. (1986). In *Highlights of astronomy*, vol 7, ed. J.-P. Swings, p. 3. Dordrecht: D. Reidel Publishing Co.

Ruderman, M.A. and Sutherland, P.G. (1975). *Astrophys. J.*, **196**, 51.

Schwarzschild, K. (1916). Sitzungsberichte, Preussische Akademie der Wissenschaften, p. 189.

Taylor, J. (1992). The Scott Lectures, Cavendish Laboratory, University of Cambridge.

Thorne, K.S., Price, R.H. and MacDonald, D.A. (1986). *Black holes : the membrane paradigm*. New Haven: Yale University Press.

Wali, K.C. (1991). *Chandra: a biography of S. Chandrasekhar*. Chicago: University of Chicago Press.

Weisskopf, V.F. (1981). *Contemporary Phys.*, **22**, 375.

Will, C. (1989). In *The new physics*, ed. P.C.W. Davies, p. 7. Cambridge: Cambridge University Press.

Woosley, S.E. (1986). *Nucleosynthesis and stellar evolution*, in *Nucleosynthesis and chemical evolution* by J. Audouze, C. Chiosi and S.E. Woosley, p. 1. Sauverny-Versoix: Geneva Observatory Publications.

van den Heuvel, E.P.J. (1987). In *The origin and evolution of neutron stars*, eds D.J. Helfand and J.H. Huang, p. 383. Dordrecht: D. Reidel Publishing Co.

Chapter 16

General texts

Batchelor, G.K. (1967). *Introduction to fluid mechanics*. Cambridge: Cambridge University Press.

Bertout, C., Collin, S., Lasota, J-P. and Tran Thnh Van, J., eds (1991). *Structure and emission properties of accretion discs*. Gif-sur-Yvette: Edition Frontiéres.

Faber, T.E. (1994). *Fluid mechanics*. Cambridge: Cambridge University Press.

Frank, J., King, A. and Raine, D. (1992). *Accretion power in astrophysics*. Cambridge: Cambridge University Press.

Landau, L.D. and Lifshitz, E.M. (1959). *Fluid mechanics*. Oxford: Pergamon Press.

Lewin, W.H.G. and van der Heuvel, E.P.J., eds (1983). *Accretion-driven stellar X-ray sources*. Cambridge: Cambridge University Press.

Lewin, W.H.G., van Paradijs, J. and van den Heuvel, E.P.J., eds (1994). *X-ray binaries*. Cambridge: Cambridge University Press (in press).

Mason, K.O., Watson, M.G. and White, N.E., eds (1986). *Physics of accretion onto compact objects*. Heidelberg: Springer-Verlag.

Shapiro, S.I. and Teukolsky, S.A. (1983). *Black holes, white dwarfs and neutrons stars: the physics of compact objects*. New York: Wiley Interscience.

Tanaka, Y., ed. (1988). *Physics of neutron stars and black holes*. Tokyo: Universal Academic Press, Inc.

References

Córdova, F.A-D. (1994). In *X-ray binaries*, eds W.H.G. Lewin, J. van Paradijs and E.P.J. van den Heuvel. Cambridge: Cambridge University Press (in press).

Feynman, R.P. (1962). *The Feynman Lectures in physics*, Vol. 2, eds R.P. Feynman, R.B. Leighton and M.L. Sands, chap. 41. Redwood City, CA: Addison-Wesley Publishing Company.

Frank, J., King, A.R. and Lasota, J.P. (1987). *Astron. Astrophys.*, **178**, 137.

Inoue, H. (1992). *Proc. Texas/ESO-CERN symposium on Relativistic astrophysics, cosmology and fundamental particles*, eds J.D. Barrow, L. Mestel and P.A. Thomas, pp. 86–103. New York: New York Academy of Sciences.

Novikov, I.D. and Thorne, K.S. (1973). *Black holes*, eds C. DeWitt and B.S. DeWitt, 1972 Les Houches Summer School, p. 343. New York: Gordon and Breach Science Publishers.

Papaloizou, J.C.B. and Pringle, J.E. (1984). *Mon. Not. R. Astron. Soc.*, **208**, 721.

Pringle, J.E. (1981). *Ann. Rev. Astron. Astrophys.*, **19**, 140.

Rees, M.J. (1984). *Ann. Rev. Astron. Astrophys.*, **22**, 471.

Sunyaev, R.A. and Shakura, N.I. (1972). *Astron. Astrophys.*, **24**, 337.

Starrfield, S. (1988). In *Multiwavelength astrophysics*, ed. F.A. Córdova, p. 159. Cambridge: Cambridge University Press.

Wandel, A. and Mushotzky, R.F. (1986). *Astrophys. J.*, **306**, L63.

White, N. (1985). In *Interacting Binaries*, eds P.P. Eggleton and J.E. Pringle, p. 249. Dordrecht: D. Reidel Publishing Co.

Woosley, S.E. (1986). *Nucleosynthesis and stellar evolution*, in *Nucleosynthesis and chemical evolution*, by J. Audouze, C. Chiosi and S.E. Woosley, p. 1. Sauverny-Versoix: Geneva Observatory Publications.

Chapter 17

General texts

Duley, W.W. and Williams, D.A. (1984). *Interstellar chemistry*. London: Academic Press.

Michalis, D. and Binney, J. (1981). *Galactic astronomy: structure and kinematics*. San Francisco: W.H. Freeman and Co.

Osterbrock, D.E. (1989). *The Astrophysics of gaseous nebulae and active galactic nuclei*. Mill Valley, CA: University Science Books.

Shu, F.H. (1991, 1992). *The physics of astrophysics*, vols 1, 2. Mill Valley, CA: University Science Books.

Spitzer, L. (1978). *Physical processes in the interstellar medium*. New York: Wiley Interscience Publications.

Wynn-Williams, C.G. (1992). *The fullness of space*. Cambridge: Cambridge University Press.

References

Adams, F.C. and Shu, F.H. (1985). *Astrophys. J.*, **296**, 655.

Audouze, J. (1986). *Nucleosynthesis of the very light (A<12) and trans-iron (A>60) nuclei*, in *Nucleosynthesis and chemical evolution*, by J. Audouze, C. Chiosi and S.E. Woosley, p. 431. Sauverny-Versoix: Geneva Observatory Publications.

Beckwith, C., Sargent, A.I., Chini, R.S. and Güsten, R. (1990). *Astron. J.*, **99**, 924.

Blitz, L., Binney, J., Lo, K.Y., Bally, J. and Ho, P. (1993). *Nature*, **361**, 417.

Bridle, A.H. and Purton, C. (1968). *Astron. J.*, **73**, 717.

Cox, D.P. and Smith, B.W. (1974). *Astrophys. J. Lett.*, **189**, L105.

Field, G.B. (1965). *Astrophys. J.*, **142**, 531.

Field, G.B., Goldsmith, D.W. and Habing, H.J. (1969). *Astrophys. J. Lett.*, **55**, L149.

Heiles, C. (1976). *Ann. Rev. Astron. Astrophys.*, **14**, 1.

Hildebrand, R.H. (1983). *Q. J. R. Astron. Soc.*, **24**, 267.

Huang, Y.-L. and Thaddeus, P. (1986). *Astrophys. J.*, **55**, L149.

Longair, M.S. (1989). *Galaxy formation*, in *Evolution of galaxies: astronomical observations*, eds I. Appenzeller, H. Habing and P. Lena, pp. 1–93. Heidelberg: Springer-Verlag.

Mezger, P.G. and Smith, L.F. (1977). In *Star formation*, eds T. de Jong and A. Maeder, p. 133. Dordrecht: D. Reidel Publishing Co.

Pengelly, R.M. (1964). *Mon. Not. R. Astron. Soc.*, **127**, 145.

Purton, C. (1966). *The spectra of radio sources and background radiation*, Ph.D. Dissertation, Cambridge University.

Reynolds, R.J. (1990). In *Low frequency astrophysics from space*, eds N.E. Kassim and K.W. Weiler, p. 121. Berlin: Springer-Verlag.

Salpeter, E.E. (1955). *Astrophys. J.*, **121**, 161.

Shu, F.H., Adams, F.C. and Lizano, S. (1987). *Ann. Rev. Astron. Astrophys.*, **25**, 23.

Solomon, P.M., Sanders, D.B. and Roivolo, A.R. (1985). *Astrophys. J. Lett.*, **292**, L19.

Spitzer, L. (1968). *Diffuse matter in space*. New York: Interscience Publishers.

Spitzer, L. (1990). *Ann. Rev. Astron. Astrophys.*, **28**, 71.

Taylor, J. and Cordes, J.M. (1993). *Astrophys. J.*, **411**, 674.

Chapter 18

General texts

Bekefi, G. (1966). *Radiation processes in plasmas*. New York: John Wiley and Sons, Inc.

Jackson, J.D. (1975). *Classical electrodynamics*, 2nd edition. New York: John Wiley and Sons, Inc.

Pacholczyk, A.G. (1970). *Radio astrophysics*. San Francisco: W.H. Freeman and Co.

Rybicki, G.B. and Lightman, A.P. (1979). *Radiative processes in astrophysics*. New York: Interscience Publishers.

References

Abramowitz, M. and Stegun, I.A. (1965). *Handbook of mathematical functions*. New York: Dover Publications.

Caswell, J.L. (1976). *Mon. Not. R. Astron. Soc.*, **177**, 601.

Ginzburg, V.L., Sasonov, V.N. and Syrovatskii, S.I. (1968). *Uspekhi Fiz. Nauk.*, **94**, 60.

Ginzburg, V.L. and Syrovatskii, S.I. (1965). *Ann. Rev. Astron. Astrophys.*, **3**, 297.

Ginzburg, V.L. and Syrovatskii, S.I. (1969). *Ann. Rev. Astron. Astrophys.*, **7**, 375.

Kniffen, D.A. and Fichtel, C.E. (1981). *Astrophys. J.*, **250**, 389.

Le Roux, E. (1961). *Ann. Astrophys.*, **24**, 71.

Legg, M.P.C. and Westfold, K.C. (1968). *Astrophys. J.*, **154**, 499.

Scheuer, P.A.G. (1966). In *Plasma astrophysics*, ed P.A. Sturrock, Proc. International School of Physics 'Enrico Fermi', vol. 39, p. 289.

Stecker, F.W. (1975). In *Origin of cosmic rays*, eds J.L. Osborne and A.W. Wolfendale, p. 267. Dordrecht: D. Reidel Publishing Co.

Webber, W. (1983). In *Composition and origin of cosmic rays*, ed. M.M. Shapiro, p. 83. Dordrecht: D. Reidel Publishing Co.

Webster, A.S. (1974). *Mon. Not. R. Astron. Soc.*, **166**, 355.

Webster, A.S. (1978). *Mon. Not. R. Astron. Soc.*, **185**, 507.

Westfold, K.C. (1959). *Astrophys. J.*, **130**, 241.

Chapter 19

General texts

Ginzburg, V.L. and Syrovatskii, S.I. (1964). *The origin of cosmic rays*. Oxford: Pergamon Press.

References

Axford, W.I., Leer, E. and Skadron, G. (1977). *Proc. 15th International Cosmic Ray Conference*, **11**, 132.

Bell, A.R. (1978). *Mon. Not. R. Astron. Soc.*, **182**, 147.

Blandford, R.D. and Ostriker, J.P. (1978). *Astrophys. J.*, **221**, L29.

Bulanov, S.V., Dogel, V.A. and Syrovatskii, S.I. (1976). *Astrophys. Space Sci.*, **44**, 267.

Burbidge, G.R. (1956). *Phys. Rev.*, **103**, 264.

Gull, S.F. (1975). *Mon. Not. R. Astron. Soc.*, **171**, 263.

Heitler, W. (1954). *The quantum theory of radiation.* Oxford: Clarendon Press.

Jackson, J.D. (1975). *Classical electrodynamics*, 2nd edition. New York: John Wiley and Sons, Inc.

Krymsky, G.F. (1977). *Dok. Acad. Nauk. USSR*, **234**, 1306.

Lawson, K.D., Mayer, C.J., Osborne, J.L. and Parkinson, M.L. (1987). *Mon. Not. R. Astron. Soc.*, **225**, 307.

Sedov, L.I. (1959). *Similarity and dimensional methods in mechanics.* New York: Academic Press.

Taylor, G.I. (1950). *Proc. Roy. Soc.*, **A201**, 159, 175.

Toller, G.N. (1990). In *The galactic and extragalactic background radiation*, eds S. Bowyer and C. Leinert, p. 21. Dordrecht: Kluwer Academic Publishers.

Chapter 20

General texts

Ginzburg, V.L. and Syrovatskii, S.I. (1964). *The origin of cosmic rays.* Oxford: Pergamon Press.

Ramana Murthy, P. and Wolfendale, A.W. (1993). *Gamma-ray astronomy*, 2nd edition. Cambridge: Cambridge University Press.

References

Baldwin, J.E. (1967). In *Radio astronomy and the galactic system*, ed. H. van Woerden, p. 337. London: Academic Press.

Burbidge, G.R. and Brecher, K. (1971). *Comments Astrophys. Space Phys.*, **3**, 140.

Cesarsky, C. (1980). *Ann. Rev. Astron. Astrophys.*, **18**, 289.

Fichtel, C.E., Özel, M.E., Stone, R.G. and Sreekumar, P. (1991). *Astrophys. J.*, **374**, 134.

Garcia-Munoz, M., Mason, G.M. and Simpson, J.A. (1977). *Astrophys. J.*, **217**, 859.

Garcia-Munoz, M., Simpson, J.A., Guzik, T.G., Wefel, J.P. and Margolis, S.H. (1987). *Astrophys. J. Suppl.*, **64**, 269.

Ginzburg, V.L. (1970). *Comments Astrophys. Space Phys.*, **2**, 1.

Hillas, A.M. (1984). *Ann. Rev. Astron. Astrophys.*, **22**, 425.

Jenkins, E.B. (1987). In *Exploring the universe with the IUE Satellite*, ed. Y. Kondo, p. 531. Dordrecht: D. Reidel Publishing Co.

Kulsrud, R.M. and Pierce, W.P. (1969). *Astrophys. J.*, **156**, 445.

Mewaldt, R.A. and Webber, W.R. (1990). *Proc. 21st International Cosmic Ray Conference, Adelaide*, **3**, 432.

Padovani, P., Ghisellini, G., Fabian, A.C. and Celotti, A. (1993). *Mon. Not. R. Astron. Soc.*, **260**, L21.

Parker, E.N. (1965). *Astrophys. J.*, **142**, 584.

Savage, B.D. and de Boer, K.S. (1979). *Astrophys. J. Lett.*, **230**, L77.

Shapiro, M.M. (1991). In *Cosmic rays, supernovae and the interstellar medium*, eds M.M. Shapiro, R. Silberberg and J.P. Wefel, p. 1. Dordrecht: Kluwer Academic Publishers.

Simpson, J. (1983). *Ann. Rev. Nucl. Particle Sci.*, **33**, 323.

Skilling, J. (1971). *Astrophys. J.*, **170**, 265.

Sreekumar, P., Bertsch, D.L., Dingus, B.L., Fichtel, C.E., Hartman, R.C., Hunter, S.D., Kanbach, G., Kniffen, D.A., Lin, Y.C., Mattox, J.R., Mayer-Hasselwander, H.A., Michelson, P.F., von Montigny, C., Nolan, P.L., Pinkau, K., Schneid, E.J. and Thomson, D.J. (1993). *Astrophys. J.*, **400**, L67.

Stecker, F. (1977). In *The structure and content of the galaxy and galactic gamma-rays*, ed. C.E. Fichtel. Greenbelt: Goddard Space Flight Center Publications.

Wdowczyk, J. and Wolfendale, A.W. (1989). *Ann. Rev. Nucl. Particle Sci.*, **39**, 43.

Webster, A.S. (1978). *Mon. Not. R. Astron. Soc.*, **185**, 507.

Wefel, J.P. (1991). In *Cosmic rays, supernovae and the interstellar medium*, eds M.M. Shapiro, R. Silberberg and J.P. Wefel, p. 44. Dordrecht: Kluwer Academic Publishers.

Wentzel, D.G. (1974). *Ann. Rev. Astron. Astrophys.*, **12**, 71.

Wolfendale, A.W. (1991). In *Cosmic rays, supernovae and the interstellar medium*, eds M.M. Shapiro, R. Silberberg and J.P. Wefel, p. 71. Dordrecht: Kluwer Academic Publishers.

Chapter 21

References

Axford, W.I., Leer, E. and Skadron, G. (1977). *Proc. 15th International Cosmic Ray Conference*, **11**, 132.

Blandford, R.D. and Ostriker, J.P. (1978). *Astrophys. J.*, **221**, L29.

Blandford, R.D. and Eichler, D. (1987). *Phys. Rep.*, **154**, 1.

Bell, A.R. (1978). *Mon. Not. R. astr. Soc.*, **182**, 147.

Drury, L.O'C. (1983). *Rep. Prog. Phys.*, **154**, 973.

Fermi, E. (1949). *Phys. Rev.*, **75**, 1169.

Ginzburg, V.L. and Syrovatskii, S.I. (1964). *The origin of cosmic rays.* Oxford: Pergamon Press.

Gunn, J.E. and Ostriker, J.P. (1969). *Phys. Rev. Lett.*, **22**, 728.

Heavens, A.F. and Meisenheimer, K. (1987). *Mon. Not. R. Astron. Soc.*, **225**, 335.

Krymsky, G.F. (1977). *Dok. Acad. Nauk. USSR*, **234**, 1306.

Lagage, P.O. and Cesarsky, C.J. (1983). *Astron. Astrophys.*, **125**, 249.

Völk, H.J. (1987). *Proc. 20th International Cosmic Ray Conference, Moscow*, **7**, 157.

Appendices

General texts

Duffett-Smith, P.J. (1988). *Practical astronomy with your calculator*, 3rd edition. Cambridge: Cambridge University Press.

Duffett-Smith, P.J. (1990). *Astronomy with your personal computer*, 2nd edition. Cambridge: Cambridge University Press.

Murray, C.A. (1983). *Vectorial astronomy.* Bristol: Adam Hilger Ltd.

Rowan-Robinson, M. (1985). *The cosmological distance ladder.* New York: W.H. Freeman and Co.

Smart, W.M. (1977). *Text-book on spherical astronomy.* Cambridge: Cambridge University Press.

Walker, G. (1987). *Astronomical observations – an optical perspective.* Cambridge: Cambridge University Press.

Reference

Rowan-Robinson, M. (1988). *Space Sci. Rev.*, **48**, 1.

Index

Items shown in bold type are physical processes or topics which are frequently encountered in high energy astrophysics. The page references in bold type indicate where the main discussions of these items are to be found.